Volume I
SENSITIVITY and UNCERTAINTY ANALYSIS
THEORY

Dan G. Cacuci

CHAPMAN & HALL/CRC

A CRC Press Company
Boca Raton   London   New York   Washington, D.C.

Library of Congress Cataloging-in-Publication Data

Cacuci, D. G.
    Sensitivity and uncertainty analysis / Dan G. Cacuci.
        p. cm.
    Includes bibliographical references and index.
    Contents: v. 1. Theory
    ISBN 1-58488-115-1 (v. 1 : alk. paper)
    1. Sensitivity theory (Mathematics) 2. Uncertainty (Information theory) 3. Mathematical models—Evaluation. I. Title.

QA402.3.C255 2003
003'.5—dc21   2003043992

This book contains information obtained from authentic and highly regarded sources. Reprinted material is quoted with permission, and sources are indicated. A wide variety of references are listed. Reasonable efforts have been made to publish reliable data and information, but the author and the publisher cannot assume responsibility for the validity of all materials or for the consequences of their use.

Neither this book nor any part may be reproduced or transmitted in any form or by any means, electronic or mechanical, including photocopying, microfilming, and recording, or by any information storage or retrieval system, without prior permission in writing from the publisher.

The consent of CRC Press LLC does not extend to copying for general distribution, for promotion, for creating new works, or for resale. Specific permission must be obtained in writing from CRC Press LLC for such copying.

Direct all inquiries to CRC Press LLC, 2000 N.W. Corporate Blvd., Boca Raton, Florida 33431.

Trademark Notice: Product or corporate names may be trademarks or registered trademarks, and are used only for identification and explanation, without intent to infringe.

Visit the CRC Press Web site at www.crcpress.com

© 2003 by Chapman & Hall/CRC

No claim to original U.S. Government works
International Standard Book Number 1-58488-115-1
Library of Congress Card Number 2003043992
Printed in the United States of America 1 2 3 4 5 6 7 8 9 0
Printed on acid-free paper

INTRODUCTION

In practice, scientists and engineers often face questions such as: how well does the model under consideration represent the underlying physical phenomena? What confidence can one have that the numerical results produced by the model are correct? How far can the calculated results be extrapolated? How can the predictability and/or extrapolation limits be extended and/or improved? Such questions are logical and easy to formulate, but quite difficult to answer quantitatively, in a mathematically and physically well-founded way. Answers to such questions are provided by sensitivity and uncertainty analyses. As computer-assisted modeling and analysis of physical processes have continued to grow and diversify, sensitivity and uncertainty analyses have become indispensable investigative scientific tools in their own rights.

Since computers operate on mathematical models of physical reality, computed results must be compared to experimental measurements whenever possible. Such comparisons, though, invariably reveal discrepancies between computed and measured results. The sources of such discrepancies are the inevitable errors and uncertainties in the experimental measurements and in the respective mathematical models. In practice, the exact forms of mathematical models and/or exact values of data are not known, so their mathematical form must be estimated. The use of observations to estimate the underlying features of models forms the objective of statistics. This branch of mathematical sciences embodies both inductive and deductive reasoning, encompassing procedures for estimating parameters from incomplete knowledge and for refining prior knowledge by consistently incorporating additional information. Thus, assessing and, subsequently, reducing uncertainties in models and data requires the combined use of statistics together with the axiomatic, frequency, and Bayesian interpretations of probability.

As is well known, a mathematical model comprises independent variables, dependent variables, and relationships (e.g., equations, look-up tables, etc.) between these quantities. Mathematical models also include parameters whose actual values are not known precisely, but may vary within some ranges that reflect our incomplete knowledge or uncertainty regarding them. Furthermore, the numerical methods needed to solve the various equations introduce themselves numerical errors. The effects of such errors and/or parameter variations must be quantified in order to assess the respective model's range of validity. Moreover, the effects of uncertainties in the model's parameters on the uncertainty in the calculated results must also be quantified. Generally speaking, the objective of *sensitivity analysis* is to quantify the effects of parameter variations on calculated results. Terms such as influence, importance, ranking by importance, and dominance are all related to sensitivity analysis. On the other hand, the objective of *uncertainty analysis* is to assess the effects of parameter uncertainties on the uncertainties in calculated results. Sensitivity and uncertainty analyses can be considered as formal methods for evaluating data and models because they are associated with the computation of specific quantitative measures that allow, in particular, assessment of variability in output variables and importance of input variables.

For many investigators, a typical approach to model evaluation involves performing computations with nominal (or base-case) parameter values, then performing some computations with parameter combinations that are expected to produce extreme responses of the output, performing computations of output differences to input differences in order to obtain rough guesses of the derivatives of the output variables with respect to the input parameters, and producing scatter plots of the outputs versus inputs. While all of these steps are certainly useful for evaluating a model, they are far from being sufficient to provide the comprehensive understanding needed for a reliable and acceptable production use of the respective model. Such a comprehensive evaluation and review of models and data are provided by systematic (as opposed to haphazard) sensitivity and uncertainty analysis. Thus, the scientific goal of sensitivity and uncertainty analysis is not to confirm preconceived notions, such as about the relative importance of specific inputs, but to discover and quantify the most important features of the models under investigation.

Historically, limited considerations of sensitivity analysis already appeared a century ago, in conjunction with studies of the influence of the coefficients of a differential equation on its solution. For a long time, however, those considerations remained merely of mathematical interest. The first systematic methodology for performing sensitivity analysis was formulated by Bode (1945) for linear electrical circuits. Subsequently, sensitivity analysis provided a fundamental motivation for the use of feedback, leading to the development of modern control theory, including optimization, synthesis, and adaptation. The introduction of state-space methods in control theory, which commenced in the late 1950's, and the rapid development of digital computers have provided the proper conditions for establishing sensitivity theory as a branch of control theory and computer sciences. The number of publications dealing with sensitivity analysis applications in this field grew enormously (see, e.g., the books by: Kokotovic, 1972; Tomovic and Vucobratovic, 1972; Cruz, 1973; Frank, 1978; Fiacco, 1984; Deif, 1986; Eslami, 1994; Rosenwasser and Yusupov, 2000). In parallel, and mostly independently, ideas of sensitivity analysis have also permeated other fields of scientific and engineering activities; notable developments in this respect have occurred in the nuclear, atmospheric, geophysical, socio-economical, and biological sciences.

As has been mentioned in the foregoing, sensitivity analysis can be performed either comprehensively or just partially, by considering selected parameters only. Depending on the user's needs, the methods for sensitivity analysis differ in complexity; furthermore, each method comes with its own specific advantages and disadvantages. Thus, the simplest and most common procedure for assessing the effects of parameter variations on a model is to vary selected input parameters, rerun the code, and record the corresponding changes in the results, or responses, calculated by the code. The model parameters responsible for the largest relative changes in a response are then considered to be the most important for the respective response. For complex models, though, the large amount of computing time needed by such recalculations severely restricts the scope of this sensitivity analysis method; in practice, therefore, the modeler who uses this method can investigate only a few parameters that he judges *a priori* to be important.

Another method of investigating response sensitivities to parameters is to consider simplified models obtained by developing fast-running approximations of complex processes. Although this method makes rerunning less expensive, the parameters must still be selected a priori, and, consequently, important sensitivities may be missed. Also, it is difficult to demonstrate that the respective sensitivities of the simplified and complex models are the same.

When the parameter variations are small, the traditional way to assess their effects on calculated responses is by using perturbation theory, either directly or indirectly, via variational principles. The basic aim of perturbation theory is to predict the effects of small parameter variations without actually calculating the perturbed configuration but rather by using solely unperturbed quantities (see, e.g., Kato, 1963; Sagan, 1969; Kokotovic et al., 1972; Nayfeh, 1973; Eckhaus, 1979; Bogaevski and Povzner, 1991; O'Malley, 1991; Holmes, 1995; Kevorkian and Cole, 1996). As a branch of applied mathematics, perturbation theory relies on concepts and methods of functional analysis. Since the functional analysis of linear operators is quite mature and well established, the regular analytic perturbation theory for linear operators is also well established. Even for linear operators, though, the results obtained by using analytic perturbation theory for the continuous spectrum are less satisfactory than the results delivered by analytic perturbation theory for the essential spectrum. This is because the continuous spectrum is less stable than the essential spectrum, as can be noted by recalling that the fundamental condition underlying analytic perturbation theory is the continuity in the norm of the perturbed resolvent operator (see, e.g., Kato, 1963; Yosida, 1971). If this analytical property of the resolvent is lost, then isolated eigenvalues need no longer remain stable to perturbations, and the corresponding series developments may diverge, have only a finite number of significant terms, or may cease to exist (e.g., an unstable isolated eigenvalue may be absorbed into the continuous spectrum as soon as the perturbation is switched on). The analysis of such divergent series falls within the scope of asymptotic perturbation theory, which comprises: (i) regular or uniform asymptotic perturbation expansions, where the respective expansion can be constructed in the entire domain, and (ii) singular asymptotic perturbation expansions, characterized by the presence of singular manifolds across which the solution behavior changes qualitatively. A nonexhaustive list of typical examples of singular perturbations include: the presence or occurrence of singularities, passing through resonances, loss of the highest-order derivative, change of type of a partial differential equation, and leading operator becoming nilpotent. A variety of methods have been developed for analyzing such singular perturbations; among the most prominent are the method of matched asymptotic expansions, the method of strained coordinates, the method of multiple scales, the WKB (Wentzel-Kramers-Brillouin, or the phase-integral) method, the KBM (Krylov-Bogoliubov-Mitropolsky) method, Whitham's method, and variations thereof. Many fundamental issues in asymptotic perturbation theory are still unresolved, and a comprehensive theory encompassing all types of operators (in particular, differential operators) does not exist yet. Actually, the problems tackled by singular perturbation theory are so diverse that this part of applied mathematics appears to the nonspecialist as a collection of almost disparate methods that often require some clever *a priori* guessing at the structure of the very answer one is looking for. The lack of a unified method for singularly perturbed problems is particularly evident for nonlinear systems, and this state of affairs is not surprising given the fact that nonlinear functional analysis is much less well developed than its linear counterpart. As can be surmised from the above arguments, although perturbation theory can be a valuable tool in certain instances for performing sensitivity analysis, it should be noted already at this stage that the aims of perturbation theory and sensitivity analysis do not coincide, and the two scientific disciplines are distinct from each other.

For models that involve a large number of parameters and comparatively few responses, sensitivity analysis can be performed very efficiently by using deterministic methods based on adjoint functions. The use of adjoint functions for analyzing the effects of small perturbations in a linear system was introduced by Wigner (1945). Specifically, he showed that the effects of perturbing the material properties in a critical nuclear reactor can be assessed most efficiently by using the adjoint neutron flux, defined as the solution of the adjoint neutron transport equation. Since the neutron transport operator is linear, its adjoint operator is straightforward to obtain. In the same report, Wigner was also the first to show that the adjoint neutron flux can be interpreted as the importance of a neutron in contributing to the detector response. Wigner's original work on the linear neutron diffusion and transport equations laid the foundation for the development of a comprehensive and efficient deterministic methodology, using adjoint fluxes, for performing systematic sensitivity and uncertainty analyses of eigenvalues and reaction rates to uncertainties in the cross sections in nuclear reactor physics problems (see, e.g., Selengut, 1959, 1963; Weinberg and Wigner, 1958; Usachev, 1964; Stacey, 1974; Greenspan, 1982; Gandini, 1987). Since the neutron transport and, respectively, neutron diffusion equations underlying problems in nuclear reactor physics are linear in the dependent variable (i.e., the neutron flux), the respective adjoint operators and adjoint fluxes are easy to obtain, a fact that undoubtedly facilitated the development of the respective sensitivity and uncertainty analysis methods. Furthermore, the responses considered in all of these works were functionals of the neutron forward and/or adjoint fluxes, and the sensitivities were defined as the derivatives of the respective responses with respect to scalar parameters, such as atomic number densities and energy-group-averaged cross sections.

As the need for systematic sensitivity/uncertainty analysis gained recognition, additional sensitivity analysis methods were developed, most notably in conjunction with applications to chemical kinetics, optimization theory, atmospheric sciences (see, e.g., Hall and Cacuci, 1983; Navon et al., 1992; Zhou et al., 1993), and nuclear engineering (see, e.g., March-Leuba et al., 1984, 1986; Ronen, 1988; Rief, 1988; Lillie et al., 1988). For example, three sensitivity analysis methods developed in conjunction with applications in chemical kinetics (see, e.g., Dickinson and Gelinas, 1976: Demiralp and Rabitz, 1981, and references therein) are the Fourier Amplitude Sensitivity Test (acronym FAST), the direct method, and the Green's function method. The FAST technique gives statistical mean values and standard deviations for model responses, but requires a very large number of calculations (despite its acronym, it is not a fast-running technique). In contrast to the FAST technique, the direct method and Green's function method are deterministic, rather than statistical, methods. The direct method is practically identical to the sensitivity analysis methods used in control theory, involving differentiation of the equations describing the model with respect to a parameter. The resulting set of equations is solved for the derivative of all the model variables with respect to that parameter. Note that the actual form of the differentiated equations depends on the parameter under consideration. Consequently, for each parameter, a different set of equations must be solved to obtain the corresponding sensitivity. Hence, all of these methods become prohibitively expensive, computationally, for large-scale nonlinear models, involving many parameters, as are usually encountered in practical situations.

The direct method, on the one hand, and the variational/perturbation methods using adjoint functions, on the other hand, were unified and generalized by Cacuci et al. (1980), who presented a comprehensive methodology, based on Frechet-derivatives, for performing systematic and efficient sensitivity analyses of large-scale continuous and/or discrete linear and/or nonlinear systems. Shortly thereafter, this methodology was further generalized by Cacuci (1981, I-II), who used methods of nonlinear functional analysis to introduce a rigorous definition of the concept of sensitivity as the first Gâteaux-differential - in general a nonlinear operator - of the system response along an arbitrary direction in the hyperplane tangent to the base-case solution in the phase-space of parameters and dependent variables. These works presented a rigorous theory not only for sensitivity analysis of functional-type responses but also for responses that are general nonlinear operators, and for responses defined at critical points. Furthermore, these works have also set apart sensitivity analysis from perturbation theory, by defining the scope of the former to be the exact and efficient calculation of all sensitivities, regardless of their use. Two procedures were developed for calculating the sensitivities, namely the Forward Sensitivity Analysis Procedure (*FSAP*) and the Adjoint Sensitivity Analysis Procedure (*ASAP*). The *FSAP* generalizes the direct method of sensitivity analysis, in that the equations obtained within the *FSAP* comprise *all* of the equations obtained within the direct method for calculating sensitivities. Just as the direct method of sensitivity analysis, the *FSAP* is advantageous to employ only if, in the problem under consideration, the number of responses of interest exceeds the number of system parameters and/or parameter variations to be considered. However, most problems of practical interest comprise a large number of parameters and comparatively few responses. In such situations, it is by far more advantageous to employ the *ASAP*. Note, though, that the *ASAP* is not easy to implement for complicated nonlinear models, particularly in the presence of structural discontinuities. Furthermore, Cacuci (1981, I-II) has underscored the fact that the adjoint functions needed for the sensitivity analysis of nonlinear systems depend on the unperturbed forward (i.e., nonlinear) solution, a fact that is in contradistinction to the case of linear systems. These works have also shown that the adjoint functions corresponding to the Gâteaux-differentiated nonlinear systems can be interpreted as *importance functions* - in that they measure the importance of a region and/or event in phase-space in contributing to the system's response under consideration; this interpretation is similar to that originally assigned by Wigner (1945) to the adjoint neutron flux for linear neutron diffusion and/or transport problems in reactor physics and shielding.

Once they become available, the sensitivities can be used for various purposes, such as for ranking the respective parameters in order of their relative importance to the response, for assessing changes in the response due to parameter variations, for performing uncertainty analysis using either the Bayesian approach or the response surface approach, or for data adjustment and/or assimilation. As highlighted by Cacuci (1981, I-II), it is necessary to define rigorously the concept of sensitivity and to separate the calculation of sensitivities from their use, in order to compare clearly, for each particular problem, the relative advantages and disadvantages of using one or the other of the competing deterministic methods, statistical methods, or Monte Carlo methods of sensitivity and uncertainty analysis.

Although the numerical computations are generally considered to be deterministic, the output (or responses) of such calculations can be considered to be random variables, since the input variables are known imprecisely or are inherently random. For this reason, in addition to the deterministic methods mentioned above for sensitivity analysis, a variety of statistical techniques have also been developed for performing sensitivity and uncertainty analysis. Most of these methods are collectively known as "response surface methods" (see, e.g., Myers, 1971), and they involve the following basic steps:

(a) selecting a small number of model parameters that are thought by the modeler to be important;

(b) selecting design points, in the space of model parameters, at which the responses of interest to sensitivity/uncertainty analysis are computed;

(c) performing response recalculations using altered parameter values, around the design points selected in (b) for the model parameters selected in (a);

(d) using the recalculated results obtained in (c) to construct a "response surface," i.e., a simplified model that approximately represents the behavior of the response as a relatively simple function of the chosen model parameters;

(e) using the response surface obtained in (d) to replace the original model for subsequent statistical studies (e.g., Monte Carlo, moment matching) to estimate sensitivities and uncertainty distributions for the responses.

Descriptions and applications of some of the modern statistical methods (including response surface methods) for sensitivity and uncertainty analysis are provided, for example, in the works of McKay, 1988; Rief, 1988; Boyack et al., 1990; Committee on the Safety of Nuclear Installations, 1994; D'Auria and Giannotti, 2000. As can be noted from these works, the application of statistical methods to sensitivity and uncertainty analysis of computer codes has the advantages of being conceptually straightforward and requiring relatively little developmental work. However, despite progress towards reducing the number of recalculations needed to construct the response surfaces, statistical methods remain expensive and relatively limited in scope, particularly because of the following practical considerations: (i) when the number of model parameters is very large, only a small subset can be selected for sensitivity/uncertainty analysis; (ii) information about data importance is required prior to initiating the analysis, but the data importance is largely unknown, so that a considerable probability of missing important effects exists; (iii) the actual parameter sensitivities can be estimated, but cannot be calculated exactly by using response surfaces. Last but not least, the sensitivities and parameter uncertainties are not separated, but amalgamated into producing the response surface. Consequently, if the ranges of parameter uncertainties change for some reason, such as obtaining more accurate parameters, the entire response surface must be calculated anew, with all the accompanying computational expenses. Similarly, the response surface would need to be calculated anew if the model is improved, since model improvements would cause the respective response sensitivities change.

With a notable exception, the main methods for sensitivity and uncertainty analysis reviewed briefly in the foregoing have been adequately covered in the books mentioned in the Reference section at the end of this volume, the exception being a systematic exposure of the powerful sensitivity analysis method based on the use of adjoint operators. Consequently, this book aims at filling this gap, and is therefore focused on the mathematical considerations underlying the development of FSAP and ASAP, and the use of deterministically obtained sensitivities for subsequent uncertainty analysis. Either or both of these methods can be applied to a much wider variety of problems than can the traditional methods of sensitivity analysis as used in modern control theory. Hence, this book complements the books on sensitivity and uncertainty analysis that have been published thus far.

To assist the reader, the first three chapters of this volume provide a brief review of the mathematical concepts of matrix and operator theory, probability and statistics, and the theory and practice of measurements and errors, respectively, as used for sensitivity and uncertainty analysis of data and models. Thus, Chapter I summarizes the salient concepts needed for sensitivity analysis from matrix theory, linear operators and their adjoints (in both finite and infinite dimensional vector spaces), and differential calculus in vector spaces. Note that the material presented in this chapter does not contain detailed illustrative examples, and is not intended to serve as a replacement for the many excellent textbooks that have been published on the various specialized mathematical topics (see, e.g., Liusternik and Sobolev, 1961; Pontryagin et al., 1962; Varga, 1962; Vainberg, 1964; Zemanian, 1965; Ostrowski, 1966; Dieudonne, 1969; Morse, 1969; Ortega, 1970; Kanwal, 1971; Rall, 1971; Yosida, 1971; Krasnosel'skii et al., 1972; Gourlay and Watson, 1973; Schechter, 1977; Kahn, 1980; Reed and Simon, 1980; Lions, 1981; Nowinski, 1981; Deif, 1982; Smith, 1984; Zwillinger, 1992; Drazin, 1994; Horn and Johnson, 1996).

Chapter II provides a summary of the main definitions and features of the mathematical and philosophical concepts of probability theory and statistics as needed for sensitivity and uncertainty analysis of data and models. From a mathematical point of view, the concepts of probability theory are best introduced by using Kolmogorov's axiomatic approach, in which probability is postulated in terms of abstract functions operating on well-defined event spaces. This axiomatic approach avoids both the mathematical ambiguities inherent to the concept of relative frequencies and the pitfalls of inadvertently misusing the concept of inductive reasoning. Nevertheless, the other two main interpretations of probability, namely that of *relative frequency* (which is used, in particular, for assigning statistical errors to measurements) and that of *subjective probability* (which is used, in particular, to quantify systematic uncertainties), are also discussed, in conjunction with their uses for sensitivity and uncertainty analysis of models and data. Thus, a systematic uncertainty analysis of practical problems inevitably requires the synergetic use of the various interpretations of probability, including the axiomatic, frequency, and Bayesian interpretations and methodologies (see, e.g., Kendal and Stuart, 1976; Box et al., 1978; Conover, 1980; Bauer, 1981; Shapiro and Gross, 1981; Jaynes, 1983; Smith, 1991; Cowan,1998; Ross, 2000).

Chapter III provides a summary of definitions and practical considerations, as needed for uncertainty analysis of data and models, regarding the theory and practice of measurements and, in particular, concerning the errors that are inevitably associated with data and models. The chapter commences with a review of the main sources and features of errors, and then describes the procedures for dealing with the types of errors and uncertainties which affect direct and indirect measurements. This presentation sets the stage for introducing one of the most basic methods for quantifying the magnitudes and effects of errors in (and on) complex measurements and computations; this method is usually referred to as the method of "propagation of errors" or, alternatively, the method of "propagation of moments" (see, e.g., Shapiro and Gross, 1981; Ronen, 1988; Smith, 1991; Cowan, 1998; Rabinovich, 2000). As highlighted in Chapter

III, the "propagation of errors" method provides a systematic way of obtaining the uncertainties in results of measurements and computations, arising not only from uncertainties in the parameters that enter the respective computational model but also from numerical approximations. The propagation of errors method combines systematically and consistently the parameter errors with the sensitivities of responses (i.e., results of measurements and/or computations) to the respective parameters, thus providing the symbiotic linchpin between the objectives of uncertainty analysis and those of sensitivity analysis.

The efficient computation of sensitivities and, subsequently, uncertainties in results produced by various models (algebraic, differential, integrals, etc.) forms the objectives of the remaining chapters in Vol. I of this book. Thus, the concepts underlying *local* sensitivity analysis of system responses around a chosen point or trajectory in the combined phase space of parameters and state variables are presented in Chapters IV and V, for both linear and nonlinear systems. Chapter IV presents the two fundamental procedures for calculating response sensitivities to system parameters, namely the *FSAP* and the *ASAP*. Although the *FSAP* is conceptually easier to develop and implement than the *ASAP*, the *FSAP* is advantageous to employ only if, in the problem under consideration, the number of different responses of interest exceeds the number of system parameters and/or parameter variations to be considered. However, most problems of practical interest comprise a large number of parameters and comparatively few responses. In such situations, it is by far more advantageous to employ the *ASAP*. It is important to note that the physical systems considered in Chapter IV are *linear in the dependent (state) variables. This important feature makes it possible to solve the adjoint equations underlying the ASAP independently of the original (forward) equations.* Several examples illustrating the uses of sensitivities are also provided.

Chapter V continues the presentation of local sensitivity analysis of system responses around a chosen point or trajectory in the combined phase space of parameters and state variables. In contradistinction to the problems considered in Chapter IV, though, the operators considered in this chapter act nonlinearly not only on the system's parameters but also on the dependent variables, and include linear and/or nonlinear feedback. Consequently, it will no longer be possible to solve the adjoint equations underlying the *ASAP* independently of the original (forward) nonlinear equations. This is the fundamental distinction between applying the *ASAP* to nonlinear versus linear systems, and the consequences of this distinction are illustrated by applications to several representative nonlinear problems. It is also noted that the *ASAP* is more difficult to implement for nonlinear problems (by comparison to linear ones), particularly if the structural flexibility of the original computer codes is limited.

Since the objective of *local sensitivity analysis* is to analyze the behavior of the system responses *locally* around a chosen point or trajectory in the combined phase space of parameters and state variables, the local sensitivities yield only the first-order contributions to the total response variation, valid only within sufficiently small neighborhoods of the nominal values of the respective parameters. In the simplest case, the local sensitivities are equivalent to the first-order partial derivatives of the system's response with respect to the system parameters, calculated at the parameters' nominal values. In principle, the *ASAP* could be extended to calculate second- and higher-order sensitivities. For most practical problems, however, it is seldom profitable to calculate higher-order sensitivities (by any method), since the respective equations become increasingly more difficult to solve. Within the *ASAP*, for example, *the adjoint system needed to calculate the first-order sensitivities is independent of the perturbation value*, but the adjoint systems needed to compute *second-* and *higher-order* sensitivities *would* depend on the value of the perturbation, and this dependence cannot be avoided. Consequently, the computation of second- and higher-order sensitivities would require solving as many adjoint systems as there are parameter perturbations. This is in contradistinction with the significant practical advantages enjoyed by the *ASAP* for calculating the first-order sensitivities, where only one adjoint system needs to be solved per response. Furthermore, even if all the second- and higher-order sensitivities could be computed, the additional information gained over that provided by the first-order sensitivities would still be about the *local, rather than the global, behavior* of the response and system around the nominal parameter values. This is because the respective operator-valued Taylor-series only provides local, rather than global, information. In summary, therefore, we note that calculating the first-order sensitivities by the *ASAP* provides a high ratio of payoff-to-effort (analytical and computational); this ratio decreases dramatically when calculating second- and higher-order sensitivities.

Chapter VI presents a *global sensitivity analysis* procedure (Cacuci, 1990), aimed at determining all of the system's critical points (bifurcations, turning points, saddle points, response extrema) in the combined phase-space formed by the parameters, forward state variables, and adjoint variables, and subsequently analyze these critical points locally by the efficient *ASAP*. The local sensitivities around a point in phase-space (including critical points) provide, in particular, information about the linear stability of the respective point. The aim of global sensitivity analysis is attained by devising a conceptual and computational framework using a global continuation method (see, e.g., Wacker, 1978), based on homotopy theory, for determining interesting features of the physical problem (i.e., various types of critical points) over the entire space of parameter variations. Furthermore, computational difficulties (such as slow convergence, small steps, or even failure to converge) are circumvented by using a "pseudo-arc-length" (i.e., distance along a local tangent) continuation method, which employs a scalar constraint in addition to the basic homotopy, thus "inflating" the original problem into one of a higher dimension. To compute the homotopy path efficiently, it is advantageous to use Newton's method because of its simplicity and superior convergence properties (quadratic or, at worst, superlinear at bifurcation points), switching at the critical points to the secant method coupled with regula falsi, to ensure rapid convergence. Finally, the use of the global computational algorithm described in the foregoing is illustrated by determining the critical and first-order bifurcation points of a paradigm nonlinear lumped parameter model describing the dynamic behavior of a boiling water nuclear reactor (March-Leuba et al., 1984, 1986).

To keep this volume to a reasonable size, methods of data adjustment and data assimilation in the presence of uncertainties, as well as various applications of sensitivity analysis to large-scale problems, have been deferred for presentation in subsequent volume(s). Last but not least, I would like to acknowledge the essential contributions made by others in bringing this book to the readers. To Dr. Mihaela Ionescu-Bujor, my close collaborator at the Forschungszentrum Karlsruhe, Germany, I owe a great deal of gratitude not only for her constructive criticism, which brought the originally very mathematical presentation to a form readable by a much wider audience, but also for her guiding me through the intricacies of desktop publishing. To Ms. Helena Redshaw of CRC Press, I am thankful not only for her holding my hand through the publication process, but also for keeping the publication schedule on track with her friendly and humorous e-mails. Words are not enough to express my thanks to Bob Stern, CRC Executive Editor; were it not for his unbounded patience, this book would certainly not have appeared in its present form.

TABLE OF CONTENTS

I.	**CONCEPTS OF MATRIX AND OPERATOR THEORY FOR SENSITIVITY AND UNCERTAINTY ANALYSIS**	**1**
A.	MATRICES: BASIC DEFINITIONS AND PROPERTIES	1
B.	OPERATORS IN VECTOR SPACES: BASIC DEFINITIONS AND PROPERTIES	12
C.	ADJOINT OPERATORS: BASIC DEFINITIONS AND PROPERTIES	27
D.	ELEMENTS OF DIFFERENTIAL CALCULUS IN NORMED SPACES	33
II.	**CONCEPTS OF PROBABILITY THEORY FOR SENSITIVITY AND UNCERTAINTY ANALYSIS**	**39**
A.	INTRODUCTION AND TERMINOLOGY	39
B.	INTERPRETATION OF PROBABILITY	48
	1. Probability as a Relative Frequency	48
	2. Subjective Probability	49
C.	MULTIVARIATE PROBABILITY DISTRIBUTIONS	51
D.	FUNDAMENTAL PARAMETERS OF PROBABILITY FUNCTIONS	55
	1. Expectations and Moments	55
	2. Variance, Standard Deviation, Covariance, and Correlation	58
E.	CHARACTERISTIC AND MOMENT GENERATING FUNCTIONS	65
F.	COMMONLY ENCOUNTERED PROBABILITY DISTRIBUTIONS	69
G.	CENTRAL LIMIT THEOREM	85
H.	BASIC CONCEPTS OF STATISTICAL ESTIMATION	88
I.	ERROR AND UNCERTAINTY	100
J.	PRESENTATION OF RESULTS OF MEASUREMENTS: RULES FOR ROUNDING OFF	101
III.	**MEASUREMENT ERRORS AND UNCERTAINTIES: BASIC CONCEPTS**	**105**
A.	MEASUREMENTS: BASIC CONCEPTS AND TERMINOLOGY	105
B.	CLASSIFICATION OF MEASUREMENT ERRORS	109
C.	NATURAL PRECISION LIMITS OF MACROSCOPIC MEASUREMENTS	114

D.	DIRECT MEASUREMENTS	116
	1. Errors and Uncertainties in Single Direct Measurements	116
	2. Errors and Uncertainties in Multiple Direct Measurements	117
E.	INDIRECT MEASUREMENTS	119
F.	PROPAGATION OF MOMENTS (ERRORS)	120
GLOSSARY		127

IV. LOCAL SENSITIVITY AND UNCERTAINTY ANALYSIS OF LINEAR SYSTEMS   129

A.	THE FORWARD SENSITIVITY ANALYSIS PROCEDURE (FSAP)	132
B.	THE ADJOINT SENSITIVITY ANALYSIS PROCEDURE (ASAP)	134
	1. System Responses: Functionals	136
	2. System Responses: Operators	137
C.	ILLUSTRATIVE EXAMPLE: SENSITIVITY ANALYSIS OF LINEAR ALGEBRAIC MODELS	142
D.	ILLUSTRATIVE EXAMPLE: LOCAL SENSITIVITY ANALYSIS OF A LINEAR NEUTRON DIFFUSION PROBLEM	146
E.	ILLUSTRATIVE EXAMPLE: SENSITIVITY ANALYSIS OF EULER'S ONE-STEP METHOD APPLIED TO A FIRST-ORDER LINEAR ODE	154
F.	ILLUSTRATIVE EXAMPLE: SENSITIVITY AND UNCERTAINTY ANALYSIS OF A SIMPLE MARKOV CHAIN SYSTEM	165
G.	ILLUSTRATIVE EXAMPLE: SENSITIVITY AND UNCERTAINTY ANALYSIS OF NEUTRON AND GAMMA RADIATION TRANSPORT	182
	1. Sensitivity Analysis	183
	2. Uncertainty Analysis	191

V. LOCAL SENSITIVITY AND UNCERTAINTY ANALYSIS OF NONLINEAR SYSTEMS   199

A.	SENSITIVITY ANALYSIS OF NONLINEAR SYSTEMS WITH FEEDBACK AND OPERATOR-TYPE RESPONSES	199
	1. The Forward Sensitivity Analysis Procedure (FSAP)	202
	2. Adjoint (Local) Sensitivity Analysis Procedure (ASAP)	202
B.	EXAMPLE: A PARADIGM CLIMATE MODEL WITH FEEDBACK	205

C. SENSITIVITY ANALYSIS OF NONLINEAR ALGEBRAIC
   MODELS: THE DISCRETE ADJOINT SENSITIVITY
   ANALYSIS PROCEDURE (DASAP) 209
   1. The Discrete Forward Sensitivity
      Analysis Procedure (DFSAP) 212
   2. The Discrete Adjoint Sensitivity
      Analysis Procedure (DASAP) 216
D. ILLUSTRATIVE EXAMPLE: A TRANSIENT
   NONLINEAR REACTOR THERMAL
   HYDRAULICS PROBLEM 217
   1. Adjoint Sensitivity Analysis Procedure (ASAP) 221
   2. Discrete Adjoint Sensitivity Analysis
      Procedure (DASAP) 224
   3. Comparative Discussion of the ASAP
      and DASAP Formalisms 227
E. ILLUSTRATIVE EXAMPLE: SENSITIVITY ANALYSIS
   OF THE TWO-STEP RUNGE-KUTTA METHOD FOR
   NONLINEAR DIFFERENTIAL EQUATIONS 231
F. ILLUSTRATIVE EXAMPLE: SENSITIVITY ANALYSIS
   OF A SCALAR NONLINEAR HYPERBOLIC PROBLEM 238

VI. GLOBAL OPTIMIZATION AND SENSITIVITY ANALYSIS   245

A. MATHEMATICAL FRAMEWORK 247
B. CRITICAL POINTS AND GLOBAL OPTIMIZATION 249
C. SENSITIVITY ANALYSIS 252
D. GLOBAL COMPUTATION OF FIXED POINTS 256
E. A CONSTRAINED OPTIMIZATION PROBLEM FOR
   TESTING THE GLOBAL COMPUTATIONAL ALGORITHM 263
F. ILLUSTRATIVE EXAMPLE: GLOBAL ANALYSIS
   OF A NONLINEAR LUMPED-PARAMETER MODEL
   OF A BOILING WATER REACTOR 264

REFERENCES 267

INDEX 273

Chapter I

CONCEPTS OF MATRIX AND OPERATOR THEORY FOR SENSITIVITY AND UNCERTAINTY ANALYSIS

This chapter is intended to provide a summary of selected definitions and concepts of matrix and operator theory that underlie sensitivity and uncertainty analysis of data and models. The topics covered for this purpose are matrices, linear operators and their adjoints in finite and infinite dimensional vector spaces, and differential calculus in vector spaces. The material covered in this chapter serves as a self-contained mathematical basis for sensitivity and uncertainty analysis, and is provided for easy reference. Note, though, that the material presented in this chapter does not contain detailed examples for illustrating the various mathematical concepts, and is not intended to serve as a replacement for the many excellent textbooks on the respective mathematical topics. A list of suggested reference mathematical textbooks is provided in the Reference Section of this book.

I.A. MATRICES: BASIC DEFINITIONS AND PROPERTIES

A *number field* $\mathcal{F}$ is defined as an arbitrary collection of numbers within which the four operations of addition, subtraction, multiplication, and division by a nonzero number can always be carried out. Examples of number fields are the set of all integer numbers, the set of all rational numbers, the set of all real numbers, and the set of all complex numbers.

A rectangular array of numbers of the field $\mathcal{F}$, composed of m rows and n columns

$$A = \begin{pmatrix} a_{11} & a_{12} \cdots a_{1n} \\ a_{21} & a_{22} \cdots a_{2n} \\ \cdots\cdots\cdots\cdots\cdots \\ a_{m1} & a_{m2} \cdots a_{mn} \end{pmatrix}, \qquad (\text{I.A.1})$$

is called a *rectangular matrix* of dimension $m \times n$. The numbers a_{ij}, $(i=1,2,...,m;\ j=1,2,...,n)$; are called the elements of A. When $m=n$, the matrix A is called square and the number m, equal to n, is called the order of

A. Note that, in the double-subscript notation for the elements a_{ij}, the order of the subscripts is important: the first subscript, i, denotes the row, while the second subscript, j, denotes the column containing the given element. Often, the notation for the matrix defined in (I.A.1) is abbreviated as

$$A = (a_{ij}), (i = 1,2,...,m; j = 1,2,...,n), \tag{I.A.2}$$

or simply A, when its elements and dimensions are understood from context.
A *column matrix* is a rectangular matrix consisting of a single column,

$$\begin{pmatrix} x_1 \\ x_2 \\ \vdots \\ x_n \end{pmatrix}, \tag{I.A.3}$$

while a *row matrix* is a rectangular matrix consisting of a single row,

$$(a_1, a_2, \cdots, a_n). \tag{I.A.4}$$

A *diagonal matrix*, D, is a square matrix whose off-diagonal elements are zero: $a_{ij} = 0$, *for* $i \neq j$. Note that the diagonal elements $a_{ii} = d_i$ may (or may not) be different from 0. Note also that

$$D = (d_i \delta_{ik}), \text{ where } (\delta_{ik}) \text{ is the Kronecker symbol: } \delta_{ik} = \begin{cases} 1, & i=k \\ 0, & i \neq k \end{cases}.$$

A *unit* (or *identity*) *matrix*, I, is a diagonal matrix with $a_{ii} = 1$, for all i, i.e., $I = (\delta_{ik})$. A *zero matrix*, $\mathbf{0}$, is a matrix with all its elements zero. A *lower triangular matrix* is a matrix with elements $a_{ij} = 0$, *for* $i < j$, while an *upper triangular matrix* is a matrix with elements $a_{ij} = 0$, *for* $i > j$. Two matrices A and B are said to be equal if their corresponding elements are equal, i.e., if $a_{ij} = b_{ij}$, for all $i = 1,...,m$ and $j = 1,...,n$. The *transpose* A^T of a matrix $A = (a_{ik})$, $(i = 1,2,...,m; k = 1,2,...,n)$, is defined as $A^T = \|a_{ki}\|$, $(i = 1,2,...,m; k = 1,2,...,n)$. The following properties of the operation of transposition are often encountered in applications: $(A + B)^T = A^T + B^T$, $(aA)^T = aA^T$, $(AB)^T = B^T A^T$. A square matrix S that coincides with its transpose, i.e.,

$a_{ij} = a_{ji}$, for all i, j, is called a *symmetric* matrix. A *skew-symmetric matrix* is a matrix with elements $a_{ij} = -a_{ji}$, for all i, j. A *hermitian matrix* is a matrix with elements $a_{ij} = \overline{a}_{ji}$, for all i, j; here, the overbar denotes complex conjugation. This definition implies that the diagonal elements of a hermitian matrix are real. A *skew-hermitian matrix* is a matrix with elements $a_{ij} = -\overline{a}_{ji}$, for all i, j; this definition indicates that the diagonal elements of a skew-hermitian matrix are purely imaginary.

If m quantities $y_1, y_2, ..., y_m$ have linear and homogeneous expressions in terms of n other quantities $x_1, x_2, ..., x_n$ such that

$$\left.\begin{aligned} y_1 &= a_{11}x_1 + a_{12}x_2 + \cdots + a_{1n}x_n \\ y_2 &= a_{21}x_1 + a_{22}x_2 + \cdots + a_{2n}x_n \\ &\cdots\cdots\cdots\cdots\cdots\cdots\cdots\cdots\cdots \\ y_m &= a_{m1}x_1 + a_{m2}x_2 + \cdots + a_{mn}x_n \end{aligned}\right\}, \quad (I.A.5)$$

then the transformation of the quantities $x_1, x_2, ..., x_n$ into the quantities $y_1, y_2, ..., y_m$ by means of the formulas (I.A.5) is called a *linear transformation*.

The coefficients of the transformation (I.A.5) form the rectangular matrix of dimension $m \times n$ defined in Eq. (I.A.1). Note that the linear transformation (I.A.5) determines the matrix (I.A.1) uniquely, and vice versa. In the next section, the properties of the linear transformation (I.A.5) will serve as the starting point for defining the basic operations on rectangular matrices.

The basic operations on matrices are: *addition (summation) of matrices, scalar multiplication of a matrix by a number*, and *multiplication* of matrices. Thus, the *sum* of two rectangular matrices $A = (a_{ik})$ and $B = (b_{ik})$, both of dimension $m \times n$, is the matrix $C = (c_{ik})$, of the same dimension, whose elements are

$$C = A + B, \text{ with } c_{ik} = a_{ik} + b_{ik}, (i = 1,...,m, \text{ and } k = 1,...,n). \quad (I.A.6)$$

Note that the matrix addition of arbitrary rectangular matrices of equal dimensions is commutative and associative, i.e.,

$$A + B = B + A, \text{ and } (A + B) + C = A + (B + C)$$

Next, the scalar multiplication of a matrix $A = (a_{ik})$ by a number β of $\mathcal{F}$ is the matrix $C = (c_{ik})$ such that

$$C = \beta A, \text{ with } c_{ik} = \beta a_{ik}, (i = 1,...,m, \text{ and } k = 1,...,n). \quad (I.A.7)$$

The *difference* $(A - B)$ of two rectangular matrices of equal dimensions is defined as $A - B = A + (-1)B$.

Matrix multiplication is the operation of forming the product of two rectangular matrices

$$A = \begin{pmatrix} a_{11} & a_{12} \cdots a_{1n} \\ a_{21} & a_{22} \cdots a_{2n} \\ \cdots\cdots\cdots\cdots\cdots \\ a_{m1} & a_{m2} \cdots a_{mn} \end{pmatrix}, \text{ and } B = \begin{pmatrix} b_{11} & b_{12} \cdots b_{1q} \\ b_{21} & b_{22} \cdots b_{nq} \\ \cdots\cdots\cdots\cdots\cdots \\ b_{n1} & b_{n2} \cdots b_{nq} \end{pmatrix}$$

to obtain the matrix

$$C = \begin{pmatrix} c_{11} & c_{12} \cdots c_{1q} \\ c_{21} & c_{22} \cdots c_{2q} \\ \cdots\cdots\cdots\cdots\cdots \\ c_{m1} & c_{m2} \cdots c_{mq} \end{pmatrix}.$$

The element c_{ij}, at the intersection of the i-th row and the j-th column of C, is obtained as the sum of the products of the elements of the i-th row of the first matrix A by the elements of the j-th column of the second matrix B. Thus

$$C = AB, \text{ where } c_{ij} = \sum_{k=1}^{n} a_{ik} b_{kj}, \quad (i = 1,2,...,m; j = 1,2,...,q). \quad \text{(I.A.8)}$$

The multiplication of two rectangular matrices can only be carried out when the number of columns of the first matrix is equal to the number of rows of the second. Matrix multiplication is *associative* and also *distributive* with respect to addition: $(AB)C = A(BC)$, $(A + B)C = AC + BC$, $A(B + C) = AB + AC$.

In particular, multiplication is always possible when both matrices are square, of the same order. Note, though, that even in this special case the multiplication of matrices does not, in general, commute. In the *special case when* $AB = BA$, the matrices A and B are called *permutable* or *commuting*.

The *trace* of a square matrix A, denoted by trA, is a scalar quantity obtained by summing up the diagonal elements of A, i.e.,

$$trA = \sum_{i=1}^{n} a_{ii}. \quad \text{(I.A.9)}$$

The *power of a square matrix* $A = (a_{ik})$ is defined as

$$A^p = \underbrace{AA \cdots A}_{p \text{ times}}, (p = 1, 2, \ldots); \text{ with } A^0 \equiv I. \qquad (\text{I.A.10})$$

Since matrix multiplication is associative, it follows that $A^p A^q = A^{p+q}$, for p and q arbitrary nonnegative integers. A square matrix A is called *nilpotent* if and only if there is an integer p such that $A^p = 0$. The smallest such integer p is called the *index of nilpotence*.

Consider now a polynomial $f(t)$ with coefficients in the field $\mathcal{F}$, namely $f(t) = \alpha_0 t^m + \alpha_1 t^{m-1} + \cdots + \alpha_m$. Then a *polynomial*, $f(A)$, *in a matrix* A, is defined to be the matrix

$$f(A) = \alpha_0 A^m + \alpha_1 A^{m-1} + \cdots + \alpha_m I. \qquad (\text{I.A.11})$$

The *determinant* of a square matrix A is a scalar quantity, denoted by $\det A$ or $|A|$, and is given by

$$|A| = \sum_{j_1, j_2, \cdots, j_n} p(j_1, j_2, \cdots, j_n) a_{1j_1} a_{2j_2} \cdots a_{nj_n}, \qquad (\text{I.A.12})$$

where $p(j_1, j_2, \cdots, j_n)$ is a permutation equal to ± 1, written in general as $p(j_1, j_2, \cdots, j_n) = \text{sign} \prod_{1 \leq s < r \leq n} (j_r - j_s)$. The determinant can also be written in the alternative forms

$$|A| = \sum_{j=1}^{n} a_{ij} C_{ij}, \quad i = \text{constant, or } |A| = \sum_{i=1}^{n} a_{ij} C_{ij}, \quad j = \text{constant}, \qquad (\text{I.A.13})$$

where C_{ij} is called the *cofactor* of the element a_{ij}. Note also that $C_{ij} = (-1)^{i+j} M_{ij}$, where M_{ij}, called the *minor*, is the determinant of the submatrix obtained from the matrix A by deleting the i-th row and j-th column. When the rows and columns deleted from A have the same indices, the resulting submatrix is located symmetrically with respect to the main diagonal of A, and the corresponding minor is called the *principal* minor of A. The minor

$$\begin{vmatrix} a_{11} & \cdots & a_{1r} \\ \vdots & & \vdots \\ a_{r1} & & a_{rr} \end{vmatrix}$$

is called the *leading* principal minor of A of order r.

For easy reference, some commonly used properties of determinants will now be listed:

1. $|A| = |A^T|$.

2. If any column or row of A is multiplied by a scalar α, then the determinant of the new matrix becomes $\alpha |A|$.

3. If any column or row of A is zero, then $|A| = 0$.

4. If two rows or columns of A are interchanged, then the determinant of the new matrix has the same absolute value but opposite sign to the original determinant.

5. If two rows or columns in A are identical, then $|A| = 0$.

6. If a matrix A has elements $a_{sj} = c_{sj} + b_{sj}$, for all $j = 1, 2, \ldots, n$, then the determinant of A is equal to the sum of two determinants such that the first determinant is obtained by replacing a_{sj} in A by c_{sj} for all $j = 1, 2, \ldots, n$, while the second determinant is obtained by replacing a_{sj} in A by b_{sj} for all $j = 1, \ldots, n$.

7. $|AB| = |A||B|$.

8. The sum of the product of the elements of a row by the cofactors of another row is equal to zero, i.e., $\sum_{j=1}^{n} a_{ij} C_{kj} = 0$, for all $i \neq k$.

9. The determinant of a matrix A does not change if the elements of any row are multiplied by a constant and then added to another row.

10. The derivative $(d|A(\lambda)|/d\lambda)$, of a determinant $|A(\lambda)| = \sum_{j_1,\ldots,j_n} p(j_1,\ldots,j_n) a_{1j_1}(\lambda) \ldots a_{nj_n}(\lambda)$, with respect to a parameter λ, is obtained as the sum of the determinants obtained by differentiating the rows of A with respect to λ one at a time, i.e.,

$$\frac{d}{d\lambda}|A| = \sum_{j_1,\ldots,j_n} p(j_1,\ldots,j_n)\frac{da_{1j_1}(\lambda)}{d\lambda} a_{2j_2}(\lambda)\ldots a_{nj_n}(\lambda)$$
$$+ \sum_{j_1,\ldots,j_n} p(j_1,\ldots,j_n)a_{1j_1}(\lambda)\frac{da_{2j_2}(\lambda)}{d\lambda}\ldots a_{nj_n}(\lambda) \qquad (\text{I.A.14})$$
$$+\ldots+ \sum_{j_1,\ldots,j_n} p(j_1,\ldots,j_n)a_{1j_1}(\lambda)\ldots a_{n-1j_{n-1}}(\lambda)\frac{da_{nj_n}}{d\lambda}(\lambda).$$

A square matrix A is called *singular* if $|A|=0$. Otherwise A is called *nonsingular*. Consider now the linear transformation

$$y_i = \sum_{k=1}^{n} a_{ik} x_k \quad (i=1,2,\ldots,n), \qquad (\text{I.A.15})$$

with $A=(a_{ik})$ being nonsingular, i.e., $|A|\neq 0$. Regarding (I.A.15) as equations for $x_1, x_2, \ldots, x_n$, and solving them in terms of $y_1, y_2, \ldots, y_n$ yields:

$$x_i = \frac{1}{|A|}\begin{vmatrix} a_{11} \cdots a_{1,i-1} & y_1 & a_{1,i+1} \cdots a_{1n} \\ a_{21} \cdots a_{2,i-1} & y_2 & a_{2,i+1} \cdots a_{2n} \\ \cdots\cdots\cdots & \cdots & \cdots\cdots\cdots \\ a_{n1} \cdots a_{n,i-1} & y_n & a_{n,i+1} \cdots a_{nn} \end{vmatrix} \equiv \sum_{k=1}^{n} a_{ik}^{(-1)} y_k \ (i=1,2,\ldots,n). \quad (\text{I.A.16})$$

The transformation represented by Eq. (I.A.16) is the "inverse" of the transformation represented by Eq. (I.A.15). The coefficient matrix

$$A^{-1} = \left(a_{ik}^{(-1)}\right) \quad \text{with} \quad a_{ik}^{(-1)} = \frac{A_{ki}}{|A|}, \quad (i,k=1,2,\ldots,n), \qquad (\text{I.A.17})$$

is called the *inverse matrix* of A, and A_{ki} is the algebraic complement (the cofactor) of the elements a_{ki} $(i,k=1,2,\ldots,n)$ in the determinant $|A|$.

The composite transformation of Eqs. (I.A.15) and (I.A.16), in either order, gives the identity transformation $AA^{-1} = A^{-1}A = I$. Note that the matrix equations $AX = I$ and $XA = I$, $|A|\neq 0$ admit only the solution $X = A^{-1}$. Note also that (I.A.16) implies

$$|A^{-1}| = \frac{1}{|A|}. \tag{I.A.18}$$

Furthermore, the inverse of the product of any two nonsingular matrices is given by $(AB)^{-1} = B^{-1}A^{-1}$. Note also the relation $(A^{-1})^T = (A^T)^{-1}$.

In many applications, it is convenient to use matrices that are partitioned into rectangular parts, or "cells" or "blocks." An arbitrary rectangular matrix $A = (a_{ik})$ $(i = 1,2,...,m; k = 1,2,...,n)$ can be partitioned into rectangular blocks by means of horizontal and vertical lines, as follows:

$$A = \begin{pmatrix} A_{11} & A_{12} & \cdots & A_{1t} \\ A_{21} & A_{22} & \cdots & A_{2t} \\ \cdots & \cdots & \cdots & \cdots \\ A_{s1} & A_{s2} & \cdots & A_{st} \end{pmatrix} \begin{matrix} \} m_1 \\ \} m_2 \\ \vdots \\ \} m_s \end{matrix}. \tag{I.A.19}$$

$$\underbrace{\phantom{A_{11}}}_{n_1} \underbrace{\phantom{A_{12}}}_{n_2} \underbrace{\phantom{A_{1t}}}_{n_t}$$

The matrix A in Eq. (I.A.19) is said to be *partitioned* into $s \times t$ blocks (or cells) $A_{\alpha\beta}$ of dimensions $m_\alpha \times n_\beta$ $(\alpha = 1,2,...,s; \beta = 1,2,...,t)$, or that it is represented in the form of a *partitioned*, or *blocked*, matrix. By analogy with the customary notation for matrices, the matrix A in Eq. (I.A.19) is written as $A = (A_{\alpha\beta})$, $(\alpha = 1,2,...,s; \beta = 1,2,...,t)$. When $s = t$, then A is a square block matrix, denoted as $A = (A_{\alpha\beta})_1^s$, or simply $A = (A_{\alpha\beta})$, when the meaning of s is clear from the context. Operations on partitioned matrices are performed using the same formal rules as used for matrices that have scalar elements instead of blocks. For example, the operation of addition of two rectangular matrices $A = (A_{\alpha\beta})$, $B = (B_{\alpha\beta})$, $(\alpha = 1,2,...,s; \beta = 1,2,...,t)$ becomes $A + B = (A_{\alpha\beta} + B_{\alpha\beta})$.

For block-multiplication of two rectangular matrices A and B, the length of the rows of the first factor A must be the same as the height of the columns of the second factor B, and the partitioning into blocks must ensure that the horizontal dimensions in the first factor are the same as the corresponding vertical dimensions in the second:

$$A = \begin{pmatrix} A_{11} & A_{12} & \cdots & A_{1t} \\ A_{21} & A_{22} & \cdots & A_{2t} \\ \cdots & \cdots & \cdots & \cdots \\ A_{s1} & A_{s2} & \cdots & A_{st} \end{pmatrix} \begin{matrix} \} m_1 \\ \} m_2 \\ \vdots \\ \} m_s \end{matrix}, \quad B = \begin{pmatrix} B_{11} & B_{12} & \cdots & B_{1u} \\ B_{21} & B_{22} & \cdots & B_{2u} \\ \cdots & \cdots & \cdots & \cdots \\ B_{t1} & B_{t2} & \cdots & B_{tu} \end{pmatrix} \begin{matrix} \} n_1 \\ \} n_2 \\ \vdots \\ \} n_t \end{matrix}.$$

The operation of block-multiplication can then be written in the customary form, namely:

$$AB = C = (C_{\alpha\beta}), \quad \text{where} \quad C_{\alpha\beta} = \sum_{\delta=1}^{t} A_{\alpha\delta} B_{\delta\beta} \quad \begin{pmatrix} \alpha = 1,2,\ldots,s \\ \beta = 1,2,\ldots,u \end{pmatrix}. \quad \text{(I.A.20)}$$

When A is a *quasi-diagonal* matrix (with $s = t$ and $A_{\alpha\beta} = 0$ when $\alpha \neq \beta$), then Eq. (I.A.20) reduces to $C_{\alpha\beta} = A_{\alpha\alpha} B_{\alpha\beta}$, $(\alpha = 1,2,\ldots,s; \beta = 1,2,\ldots,u)$. Similarly, when B is a quasi-diagonal matrix (with $t = u$ and $B_{\alpha\beta} = 0$ for $\alpha \neq \beta$), then Eq. (I.A.20) becomes $C_{\alpha\beta} = A_{\alpha\beta} B_{\beta\beta}$. The partitioned matrix in Eq. (I.A.20) is called *upper (lower) quasi-triangular* if $s = t$ and $A_{\alpha\beta} = 0$ for $\alpha > \beta$ ($\alpha < \beta$). A quasi-diagonal matrix is a special case of a quasi-triangular matrix. Note from Eq. (I.A.20) that the product of two upper (lower) quasi-triangular matrices is itself an upper (lower) quasi-triangular matrix. Furthermore, the determinant of a quasi-triangular matrix (or, in particular, a quasi-diagonal matrix) is equal to the product of the determinant of the diagonal cells: $|A| = |A_{11}||A_{22}|\cdots|A_{ss}|$.

Consider that the block A_{11} of the partitioned matrix A in Eq. (I.A.19) is square and nonsingular $(|A_{11}| \neq 0)$. The block A_{21} can be eliminated by multiplying the first row of A on the left by $-A_{\alpha 1} A_{11}^{-1}$ ($\alpha = 2,\ldots,s$) and adding the result to the α-th row of A. If the matrix $A_{22}^{(1)}$ is square and nonsingular, the process can be continued. This procedure is called Gauss elimination for block-matrices, and provides a method for computing the determinant of a partitioned matrix:

$$|A| = |B_1| = |A_{11}| \begin{vmatrix} A_{22}^{(1)} \cdots A_{2t}^{(1)} \\ \cdots\cdots\cdots\cdots \\ A_{s2}^{(1)} \cdots A_{st}^{(1)} \end{vmatrix}. \quad \text{(I.A.21)}$$

The formula represented by Eq. (I.A.21) reduces the computation of the determinant $|A|$, consisting of $s \times t$ blocks, to the computation of a determinant of lower order, consisting of $(s-1) \times (t-1)$ blocks. If $A_{22}^{(1)}$ is a square matrix and $|A_{22}^{(1)}| \neq 0$, then this determinant of $(s-1)(t-1)$ blocks can again be subjected to the same procedure as above. As a particular example, consider a determinant

Δ partitioned into four blocks, namely: $\Delta = \begin{vmatrix} A & B \\ C & D \end{vmatrix}$, where A and D are square matrices, with $|A| \neq 0$. Multiplying the first row on the left by CA^{-1} and subtracting it from the second row yields

$$\Delta = \begin{vmatrix} A & B \\ 0 & D - CA^{-1}B \end{vmatrix} = |A||D - CA^{-1}B|.$$

On the other hand, if $|D| \neq 0$, then the Gauss elimination procedure can be applied to the second row of Δ, leading to

$$\Delta = \begin{vmatrix} A - BD^{-1}C & 0 \\ C & D \end{vmatrix} = |A - BD^{-1}C||D|.$$

In the special case in which all four matrices, A, B, C, and D, are square matrices of order n, then these equations become $\Delta = |AD - ACA^{-1}B|$, when $|A| \neq 0$ or $\Delta = |AD - BD^{-1}CD|$, when $|D| \neq 0$, respectively. These results are known as the *formulas of Schur*, and they reduce the computation of a determinant of order $2n$ to the computation of a determinant of order n.

Two matrices A and B connected by the relation $B = T^{-1}AT$, where T is a nonsingular matrix, are called *similar*, and have the following three properties: *reflexivity* (a matrix A is always similar to itself); *symmetry* (if A is similar to B, then B is similar to A); and *transitivity* (if A is similar to B, and B is similar to C, then A is similar to C). Although two similar matrices have the same determinant, since $|B| = |T|^{-1}|A||T| = |A|$, the equation $|B| = |A|$ is a necessary, but not a sufficient condition for the similarity of the matrices A and B.

The vectors x for which

$$Ax = \lambda x, \quad \lambda \in F, x \neq 0 \qquad (I.A.22)$$

are called *characteristic vectors*, and the numbers λ corresponding to them are called *characteristic values* or *characteristic roots* of the matrix A. (Other terms in use for the characteristic vectors are: proper vector, latent vector, or eigenvector; other terms for the characteristic values are: proper values, latent values, latent roots, latent numbers, characteristic numbers, eigenvalues, etc.). The characteristic values of $A = (a_{ik})$ are the nonzero solutions of the system

$$\Delta(\lambda) \equiv |A - \lambda I| \equiv \begin{vmatrix} (a_{11} - \lambda)x_1 + a_{12}x_2 + \ldots + a_{1n}x_n = 0 \\ a_{21}x_1 + (a_{22} - \lambda)x_2 + \ldots + a_{2n}x_n = 0 \\ \ldots\ldots\ldots\ldots\ldots\ldots\ldots\ldots\ldots\ldots\ldots\ldots\ldots\ldots \\ a_{n1}x_1 + a_{n2}x_2 + \ldots + (a_{nn} - \lambda)x_n = 0 \end{vmatrix} = 0. \quad \text{(I.A.23)}$$

The equation above is called the *characteristic equation* or the *secular equation* of the matrix $A = (a_{ik})$; the left-side of Eq. (I.A.23) is called the *characteristic polynomial of* A. The *Hamilton-Cayley Theorem* states that every square matrix A satisfies its characteristic equation, $\Delta(A) = 0$.

In general, Eq. (I.A.22) admits three kinds of eigenvalues: (a) distinct eigenvalues with corresponding linearly independent eigenvectors, (b) multiple eigenvalues with corresponding linearly independent eigenvectors, and (c) multiple eigenvalues with linearly dependent eigenvectors. The first two kinds are called *semi-simple* eigenvalues and the last ones are called *nonsemi-simple* eigenvalues. The following figure shows all possibilities for a matrix A:

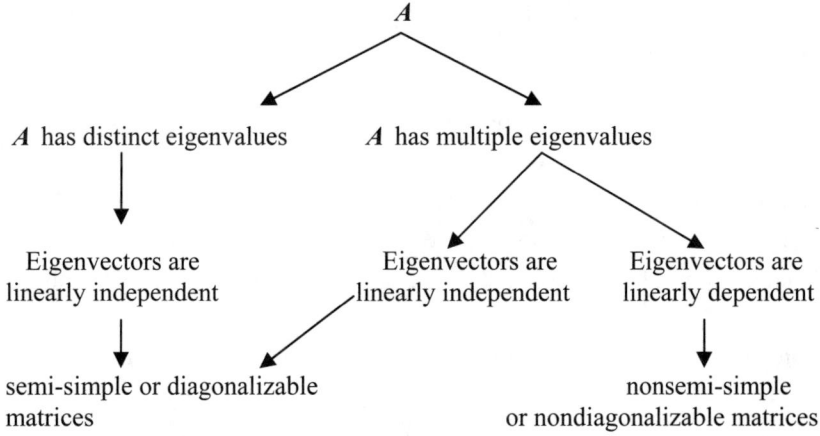

If $\lambda_1, \lambda_2, \ldots, \lambda_n$ are all the characteristic values (with the proper multiplicities) of a matrix A, and if $g(\mu)$ is a scalar polynomial, then $g(\lambda_1), g(\lambda_2), \ldots, g(\lambda_n)$ are the characteristic values of $g(A)$. In particular, if A has the characteristic values $\lambda_1, \lambda_2, \ldots, \lambda_n$, then A^k has the characteristic values $\lambda_1^k, \lambda_2^k, \ldots, \lambda_n^k$, $(k = 0, 1, 2, \ldots)$.

Under certain conditions imposed on a matrix A, it is possible to define elementary functions of A, just as one defines functions of a scalar. The matrix functions most commonly encountered in practical applications are:

$$e^A = I + A + \frac{A^2}{2!} + \ldots; \quad \sin A = A - \frac{A^3}{3!} + \frac{A^5}{5!} - \ldots; \quad \cos A = I - \frac{A^2}{2!} + \frac{A^4}{4!} - \ldots;$$

$$\sinh A = A + \frac{A^3}{3!} + \frac{A^5}{5!} + \ldots; \quad \cosh A = I + \frac{A^2}{2!} + \frac{A^4}{4!} + \ldots;$$

$$\tan A = (\cos A)^{-1} \sin A; \quad \tanh A = (\cosh A)^{-1} \sinh A;$$

$$\log(I + A) = A - \frac{A^2}{2} + \frac{A^3}{3} - \ldots; \quad (I - A)^{-1} = I + A + A^2 + \ldots.$$

A matrix A has a generalized inverse A^i, if and only if (**iff**) $AA^iA = A$, where $A^i = P \begin{bmatrix} I_\rho & U \\ V & W \end{bmatrix} R$, where: (a) ρ is the rank of A, (b) P and R are selected such that $RAP = \begin{bmatrix} I_\rho & 0 \\ 0 & 0 \end{bmatrix}$, and (c) U, V, and W are arbitrary matrices. The generalized inverse A^i is used to solve a set of linear simultaneous equations $Ax = b$, where A is of order (m,n). The solution is given by $x = A^i b$, or, equivalently,

$$x = A^t b + (I - A^t A)z, \tag{I.A.24}$$

where z is an arbitrary vector (which accounts for U, V, and W), and A^t is the part of A^i that does not depend on U, V, and W. In general, A^t is obtained from A^i by setting U, V, and W to zero, or, equivalently, by partitioning P and R in the form $P = [P_1 \;\vdots\; P_2]$, $R = \begin{bmatrix} R_1 \\ \cdots \\ R_2 \end{bmatrix}$, where P_1 and R_1 are respectively of order (n, ρ) and (ρ, m), so that $A^t = P_1 R_1$.

I.B. OPERATORS IN VECTOR SPACES: BASIC DEFINITIONS AND PROPERTIES

A nonempty set $\mathcal{V}$ is called a *vector space* if any pair of elements $f, g \in \mathcal{V}$ can be:

(i) added together by an operation called addition to give an element $f + g$ in $\mathcal{V}$, such that, for any $f, g, h \in \mathcal{V}$, the following properties hold:

$$f + g = g + f;$$
$$f + (g + h) = (f + g) + h;$$

There is a unique element 0 in $\mathcal{V}$ such that $f + 0 = f$ for all $f \in \mathcal{V}$;
For each $f \in \mathcal{V}$ there is a unique element $(-f)$ in $\mathcal{V}$ such that $f + (-f) = 0$.
 (ii) multiplied by any scalar α of a field $\mathcal{F}$ to give an element αf in $\mathcal{V}$; furthermore, for any scalars α, β, the following properties must hold:

$$\alpha(f + g) = \alpha f + \alpha g;$$
$$(\alpha + \beta)f = \alpha f + \beta f;$$
$$(\alpha\beta)f = \alpha(\beta f);$$
$$1 \times f = f.$$

The space $\mathcal{V}$ is called a complex vector space if the scalar field is $\mathbb{C}$, or a real vector space if the scalar field is $\mathbb{R}$. The members f, g, h of $\mathcal{V}$ are called points, elements, or vectors depending on the respective context.

A finite set $\mathcal{S} = \{f_j\}_1^n$ of vectors in $\mathcal{V}$ is called *linearly dependent* iff there are scalars $\alpha_1, \ldots, \alpha_n$, not all of which are zero, such that $\sum \alpha_j f_j = 0$; otherwise, $\mathcal{S}$ is called *linearly independent*. An arbitrary set $\mathcal{S}$ of vectors in $\mathcal{V}$ is linearly independent iff every finite nonempty subset of $\mathcal{S}$ is linearly independent; otherwise, the set $\mathcal{S}$ is linearly dependent. If there exists a positive integer n such that $\mathcal{V}$ contains n, but not $n+1$ linearly independent vectors, then $\mathcal{V}$ is said to be *finite dimensional* with dimension n; $\mathcal{V}$ is *infinite dimensional* iff it is not finite dimensional. The finite set $\mathcal{S}$ of vectors in an n-dimensional space $\mathcal{V}$ is called a *basis* of $\mathcal{V}$ iff $\mathcal{S}$ is linearly independent, and each element of $\mathcal{V}$ may be written as $\sum_1^n \alpha_j f_j$ for some $\alpha_1, \ldots, \alpha_n \in \mathbb{C}$ and $f_1, \ldots, f_n \in \mathcal{S}$. An infinite-dimensional space is obtained by generalizing $\mathbb{R}^n$ or $\mathbb{C}^n$, and taking infinite sequences $f = (f_n)$ as the elements of the space. This space is known as a *sequence space* and is usually denoted by ℓ.

Consider that $\mathcal{S}$ is a subset of $\mathcal{V}$, and define its complement $\mathcal{V}/\mathcal{S}$ as the set of all elements in $\mathcal{V}$ that do not belong to $\mathcal{S}$. Next, define a new subset $\overline{\mathcal{S}} \subset \mathcal{V}$, called the closure of $\mathcal{S}$, by requiring that $f \in \overline{\mathcal{S}}$ iff there is a sequence of (not necessarily distinct) points of $\mathcal{S}$ converging to f. The set $\mathcal{S}$ is called closed iff $\mathcal{S} = \overline{\mathcal{S}}$. A subset $\mathcal{S}$ of $\mathcal{V}$ is said to be open iff its complement $\mathcal{V}/\mathcal{S}$ is closed. If $\mathcal{S}_1 \subset \mathcal{S}_2 \subset \mathcal{V}$, then $\mathcal{S}_1$ is said to be *open* in

$\mathcal{S}_2$ iff it is the intersection of an open set with $\mathcal{S}_2$. A *neighborhood* of a point is any set that contains an open set that itself contains the point. A point f is an *interior point* of $\mathcal{S} \subset \mathcal{V}$ iff there is a neighborhood of f contained in $\mathcal{S}$. The *interior* $\mathcal{S}^o$ of $\mathcal{S}$ is the set of interior points of $\mathcal{S}$ (and is open). A point f is a *boundary point* of $\mathcal{S}$ iff every neighborhood of f contains points of both $\mathcal{S}$ and its complement $\mathcal{V}/\mathcal{S}$. The *boundary* $\partial \mathcal{S}$ of $\mathcal{S}$ is the set of boundary points of $\mathcal{S}$.

The *norm of a vector* $f \in \mathcal{V}$, denoted by $\|f\|$, is a nonnegative real scalar that satisfies the following relations, for all $f, g \in \mathcal{V}$:

$\|f\| = 0$ iff $f = 0$;

$\|\alpha f\| = |\alpha| \|f\|$ for any scalar α;

$\|f + g\| \le \|f\| + \|g\|$ (the triangle inequality).

A vector space $\mathcal{V}$ endowed with a norm as defined above is called a *normed vector space*; the notation $\|\bullet\|$ highlights the role of the norm as a generalization of the customary distance in $\mathbb{R}^3$.

For finite dimensional spaces, the most frequently used vector norm is the *Hölder* or ℓ_p-*norm*, defined as

$$\|x\|_p = \left(\sum_{i=1}^{n} |x_i|^p \right)^{1/p}, \quad p > 0. \tag{I.B.1}$$

The most frequently used values of the *Hölder norm* are

$$\|x\|_1 = \sum_{i=1}^{n} |x_i|, \|x\|_2 = \sqrt{\sum_{i=1}^{n} |x_i|^2}, \text{ and } \|x\|_\infty = \max_i |x_i|.$$

The *norm of a square matrix* A, denoted by $\|A\|$, is a nonnegative real scalar that satisfies the following relations:

1. $\|A\| > 0$, if $A \ne \mathbf{0}$ (nonnegative);
2. $\|\alpha A\| = |\alpha| \|A\|$, for any $\alpha \in \mathbb{C}$ and any $A \in \mathbb{R}^n$ (homogeneous);
3. $\|A + B\| \le \|A\| + \|B\|$, for any $A, B \in \mathbb{R}^n$ (triangle inequality);
4. $\|AB\| \le \|A\| \|B\|$, for any $A, B \in \mathbb{R}^n$ (sub-multiplicative).

There are many matrix norms that satisfy the above relation; the most frequently used norm is the *matrix-Hölder norm induced by (or subordinate to) the Hölder vector norm*, defined as

$$\|A\|_p \equiv \max_x \frac{\|Ax\|_p}{\|x\|_p}. \qquad (\text{I.B.2})$$

The most widely used matrix-Hölder norms are those obtained for the following values of p:

(a) $p = 1$, the *maximum column sum matrix norm*, induced by the ℓ_1 vector norm: $\|A\|_1 \equiv \max_x \frac{\|Ax\|_1}{\|x\|_1} = \max_j \sum_{i=1}^n |a_{ij}|$;

(b) $p = 2$, the *spectral norm*, induced by the ℓ_2 vector norm:

$\|A\|_2 \equiv \max_x \frac{\|Ax\|_2}{\|x\|_2} = \max_\lambda \{\sqrt{\lambda}\}$, where λ is an eigenvalue of $(A^*)A$;

(c) $p = \infty$, the *maximum row sum matrix norm*, induced by the ℓ_∞ vector norm: $\|A\|_\infty \equiv \max_x \frac{\|Ax\|_\infty}{\|x\|_\infty} = \max_i \sum_{j=1}^n |a_{ij}|$.

The *condition number* of a square matrix A is a nonnegative real scalar defined in connection with the Hölder matrix and vector norms as

$$\gamma_p(A) \equiv \max_{u,v} \frac{\|Au\|_p}{\|Av\|_p}, \quad \|u\|_p = \|v\|_p = 1. \qquad (\text{I.B.3})$$

A geometrical interpretation of the condition number $\gamma_p(A)$ can be envisaged by considering that the surface $\|x\| = 1$ is mapped by the linear transformation $y = Ax$ onto some surface S. Then, the condition number $\gamma_p(A)$ is the ratio of the largest to the smallest distances from the origin to points on S. This interpretation indicates that $\gamma_p(A) \geq 1$, when A is nonsingular, and $\gamma_p(A) = \infty$, when A is singular.

Rearranging the above definition leads to $\gamma_p(A) = \|A^{-1}\|\|A\|$, which implies that $\gamma_p(A)$ becomes large when $|A|$ is small; in such cases, the matrix A is called *ill-conditioned*. Ill-conditioning may be regarded as an approach towards singularity, a situation that causes considerable difficulties when solving linear simultaneous equations, since a small change in the coefficients of the equations causes a large displacement in the solution, leading to loss of solution accuracy due to the loss of significant figures during the computation.

In the *sequence space* ℓ_p, the quantity $\|\bullet\|_p$, defined as

$$\|f\|_p = \left\{\sum |f_n|^p\right\}^{1/p}, \quad (1 \le p < \infty),$$
$$\|f\|_\infty = \sup |f_n| \quad\quad\quad\quad\quad\quad\quad\quad (I.B.4)$$

is a norm on the subset of those f where $\|f\|_p$ is finite. Two numbers, p, q, with $1 \le p, q \le \infty$, are called *conjugate indices* iff $p^{-1} + q^{-1} = 1$; if $p = 1$, then $q = \infty$. When $1 \le p \le \infty$, and q is the conjugate index, the following inequalities hold for any $f, g \in \ell_p$ (infinite values being allowed):

$\|fg\|_1 \le \|f\|_p \|g\|_q$ (Hölder's inequality);

$\|f + g\|_p \le \|f\|_p + \|g\|_p$ (Minkowski's inequality).

A complex valued function f, defined on $\Omega \in \mathbb{R}^n$, is called *continuous at the point* $x_0 \in \Omega$ iff for each $\varepsilon > 0$ there exists $\delta > 0$ such that $|f(x) - f(x_0)| < \varepsilon$ whenever $x \in \Omega$ and $|x - x_0| < \delta$. The function f is said to be *uniformly continuous* on Ω iff for each $\varepsilon > 0$ there exists a $\delta > 0$ such that $|f(x) - f(x_0)| < \varepsilon$, whenever $x, x_0 \in \Omega$ and $|x - x_0| < \delta$.

Thus, when Ω is the finite interval $[a, b]$, continuity at a or b is to be interpreted as continuity from the right or left, respectively. If f is continuous on $[a, b]$, then it is bounded, but f need not be bounded if it is continuous only on (a, b). For functions of several variables, continuity in the sense of the above definition is sometimes referred to as "joint continuity" to distinguish it from "separate continuity;" the latter terminology means that the function is continuous in each variable in turn when the other variables are fixed. For example, if f is a function of two variables, separate continuity requires only that $f(x, \bullet)$ and $f(\bullet, y)$ should be continuous for fixed x and y, respectively. (The notation $f(x, \bullet)$ indicates that x is fixed, and the function is regarded as a function of its second argument only.) Uniform continuity is in general stronger than continuity, since the same δ must serve for every $x_0 \in \Omega$, but if $\Omega \subset \mathbb{R}^n$ is closed and bounded, then these concepts are equivalent.

The vector space of bounded continuous complex valued functions defined on $\Omega \subset \mathbb{R}^n$ is denoted by $\mathscr{C}(\Omega)$. The space $\mathscr{C}(\Omega)$ may be normed in several ways. The *sup norm* $\|\bullet\|$, defined as

$$\|f\| = \sup_{x \in \Omega} |f(x)|, \tag{I.B.5}$$

is most often used in practical applications. Generalizations of $\mathscr{C}(\Omega)$ are often used for treating differential equations. Thus, $\mathscr{C}(\Omega, \mathbb{R}^m)$ denotes the normed vector space of $\mathbb{R}^m$-valued functions equipped with the sup norm

$$\|f\| = \max_{1 \le j \le m} \sup_{x \in \Omega} |f_j(x)|. \tag{I.B.6}$$

The vector space denoted by $\mathscr{C}^k(\Omega, \mathbb{R}^m)$ consists of all $\mathbb{R}^m$-valued functions defined on Ω such that all partial derivatives up to and including those of order $k > 0$ of all components are bounded and continuous. The vector space $\mathscr{C}^\infty(\Omega, \mathbb{R}^m)$ consists of functions in $\mathscr{C}^k(\Omega, \mathbb{R}^m)$ such that $\mathscr{C}^\infty(\Omega, \mathbb{R}^m) = \bigcap_{k=1}^{\infty} \mathscr{C}^k(\Omega, \mathbb{R}^m)$. The space $\mathscr{C}^k(\overline{\Omega}, \mathbb{R}^m)$ consists of those continuous functions defined on $\overline{\Omega}$ which on Ω have bounded and *uniformly* continuous partial derivatives up to and including those of order k. (For $n > 1$, this convention avoids difficulties with the definition of derivatives on $\partial\Omega$, which is not necessarily a smooth set). It also follows that $\mathscr{C}^\infty(\overline{\Omega}, \mathbb{R}^m) = \bigcap_{k=1}^{\infty} \mathscr{C}^k(\overline{\Omega}, \mathbb{R}^m)$.

The spaces $\mathscr{C}_0^\infty(\Omega', \mathbb{R}^m)$, where $\Omega \subset \Omega' \subset \overline{\Omega}$, are used when it is convenient to exclude the boundary from consideration. These spaces consist of those functions in $\mathscr{C}^k(\Omega, \mathbb{R}^m)$ that have bounded support contained in the interior of Ω'. (Recall that the support of a function is the closure of the set on which the function is nonzero; the support may vary from function to function).

For example, for the finite interval $[a,b]$, a function is in $\mathscr{C}^1([a,b])$ iff it has a continuous derivative on (a,b) and has left and right derivatives at b and a, respectively, which are the limits of the derivatives in the interior. Another possibility is to take $\Omega = (a,b)$ and to set

$$\|f\|_{\mathscr{C}^k} = \sum_{j=0}^{k} \sup_{x \in [a,b]} \left| f^{(j)}(x) \right|, \qquad (\text{I.B.7})$$

where $f^{(j)}$ denotes the jth derivative of f. The above norm is often used as a basis for analysis in $\mathscr{C}^k(\overline{\Omega})$. Corresponding norms may be defined when Ω is a subset of $\mathbb{R}^n$ for $n > 1$ by summing over the partial derivatives.

A comparison of the concept of "closeness" for two functions f and g in the sup norm and in $\|\bullet\|_1$ shows that the sup norm bounds the difference $|f(x) - g(x)|$ for every x, whereas $\|\bullet\|_1$ only restricts the average value of this difference. Convergence in sup norm is a very strong condition and implies convergence in $\|\bullet\|_1$; the converse implication is not true.

A sequence of functions (f_n) in a normed vector space $\mathcal{V}$ is called *Cauchy* iff $\lim_{m,n \to \infty} \|f_n - f_m\| = 0$. A set $\mathcal{S}$ in a normed vector space $\mathcal{V}$ is said to be *complete* iff each Cauchy sequence in $\mathcal{S}$ converges to a point of $\mathcal{S}$. A complete normed vector space $\mathcal{V}$ is usually called a *Banach space*.

An *inner product*, denoted by $(\bullet, \bullet)$, on a normed vector space $\mathcal{V}$ is a complex (respectively, real) valued function on $\mathcal{V} \times \mathcal{V}$ such that for all $f, g, h \in \mathcal{V}$ and $\alpha \in \mathbb{C}$ (respectively, $\alpha \in \mathbb{R}$) the following properties hold:

1. $(f, f) \geq 0$; the equality $(f, f) = 0$ holds iff $f = 0$;
2. $(f, g + h) = (f, g) + (f, h)$;
3. $(f, g) = \overline{(g, f)}$, where the overbar denotes complex conjugation;
4. $(\alpha f, g) = \alpha (f, g)$.

A space $\mathcal{V}$ equipped with an inner product is called a *pre-Hilbert* or *inner product space*. If $\mathcal{V}$ is a real vector space, and the inner product is real-valued, then the respective space is called a *real pre-Hilbert space*.

A pre-Hilbert space that is *complete with respect to the norm* is called a *Hilbert space*, and is usually denoted as $\mathcal{H}$. The spaces $\mathbb{R}^n$ and $\mathbb{C}^n$ with the usual inner product are Hilbert spaces, and so is the infinite sequence space equipped with the inner product

$$(f, g) = \sum f_n \overline{g}_n, \qquad (\text{I.B.8})$$

where $f = (f_1, f_2, \ldots)$, $g = (g_1, g_2, \ldots)$.

A set $\mathcal{K}$ of vectors in $\mathcal{H}$ is said to be *complete* iff $(f,\varphi)=0$ for all $\varphi \in \mathcal{K}$ implies that $f=0$. A countable set $\mathcal{K} = \{\varphi_n\}_{n=1}^{n=\infty}$ is called *orthonormal* iff $(\varphi_n, \varphi_m) = \delta_{nm}$ for all $m, n \geq 1$. The numbers (f, φ_n) are called the *Fourier coefficients* of f (with respect to $\mathcal{K}$), and the *Fourier series* of f is the formal series $\sum_n (f, \varphi_n) \varphi_n$. An orthonormal set $\mathcal{K} = \{\varphi_n\}$ is called an *orthonormal basis* of $\mathcal{H}$ iff every $f \in \mathcal{H}$ can be represented in the *Fourier series*

$$f = \sum_n (f, \varphi_n) \varphi_n. \tag{I.B.9}$$

A sequence of vectors $x_1, x_2, \ldots$ is called *nondegenerate* if, for every p, the vectors $x_1, x_2, \ldots, x_p$ are linearly independent. A sequence of vectors is called *orthogonal* if any two vectors of the sequence are orthogonal; *orthogonalization* of a sequence of vectors is the process of replacing the sequence by an equivalent orthogonal sequence. Every nondegenerate sequence of vectors can be orthogonalized. The orthogonalization process leads to vectors that are uniquely determined to within scalar multiples.

Every finite-dimensional subspace $\mathcal{S}$ (and, in particular, the whole space $\mathbb{R}$ if it is finite-dimensional) has an orthonormal basis; in such a basis, the coordinates of a vector are equal to its projections onto the corresponding basis vectors:

$$x = \sum_{i=1}^{n} (x, e_k) e_k. \tag{I.B.10}$$

Consider that $x_1, x_2, \ldots, x_n$ and $x'_1, x'_2, \ldots, x'_n$ are the coordinates of the same vector x in two different orthonormal bases $e_1, e_2, \ldots, e_n$ and $e'_1, e'_2, \ldots, e'_n$ of a unitary space $\mathbb{R}$. The transformation between the two coordinate systems is given by

$$x_i = \sum_{k=1}^{n} u_{ik} x'_k \quad (i = 1, 2, \ldots, n), \tag{I.B.11}$$

where the coefficients $u_{1k}, u_{2k}, \ldots, u_{nk}$ satisfy the relations

$$\sum_{i=1}^{n} u_{ik}\bar{u}_{il} = \delta_{kl} = \begin{cases} 1, & \text{for } k = l, \\ 0, & \text{for } k \neq l. \end{cases} \quad \text{(I.B.12)}$$

The transformation (I.B.11), in which the coefficients satisfy the conditions (I.B.12), is called *unitary* and the corresponding matrix $U = (u_{ik})$ is called a *unitary matrix*. Thus, in an n-dimensional unitary space the transition from one orthonormal basis to another is effected by a unitary coordinate transformation. In an n-dimensional Euclidean space $\mathbb{R}$, the transition from one orthonormal basis to another is effected by an *orthogonal* coordinate transformation, and the matrix underlying an orthogonal transformation is called an *orthogonal matrix*.

For example, orthogonalization of the sequence of powers $1, x, x^2, x^3, \ldots$, using the inner product $(f, g) = \int_a^b f(x) g(x) \tau(x) dx$, where $\tau(x) \geq 0$ and $a \leq x \leq b$, yields sequences of orthogonal polynomials. For example, using $a = -1$, $b = 1$, and $\tau(x) = 1$ leads to the Legendre polynomials

$$P_o(x) = 1, \quad P_m(x) = \frac{1}{2^m m!} \frac{d^m (x^2 - 1)^m}{dx^m} \quad (m = 1, 2, \ldots).$$

On the other hand, using $a = -1$, $b = 1$, and $\tau(x) = 1/\sqrt{1 - x^2}$ leads to the Tchebyshev (Chebyshev) polynomials $T_n(x) = \cos(n \arccos x)/(2^{n-1})$. Furthermore, using $a = -\infty$, $b = +\infty$, and $\tau(x) = e^{-x^2}$ leads to the Hermite polynomials, and so on.

Consider an arbitrary vector x, an orthonormal sequence of vectors $z_1, z_2, \ldots, z_p, \ldots$, with $p = (1, 2, \ldots)$; consider also the sequence $\xi_p \equiv (x, z_p)$. If the limit

$$\lim_{p \to \infty} \left(\|x\| - \sum_{k=1}^{p} |\xi_k|^2 \right)^{1/2} = 0 \quad \text{(I.B.13)}$$

holds, then the series $\sum_{k=1}^{\infty} \xi_k z_k$ is said to converge in the mean (or with respect to the norm) to the vector x, and the norm of x is given by:

$$\|x\| = |x|^2 = \sum_{k=1}^{\infty} |\xi_k|^2. \quad \text{(I.B.14)}$$

If, for every vector x of $\mathbb{R}$, the series $\sum_{k=1}^{\infty} \xi_k z_k$ converges in the mean to x, then the orthonormal sequence of vectors $z_1, z_2, \ldots, z_p, \ldots$ is complete. In terms of such a complete orthonormal sequence z_k, the scalar product of two vectors x and y in $\mathbb{R}$ becomes

$$(x, y) = \sum_{k=1}^{\infty} \xi_k \overline{\eta_k} \quad \left(\xi_k = \langle x, z_k \rangle, \eta_k = \langle y, z_k \rangle, k = 1, 2, \ldots \right). \quad \text{(I.B.15)}$$

As a further example, consider the space of all complex functions $f(t)$, where t is a real variable, that are piecewise continuous in the closed interval $[0, 2\pi]$, with the norm and the scalar product defined, respectively, as $\|f\| = \int_0^{2\pi} |f(t)|^2 dt$, and $(f, g) = \int_0^{2\pi} f(t) \overline{g(t)} dt$. Consider now the infinite sequence of orthogonal and complete functions $e^{ikt}/\sqrt{2\pi}$, $(k = 0, \pm 1, \pm 2, \ldots)$. Then, the *Fourier series* of $f(t)$, namely $\sum_{k=-\infty}^{\infty} f_k e^{ikt}$, with *Fourier coefficients* $f_k = \left(\int_0^{2\pi} f(t) e^{-ikt} dt \right) / 2\pi$, $(k = 0, \pm 1, \pm 2, \ldots)$, converges in the mean to $f(t)$ in the interval $[0, 2\pi]$. The condition of completeness for the function f is called *Parseval's equality*:

$$\int_0^{2\pi} f(t) \overline{g(t)} dt = \sum_{k=-\infty}^{+\infty} \frac{1}{2\pi} \int_0^{2\pi} f(t) e^{-ikt} dt \int_0^{2\pi} \overline{g(t)} e^{ikt} dt.$$

If $f(t)$ is a real function, then f_0 is real, and f_k and f_{-k} are conjugate complex numbers. Therefore, for a real function $f(t)$, the Fourier series becomes $f(t) = \frac{a_0}{2} + \sum_{k=1}^{\infty} (a_k \cos kt + b_k \sin kt)$, where $a_k = \left(\int_0^{2\pi} f(t) \cos kt \right) / \pi$ and $b_k = \left(\int_0^{2\pi} f(t) \sin kt \right) / \pi$, for $k = 0, 1, 2, \ldots$.

The space $\mathcal{L}_p$ (or $\mathcal{L}_p(X)$, if X needs emphasis) is defined to be the set of measurable functions f such that $\|f\|_p < \infty$, where

$$\|f\|_p = \left\{ \int |f|^p \, d\mu \right\}^{1/p}, \quad (1 \le p < \infty),$$
$$\|f\|_\infty = \text{ess sup } |f|.$$
(I.B.16)

Frequently, X will be an interval of the real line with end points a and b, while μ will be the Lebesgue measure; in this case, the space will be denoted by $\mathscr{L}_p(a,b)$. Two functions f and g in $\mathscr{L}_p(X)$ are said to be equal iff $f = g$ almost everywhere. Thus, all functions in $\mathscr{L}_p(X)$ are finite almost everywhere. When $p = \infty$, then $\|f\|_\infty = \sup |f(x)|$, for f continuous and μ representing the Lebesgue measure. Thus, the continuous bounded functions with the sup norm form a closed subspace of $\mathscr{L}_\infty$. If a function f in $\mathscr{L}_\infty$ is not continuous, the measure of the set for which $|f(x)| > \|f\|_\infty$ is zero. The space $\mathscr{L}_p$ may be thought of as the continuous analogue of the sequence space, with integration replacing summation; for $p \ge 1$, $\mathscr{L}_p$ is a Banach space under the norm $\|\bullet\|_p$. Furthermore, $\mathscr{L}_p$ is separable if $1 \le p < \infty$; in particular, $\mathscr{L}_2$ is a Hilbert space when equipped with the inner product $(f, g) = \int f \bar{g} d\mu$.

The set $\mathscr{D}(A)$ (sometimes denoted just by $\mathscr{D}$ if there is only one mapping under consideration) is called the *domain* of a mapping A. For $f \in \mathscr{D}(A)$, the element Af is called the *image* of f. Likewise, the image $A(\mathscr{S})$ of a set $\mathscr{S} \subset \mathscr{D}(A)$ is the set of the images of all the elements of $\mathscr{S}$. In particular, the image of $\mathscr{D}(A)$ is called the *range* of A and is written as $\mathscr{R}(A)$. The *preimage* of a set $\mathscr{S}_1 \subset \mathscr{W}$ is the set $A^{-1}(\mathscr{S}_1) = \{f : f \in \mathscr{D}(A), Af \in \mathscr{S}_1\}$.

The mapping A is called an *operator* or a *function* from $\mathscr{V}$ into $\mathscr{W}$. The notation $A: \mathscr{S} \to \mathscr{W}$ indicates that A is an operator with domain $\mathscr{S}$ and range in $\mathscr{W}$, or, equivalently, that A maps $\mathscr{S}$ into $\mathscr{W}$. Note that an operator is always single-valued, in that it assigns exactly one element of its range to each element in its domain. Furthermore, although there is no strict distinction between "operator" and "function," it is customary to reserve "function" for the case when $\mathscr{V}$ and $\mathscr{W}$ are finite dimensional and to use "operator" otherwise. In view of its importance, one particular type of operator is given a name of its own: the operator from $\mathscr{V}$ into the field $\mathscr{F}$ of scalars (real or complex) is called a *functional*.

An operator A from $\mathscr{V}$ into $\mathscr{W}$ is called *injective* iff for each $g \in \mathscr{R}(A)$, there is exactly one $f \in \mathscr{D}(A)$ such that $Af = g$; A is called *surjective* iff $\mathscr{R}(A) = \mathscr{W}$, and A is called *bijective* iff it is both injective and surjective. The terms "one-to-one," "onto," "one-to-one and onto," respectively, are common alternatives in the literature. An operator A from $\mathcal{V}$ into $\mathcal{W}$ is *continuous at the point* $f_0 = \mathcal{D}(A)$ iff for each $\varepsilon > 0$, there is a $\delta > 0$ such that $\|Af - Af_0\| < \varepsilon$ if $f \in \mathcal{D}(A)$ and $\|f - f_0\| < \delta$; A is said to be *continuous* iff it is continuous at every point of $\mathcal{D}(A)$.

An operator L from $\mathcal{V}$ into $\mathcal{W}$ with domain $\mathcal{D}(L)$ is called *linear* iff $L(\alpha f + \beta g) = \alpha Lf + \beta Lg$ for all $\alpha, \beta \in \mathbb{C}$ (or $\alpha, \beta \in \mathbb{R}$, if $\mathcal{V}$ and $\mathcal{W}$ are real spaces), and all $f, g \in \mathcal{D}(L)$. A linear operator is the vector space analogue of a function in one dimension represented by a straight line through the origin, that is, a function $\varphi : \mathbb{R} \to \mathbb{R}$ where $\varphi(x) = \lambda x$ for some $\lambda \in \mathbb{R}$. In particular, the *identity* operator, denoted by I, is the operator from $\mathcal{V}$ onto itself such that $If = f$ for all $f \in \mathbb{R}$.

A wide variety of equations may be written in the form $Lf = g$, with L a linear operator. For example, the simultaneous algebraic equations

$$\sum_{j=1}^{n} \alpha_{ij} f_j = g_i, \quad (i = 1, \ldots, m),$$

define the operator L via the relation

$$(Lf)_i = \sum_{j=1}^{n} \alpha_{ij} f_j, \quad (i = 1, \ldots, m).$$

Then $L : \mathbb{C}^n \to \mathbb{C}^m$ is a linear operator and the above equation can be written as $Lf = g$. Conversely, every linear operator $\mathbb{C}^n \to \mathbb{C}^m$ may be expressed in the above form by choosing bases for $\mathbb{C}^n$ and $\mathbb{C}^m$. The above equations may also be put in matrix form, but note that there is a distinction between the matrix, which depends on the bases chosen, and the operator L, which does not.

As a further example, the Fredholm integral equation $f(x) - \int_0^1 k(x,y) f(y) dy = g(x)$, $0 \le x \le 1$, where k and g are given and f is the unknown function, can also be written in operator form as $f - Kf = g$, where K is a linear operator. Similarly, the differential equation $a_0 f''(x) + a_1 f'(x) + a_2 f(x) = g(x)$, $0 \le x \le 1$, where $a_0, a_1, a_2 \in \mathbb{C}$, and g is a given continuous function, can be written in the form $Lf = g$, with $L : \mathbb{C}^2([0,1]) \to \mathbb{C}([0,1])$ being a linear operator.

The operator equation $Lf = g$, where L is a linear operator from $\mathcal{V}$ into $\mathcal{W}$ with domain $\mathcal{D}(L)$, may or may not have a solution f for every $g \in \mathcal{R}(L)$, depending on the following possibilities:

(i) *L is not injective,* in which case a reasonable interpretation of the inverse operator L^{-1} is not possible. The equation $Lf = g$ always has more than one solution if $g \in \mathcal{R}(L)$.

(ii) *L is injective but not surjective.* In this case, the equation $Lf = g$ has exactly one solution if $g \in \mathcal{R}(L)$, but no solution otherwise. The inverse L^{-1} is the operator with domain $\mathcal{R}(L)$ and range $\mathcal{D}(L)$ defined by $f = L^{-1}g$. The set $\mathcal{N}(L) \subset \mathcal{D}(L)$ of solutions of the equation $Lf = 0$ is called the *null space* of L. Note that $\mathcal{N}(L)$ is a linear subspace, and $\mathcal{N}(L) = 0$ iff L is injective.

(iii) *L is bijective.* In this case, L^{-1} is a linear operator with domain $\mathcal{W}$, and $Lf = g$ has exactly one solution for each $g \in \mathcal{W}$.

If a linear operator L from $\mathcal{V}$ into $\mathcal{W}$ is continuous at some point $f \in \mathcal{D}(L)$, then L is continuous everywhere. A linear operator L is *bounded* on $\mathcal{D}(L)$ iff there is a finite number m such that $\|Lf\| \leq m\|f\|$, with $f \in \mathcal{D}(L)$. If L is not bounded on $\mathcal{D}(L)$, it is said to be *unbounded*. The infimum of all constants m such that this inequality holds is denoted by $\|L\|$, and is called the operator norm of L. Note that $\|Lf\| \leq \|L\|\|f\|$; this relationship may be compared to the relation $|\varphi(x)| = |\lambda x| = |\lambda||x|$ for a linear operator $\varphi(x) = \lambda x$, $\varphi: \mathbb{R} \to \mathbb{R}$. Since $|\lambda|$ is a measure of the gradient of φ, the norm of the operator L may therefore be thought of as its maximum gradient.

Suppose that L is a (possibly unbounded) linear operator from a Banach space $\mathcal{B}$ into $\mathcal{B}$. The set $\rho(L)$ of complex numbers for which $(\lambda I - L)^{-1}$ belongs to the space of linear operators on $\mathcal{B}$ is called the *resolvent set* of L. For $\lambda \in \rho(L)$, the operator $R(\lambda; L) \equiv (\lambda I - L)^{-1}$ is known as the *resolvent* of L. The complement $\sigma(L)$ in $\mathcal{C}$ of $\rho(L)$ is the *spectrum* of L. A complex number λ is called an *eigenvalue (characteristic value)* of L iff the equation $\lambda f - Lf = 0$ has a nonzero solution. The corresponding nonzero solutions are called *eigenfunctions (characteristic functions),* and the linear subspace spanned by these is called the *eigenspace (characteristic space)* corresponding to λ. The set $\sigma_p(L)$ of eigenvalues is known as the *point spectrum* of L. The set consisting of those $\lambda \in \sigma(L)$ for which $(\lambda I - L)$ is injective and $R(\lambda I - L)$ is dense (respectively, not dense) in $\mathcal{B}$ is called the *continuous spectrum* (respectively, the *residual spectrum*). Thus, $\lambda \in \rho(L)$ iff $\lambda I - L$ is bijective; in this case $\sigma(L)$

is the union of the point, continuous and residual spectra, which are disjoint sets. If $\mathscr{B}$ is finite dimensional, then $\sigma(L) = \sigma_p(L)$.

The operator

$$l = \sum_{r=0}^{n} p_r(x)\left(\frac{d}{dx}\right)^r, \tag{I.B.17}$$

where p_r $(r = 0, 1, \ldots, n)$ are given functions on $\mathbb{R}$, is called a *formal ordinary differential operator* of order n. In a Hilbert space, it is convenient to refer loosely to any operator L obtained from l by setting $Lf = lf(x) = \sum_{r=0}^{n} p_r(x) f^{(r)}(x)$, for f in some specified domain, as a *differential operator*. For a general *partial differential equation* in n dimensions, the notational complexity can be reduced considerably by the use of multi-indices, which are defined as follows: a *multi-index* α is an n-tuple $(\alpha_1, \ldots, \alpha_n)$ of nonnegative integers. It is also convenient to use the notation $|\alpha| = \alpha_1 + \ldots + \alpha_n$ for a multi-index; even though this notation conflicts with the notation for the Euclidean distance $\mathbb{R}^n$, the meaning will always be clear from the context.

In the following, multi-indices will be denoted by α and β, and a point in $\mathbb{R}^n$ will be denoted as $x = (x_1, \ldots, x_n)$, with $|x|^2 = \sum x_j^2$ and $x^\alpha = x_1^\alpha \ldots x_n^\alpha$. The notation used for derivatives is $D_j = \partial/\partial x_j$ and $D^\alpha = D_1^{\alpha_1} \ldots D_n^{\alpha_n}$. Consider that $p_{\alpha\beta} \neq 0$ are complex-valued variable coefficients such that $p_{\alpha\beta} \in \mathscr{C}^\infty(\Omega)$, for multi-indices α and β, with $|\alpha| = |\beta| = m$. Consider, in addition, a function $\phi \in \mathscr{C}^{2m}$; then, the *formal partial differential operator* l of order $2m$ is defined as

$$l\phi \equiv \sum_{|\alpha|,|\beta| \leq m} (-1)^{|\alpha|} D^\alpha \left(p_{\alpha\beta} D^\beta \phi\right). \tag{I.B.18}$$

The operator l_P, defined as

$$l_P \phi \equiv (-1)^m \sum_{|\alpha| = |\beta| = m} D^\alpha \left(p_{\alpha\beta} D^\beta \phi\right), \tag{I.B.19}$$

is called the *principal part* of the formal partial differential operator l. Furthermore, the operator l^+ defined as

$$l^+\phi \equiv \sum_{|\alpha|,|\beta|\leq m}(-1)^{|\alpha|}D^\alpha\left(\overline{p}_{\beta\alpha}D^\beta\phi\right) \quad \left(\phi \in \mathscr{C}^{2m}\right) \tag{I.B.20}$$

is called the *formal adjoint* of l. Iff $l = l^+$, then l is called *formally self-adjoint*.

Define $\mathscr{L}_p^{loc}$ to be the set of all functions that lie in $\mathscr{L}_p(\mathscr{S})$ for every set $\mathscr{S}$ bounded and closed in $\mathbb{R}^n$. Then, a function f in $\mathscr{L}_2^{loc}$ has a α-th *weak derivative* iff there exists a function $g \in \mathscr{L}_0^{loc}$ such that

$$\int_\Omega g\phi \, dx = (-1)^{|\alpha|}\int_\Omega f \cdot \left(D^\alpha \phi\right) dx \tag{I.B.21}$$

for all $\phi \in \mathscr{C}_0^\infty$; the function g is called the α-th *weak derivative* of f, and we write $D^\alpha f = g$. Weak derivatives are unique in the context of $\mathscr{L}_2$ spaces, in the sense that if g_1 and g_2 are both weak α-th derivatives of f, then

$$\int_\Omega (g_1 - g_2)\phi \, dx = 0 \quad \left(\phi \in \mathscr{C}_0^\infty\right),$$

which implies that $g_1 = g_2$ almost everywhere in Ω. The above relation also implies that if a function has an α-th derivative g in the ordinary sense in $\mathscr{L}_2^{loc}$, then g is the weak α-th derivative of f. The weak derivatives may be thought of as averaging out the discontinuities in f. Consequently, it is permissible to exchange the order of differentiation: $D_iD_jf = D_jD_if$. In one dimension, a function has a weak first derivative iff it is absolutely continuous and has a first derivative in $\mathscr{L}_2^{loc}$. However, note that f may have an ordinary derivative almost everywhere without having a weak derivative. For example, if $f(x)=1$ for $x > 0$ and $f(x) = 0$ for $x < 0$, then $\int_{-1}^1 f \phi' \, dx = \int_0^1 \phi' \, dx = -\phi(0)$, but f does not have a weak derivative since there is no $g \in \mathscr{L}_0^{loc}$ such that $\phi(0) = \int_{-1}^1 g\phi \, dx$ for all $\phi \in \mathscr{C}_0^\infty$.

The analysis of differential operators is carried out in special Hilbert spaces, called *Sobolev spaces*. The Sobolev space $\mathscr{H}^m$ [or $\mathscr{H}^m(\Omega)$ if the domain requires emphasis] of order m, where m is a nonnegative integer, consists of the set of functions f such that for $0 \leq |\alpha| \leq m$, all the weak derivatives $D^\alpha f$

exist and are in $\mathscr{L}_2$; furthermore, $\mathscr{H}^m$ is equipped with an inner product and a norm defined, respectively, as:

$$(f,g)_m = \sum_{|\alpha|\leq m} \int_\Omega D^\alpha f \cdot \overline{D^\alpha g}\, dx, \qquad (\text{I.B.22})$$

$$\|f\|_m^2 = (f,f)_m = \sum_{|\alpha|\leq m} \int_\Omega |D^\alpha f|^2\, dx. \qquad (\text{I.B.23})$$

The above norm may be regarded as measuring the average value of the weak derivatives. Note that $\mathscr{H}^m$ is a proper subset of the set of functions with m-th weak derivatives, since the derivatives $D^\alpha f$ are required to be in $\mathscr{L}_2$, and not merely in $\mathscr{L}_2^{loc}$. The closure of $\mathscr{C}^\infty$ in $\|\cdot\|_m$ is $\mathscr{H}^m$, and the closure in $\mathscr{H}^m$ of $\mathscr{C}_0^\infty$ is the *Sobolev space* $\mathscr{H}_0^m$, of order m. The following chains of inclusions hold for the higher order Sobolev spaces:

$$\mathscr{C}^\infty \subset \ldots \subset \mathscr{H}^{m+1} \subset \mathscr{H}^m \subset \ldots \subset \mathscr{H}^0 = \mathscr{L}_2, \qquad (\text{I.B.24})$$

$$\mathscr{C}_0^\infty \subset \ldots \subset \mathscr{H}_0^{m+1} \subset \mathscr{H}_0^m \subset \ldots \subset \mathscr{H}_0^0 = \mathscr{L}_2. \qquad (\text{I.B.25})$$

The *bilinear form*, $B[f,\phi]$, associated with the formal differential operator l is defined for all functions $f, \phi \in \mathscr{H}_0^m$ as

$$B[f,\phi] = \sum_{|\alpha|,|\beta|\leq m} (p_{\alpha\beta} D^\alpha f, D^\beta \phi)_0. \qquad (\text{I.B.26})$$

Note that $B[f,\phi]$ is a bounded bilinear form on $\mathscr{H}_0^m \times \mathscr{H}_0^m$. The problem of finding $f \in \mathscr{H}_0^m$ such that $B[f,\phi] = (g,\phi)_0$ for all $\phi \in \mathscr{H}_0^m$ and $g \in \mathscr{L}_2$ is called the *generalized Dirichlet problem*.

I.C. ADJOINT OPERATORS: BASIC DEFINITIONS AND PROPERTIES

The space of continuous linear functionals on a Banach space $\mathscr{B}$ is called the *dual* of $\mathscr{B}$, and is denoted here by $\mathscr{B}^+$. For a bounded linear operator $L: \mathscr{B} \to \mathscr{C}$, the relation

$$g^+(Lf) = L^+ g^+(f), \qquad (\text{I.C.1})$$

required to hold for all $f \in \mathscr{B}$ and all $g^+ \in \mathscr{C}^+$, defines an operator L^+ from $\mathscr{C}^+$ into $\mathscr{B}^+$, called the *adjoint* of L. For example, the adjoint of an integral operator of the form $Kf(x) = \int_0^1 k(x,y) f(y) dy$, for a continuous kernel $k:[0,1] \times [0,1] \to \mathbb{R}$, is the operator $K^+ : \mathscr{L}_\infty(0,1) \to \mathscr{L}_\infty(0,1)$ defined as $K^+ g^+(y) = \int_0^1 k(x,y) g^+(x) dx$.

As the next example, consider the finite dimensional operator $L : \ell_1^{(n)} \to \ell_1^{(n)}$ corresponding to the matrix (α_{ij}), namely

$$(Lf)_i = \sum_{j=1}^n \alpha_{ij} f_j, \qquad (i=1,\ldots,n).$$

The corresponding adjoint operator, $L^+ : \ell_\infty^{(n)} \to \ell_\infty^{(n)}$, is represented by the transposed matrix $(\alpha_{ij})^T$, as can be seen by applying the definition of the adjoint operator:

$$g^+(Lf) = \sum_{i=1}^n (Lf)_i g_i^+ = \sum_{i=1}^n g_i^+ \sum_{j=1}^n \alpha_{ij} f_j = \sum_{j=1}^n f_j \sum_{i=1}^n \alpha_{ij} g_i^+ = (L^+ g^+)(f).$$

Thus, even in finite dimensions, the adjoint operator depends on the norm of the space and has more significance than the algebraic "transpose." For example, the relation $\|L\| = \|L^+\|$ will only hold if the dual space is correctly selected.

In an *n*-dimensional unitary space $\mathbb{R}$, the linear operator A^+ is called *adjoint* to the operator A iff, for any two vectors $\mathbf{x}$, $\mathbf{y}$ of $\mathbb{R}$, the following relationship holds

$$(A\mathbf{x}, \mathbf{y}) = (\mathbf{x}, A^+ \mathbf{y}). \qquad (\text{I.C.2})$$

In an orthonormal basis $e_1, e_2, \ldots, e_n$ in $\mathbb{R}$, the adjoint operator A^+ can be represented *uniquely* in the form

$$A^+ \mathbf{y} = \sum_{k=1}^n \langle \mathbf{y}, A e_k \rangle e_k, \qquad (\text{I.C.3})$$

for any vector y of $\mathbb{R}$. Consider that A is a linear operator in a unitary space and that $A = \|a_{ik}\|_1^n$ is the corresponding matrix that represents A in an orthonormal basis $e_1, e_2, \ldots, e_n$. Then, the matrix A^+ corresponding to the representation of the adjoint operator A^+ in the same basis is the complex conjugate of the transpose of A, i.e.,

$$A^+ = \overline{A}^T. \qquad (\text{I.C.4})$$

The matrix A^+ given by (I.C.4) is called the *adjoint* of A. Thus, *in an orthonormal basis, adjoint matrices correspond to adjoint operators.*

The adjoint operator has the following properties:

1. $(A^+)^+ = A$,
2. $(A+B)^+ = A^+ + B^+$,
3. $(\alpha A)^+ = \overline{\alpha} A^+$ (α a scalar),
4. $(AB)^+ = B^+ A^+$.
5. If an arbitrary subspace $\mathcal{S}$ of $\mathbb{R}$ is invariant with respect to A, then the orthogonal complement $\mathcal{E}$ of the subspace $\mathcal{S}$ is invariant with respect to A^+.

Two systems of vectors $x_1, x_2, \ldots, x_n$ and $y_1, y_2, \ldots, y_n$ are by definition *bi-orthogonal* if

$$\langle x_i, y_k \rangle = \delta_{ik}, \quad (i, k = 1, 2, \ldots, m), \qquad (\text{I.C.5})$$

where δ_{ik} is the Kronecker symbol.

If A is a linear operator of simple structure, then the adjoint operator A^+ is also of simple structure; therefore, complete systems of characteristic vectors $x_1, x_2, \ldots, x_n$ and $y_1, y_2, \ldots, y_n$ of A and A^+, respectively, can be chosen such that they are bi-orthogonal:

$$A x_i = \lambda_i x_i, \qquad A^+ y_i = \overline{\lambda_i} y_i, \qquad \langle x_i, y_k \rangle = \delta_{ik}, \quad (i, k = 1, 2, \ldots, n). \; (\text{I.C.6})$$

Furthermore, if the operators A and A^+ have a common characteristic vector, then the corresponding characteristic values are complex conjugates. A linear operator A is called *normal* if it commutes with its adjoint:

$$AA^+ = A^+A. \tag{I.C.7}$$

A linear operator H is called *hermitian* if it is equal to its adjoint:

$$H^+ = H. \tag{I.C.8}$$

A linear operator U is called *unitary* if it is inverse to its adjoint:

$$UU^+ = I. \tag{I.C.9}$$

From the above definitions, it follows that (i) the product of two unitary operators is itself a unitary operator, (ii) the unit operator I is unitary, and (iii) the inverse of a unitary operator is also unitary. Therefore the set of all unitary operators forms a group, called the *unitary group*. Note that hermitian and unitary operators are special cases of a normal operator. Note also that a hermitian operator H is called *positive semidefinite* if the inequality $\langle Hx, x \rangle \geq 0$ holds for every vector x of $\mathbb{R}$; H is called *positive definite* if the *strict* inequality $\langle Hx, x \rangle > 0$ holds for every vector $x \neq \mathbf{0}$ of $\mathbb{R}$.

Just as in the case of operators, a matrix is called *normal* if it commutes with its adjoint, *hermitian* if it is equal to its adjoint, and *unitary* if it is inverse to its adjoint. Thus, in an orthonormal basis, a normal (hermitian, unitary) operator corresponds to a normal (hermitian, unitary) matrix. Note that every characteristic vector of a normal operator A is a characteristic vector of the adjoint operator A^+, i.e., if A is a normal operator, then A and A^+ have the same characteristic vectors. Furthermore, a linear operator is normal if and only if it has a complete orthonormal system of characteristic vectors. If A is a normal operator, then each of the operators A and A^+ can be represented as a polynomial in the other; these two polynomials are determined by the characteristic values of A.

Hermitian and unitary operators can also be characterized in terms of their respective spectra. Thus, a linear operator is hermitian iff it has a complete orthonormal system of characteristic vectors with real characteristic values. Furthermore, a linear operator is unitary iff it has a complete orthonormal system of characteristic vectors with characteristic values of modulus 1. Consequently, a matrix A is normal iff it is unitarily similar to a diagonal matrix,

$$A = U \|\lambda_i \delta_{ik}\|_1^n U^{-1}, \quad U^+ = U^{-1}. \tag{I.C.10}$$

A matrix $\boldsymbol{H}$ is hermitian iff it is unitarily similar to a diagonal matrix with real diagonal elements:

$$H = U\|\lambda_i \delta_{ik}\|_1^n U^{-1}, \quad U^+ = U^{-1}, \quad \lambda_i = \overline{\lambda}_i, \quad (i=1,2,\ldots,n). \qquad \text{(I.C.11)}$$

Finally, a matrix U is unitary iff it is unitarily similar to a diagonal matrix with diagonal elements of modulus 1:

$$U = U_1\|\lambda_i \delta_{ik}\|_1^n U_1^{-1}, \quad U_1^+ = U_1^{-1}, \quad |\lambda_i| = 1, \quad (i=1,2,\ldots,n). \qquad \text{(I.C.12)}$$

Note also that a hermitian operator is positive semidefinite (positive definite) iff all its characteristic values are nonnegative (positive).

In a Hilbert space $\mathcal{H}$, Eq. (I.C.1) becomes

$$(Lf, g) = (f, L^+ g), \qquad \text{(I.C.13)}$$

for all f and $g \in \mathcal{H}$; thus, Eq. (I.C.13) defines the bounded linear operator L^+, called the (*Hilbert space*) *adjoint* of L. The *Riesz Representation Theorem* ensures that for every element g^+ of the dual $\mathcal{H}^+$ of a Hilbert space $\mathcal{H}$, there is a unique element g of $\mathcal{H}$ such that $g^+(f) = (f, g)$ for all $f \in \mathcal{H}$. The equality $\|g^+\| = \|g\|$ also holds.

The class of bounded operators from a Hilbert space $\mathcal{H}$ into itself is particularly important. Thus, if L is a linear operator on the Hilbert space $\mathcal{H}$, then L is said to be *self-adjoint* iff $L = L^+$. For example, for $\sum\sum |\alpha_{ij}|^2 < \infty$, the operator $L: \ell_2 \to \ell_2$ defined as $(Lf)_i = \sum_{j=1}^\infty \alpha_{ij} f_j$ is self-adjoint iff $\alpha_{ij} = \overline{\alpha}_{ji}$, $(i,j=1,2,\ldots)$, that is, iff the infinite matrix (α_{ij}) is *hermitian*, since $(L^+ g)_j = \sum_{i=1}^\infty \overline{\alpha}_{ij} g_i$. The adjoint $L^+ : \ell_2 \to \ell_2$ is therefore represented by the conjugate transpose of the infinite matrix (α_{ij}). As another example, the integral operator $Kf(x) = \int_0^1 k(x,y) f(y) dy$, $K: \mathcal{L}_2(0,1) \to \mathcal{L}_2(0,1)$ is bounded and is self-adjoint iff the kernel is, in the complex case, *hermitian*: $k(x,y) = \overline{k(y,x)}$, or, in the real case, *symmetric*: $k(x,y) = k(y,x)$.

The existence of solutions for operator equations involving linear compact operators is elucidated by the *Fredholm Alternative Theorem* (in a Hilbert space $\mathcal{H}$), which can be formulated as follows: consider that $L: \mathcal{H} \to \mathcal{H}$ is a linear compact operator, and consider the equation

$$(\lambda I - L) = g, \quad g \in \mathcal{H}, \quad \lambda \neq 0. \qquad \text{(I.C.14)}$$

Then, one of the following alternatives hold:

(a) The homogeneous equation has only the zero solution; in this case, $\lambda \in \rho(L)$, where $\rho(L)$ denotes the *resolvent set* of L (thus, λ cannot be an eigenvalue of L); furthermore, $(\lambda I - L)^{-1}$ is bounded, and the inhomogeneous equation has exactly one solution $f = (\lambda I - L)^{-1} g$, for each $g \in \mathcal{H}$

(b) The homogeneous equation has a nonzero solution; in this case, the inhomogeneous equation has a solution, necessarily nonunique, iff $\langle g, \varphi^+ \rangle = 0$, for every solution φ^+ of the adjoint equation $\lambda \varphi^+ = L^+ \varphi^+$, where L^+ denotes the operator adjoint to L, and $\langle , \rangle$ denotes the inner product in the respective Hilbert space $\mathcal{H}$.

Consider now that L denotes an *unbounded* linear operator from $\mathcal{H}$ into $\mathcal{H}$, with domain $\mathcal{D}(L)$ dense in $\mathcal{H}$. Recall that the specification of a domain is an essential part of the definition of an unbounded operator. Define $\mathcal{D}(L^+)$ to be the set of elements g such that there is an h with $(Lf, g) = (f, h)$ for all $f \in \mathcal{D}(L)$. Let L^+ be the operator with domain $\mathcal{D}(L^+)$ and with $L^+ g = h$ on $\mathcal{D}(L^+)$, or, equivalently, consider that L^+ satisfies the relation

$$(Lf, g) = (f, L^+ g), \qquad f \in \mathcal{D}(L), \quad g \in \mathcal{D}(L^+). \tag{I.C.15}$$

Then the operator L^+ is called the *adjoint* of L. Furthermore, a densely defined linear operator L from a Hilbert space into itself is called *self-adjoint* iff $L = L^+$. Note that necessarily $\mathcal{D}(L) = \mathcal{D}(L^+)$ for self-adjoint operators.

It is important to note that the operator theoretic concept of "adjoint" defined in Eq. (I.C.15) involves the boundary conditions in an essential manner, and is therefore a more comprehensive concept than the *formal adjoint operator* (defined in Eq. I.B.20), which merely described the coefficients of a certain differential operator. To illustrate the difference between the (operator theoretic) adjoint of an unbounded operator and the underlying formal adjoint operator, consider the formal differential operator $l = id/dx$ on the interval $[0,1]$, and consider the operator $Lf = if$. Furthermore, denote by $\mathcal{A}$ the linear subspace of $\mathcal{H} = \mathcal{L}_2(0, 1)$ consisting of absolutely continuous functions with derivatives in $\mathcal{L}_2(0, 1)$. Suppose first that no boundary conditions are specified for L, so that $\mathcal{D}(L) = \mathcal{A}$; in this case, an application of Eq. (I.C.15) shows that the proper adjoint boundary conditions are $\mathcal{D}(L^+) = \{g : g \in \mathcal{A}, g(0) = g(1) = 0\}$. On the other hand, suppose that the boundary conditions imposed on L are

$\mathcal{D}(L) = \{f : f \in \mathcal{A}, f(0) = f(1) = 0\}$; in this case, an application of Eq. (I.C.15) shows that the proper adjoint boundary conditions are $\mathcal{D}(L^+) = \mathcal{A}$. Note that L is not equal to L^+, in either of these two examples, although l is "self-adjoint" in the sense of classical differential equation theory. To construct a self-adjoint operator from the formal operator $l = id/dx$ on $[0,1]$, it is necessary to choose boundary conditions so as to ensure that $\mathcal{D}(L) = \mathcal{D}(L^+)$. Applying again Eq. (I.C.15) yields the proper domain as $\mathcal{D}(L) = \{f : f \in \mathcal{A}, f(1) = e^{i\theta} f(0), \theta \in \mathbb{R}\}$, which means that there are infinitely many self-adjoint operators based on the formal operator id/dx on the interval $[0,1]$.

I.D. ELEMENTS OF DIFFERENTIAL CALCULUS IN NORMED SPACES

Consider that I is an open interval of the real line and $\mathcal{V}$ is a normed real space. The *derivative* of a mapping (or operator) $\phi : I \to \mathcal{V}$ at $t_0 \in I$ is defined as $\lim_{t \to t_0} \dfrac{\phi(t) - \phi(t_0)}{t - t_0}$ if this limit exists, and is denoted by $\phi'(t_0)$. The limit is to be understood in the sense of the norm in $\mathcal{V}$, namely $\left\| \dfrac{\phi(t) - \phi(t_0)}{t - t_0} - \phi'(t_0) \right\| \to 0$ as $t \to 0$. Consider next that $\mathcal{V}$ and $\mathcal{W}$ are normed real spaces and $\mathcal{D}$ is an open subset of $\mathcal{V}$. Consider that $x_0 \in \mathcal{D}$ and that h is a fixed nonzero element in $\mathcal{V}$. Since $\mathcal{D}$ is open, there exists an interval $I = (-\tau, \tau)$ for some $\tau > 0$ such that if $t \in I$, then $x_0 + th \in \mathcal{D}$. If the mapping $\phi : I \to \mathcal{V}$ defined by $\phi(t) = F(x_0 + th)$ has a derivative at $t = 0$, then $\phi'(0)$ is called the *Gâteaux variation* of F at x_0 with increment h, and is denoted by $\delta F(x_0; h)$, i.e.,

$$\delta F(x_0; h) \equiv \frac{d}{dt} F(x_0 + th) \bigg|_{t=0} = \lim_{t \to 0} \frac{1}{t} \{F(x_0 + th) - F(x_0)\}. \qquad (I.D.1)$$

Note that Eq. (I.D.1) may be used to define $\delta F(x_0; h)$ when $\mathcal{V}$ is any linear space, not necessarily normed. When $\delta F(x_0; h)$ exists, it is homogeneous in h of degree one, i.e., for each real number λ, $\delta F(x_0; \lambda h)$ exists and is equal to $\lambda \delta F(x_0; h)$. The Gâteaux variation is a generalization of the notion of the directional derivative in calculus and of the notion of the first variation arising in the calculus of variations. The existence of the Gâteaux variation at $x_0 \in \mathcal{D}$ provides a local approximation property in the following sense:

$$F(x_0+h)-F(x_0)=\delta F(x_0;h)+r(x_0;h), \text{ where } \lim_{t\to 0}\frac{r(x_0;th)}{t}=0. \quad \text{(I.D.2)}$$

The existence of $\delta F(x_0;h)$ implies the *directional continuity* of F at x_0, i.e.,

$$\|F(x_0+th)-F(x_0)\|\to 0 \text{ as } t\to 0 \text{ for fixed } h, \quad \text{(I.D.3)}$$

but does not imply that F is continuous at x_0. This is equivalent to saying that, in general, Eq.(I.D.3) does not hold uniformly with respect to h on the bounded set $\{h:\|h\|=1\}$. Note also that the operator $h\to \delta F(x_0;h)$ is not necessarily linear or continuous in h.

If F and G have a Gâteaux variation at x_0, then so does the operator $T=\alpha F+\beta G$, where α,β are real numbers, and the relation $\delta T(x_0;h)=\alpha\,\delta F(x_0;h)+\beta\,\delta G(x_0;h)$ holds. However, the chain rule for the differentiation of a composite function does not hold in general.

An operator F has a *Gâteaux differential* at x_0 if $\delta F(x_0;\bullet)$ is linear and continuous; in this case, $\delta F(x_0;\bullet)$ is denoted by $DF(x_0)$ and is called the *Gâteaux derivative*. The necessary and sufficient condition for $\delta F(x_0;h)$ to be linear and continuous in h is that F satisfies the following two relations:

(a) To each h, there corresponds $\delta(h)$ such that $|t|\leq\delta$ implies

$$\|F(x_0+th)-F(x_0)\|\leq M\|th\|, \text{ where } M \text{ does not depend on } h; \quad \text{(I.D.4a)}$$

(b) $F(x_0+th_1+th_2)-F(x_0+th_1)-F(x_0+th_2)+F(x_0)=o(t)$.

$$\quad \text{(I.D.4b)}$$

Note that the chain rule does not necessarily hold for Gâteaux derivatives. Note also that if $\delta F(x_0;\bullet)$ is additive, then $\delta F(x_0;h)$ is directionally continuous in h, i.e.,

$$\lim_{\tau\to 0}\delta F(x_0;h+\tau k)=\delta F(x_0;h). \quad \text{(I.D.5)}$$

An operator $F:\mathcal{D}\to\mathcal{W}$, where $\mathcal{D}$ is an open subset of $\mathcal{V}$, and $\mathcal{V}$ and $\mathcal{W}$ are normed real linear spaces, is called *Fréchet differentiable* at $x_0\in\mathcal{D}$ if there exists a continuous linear operator $L(x_0):\mathcal{V}\to\mathcal{W}$ such that the following representation holds for every $h\in\mathcal{V}$ with $x_0+h\in\mathcal{D}$:

$$F(x_0 + h) - F(x_0) = L(x_0)h + r(x_0;h), \text{ with } \lim_{h \to 0} \frac{\|r(x_0;h)\|}{\|h\|} = 0. \quad \text{(I.D.6)}$$

The unique operator $L(x_0)h$ in Eq. (I.D.6) is called the *Fréchet differential* of F at x_0 and is usually denoted by $dF(x_0;h)$. The linear operator $F'(x_0): \mathcal{V} \to \mathcal{W}$ defined by $h \to dF(x_0;h)$ is called the *Fréchet derivative* of F at x_0, and $dF(x_0;h) = F'(x_0)h$. An operator F is Fréchet differentiable at x_0 iff (a) F is Gâteaux differentiable at x_0 and (b) Eq.(I.D.2) holds uniformly with respect to h on the set $\mathcal{S} = \{h : \|h\| = 1\}$. In this case the Gâteaux and Fréchet differentials coincide.

The Fréchet differential has the usual properties of the classical differential of a function of one or several variables. In particular, the chain rule holds for $F \bullet G$ if F has a Fréchet differential and G has a Gâteaux differential. Note that the chain rule may not hold for $F \bullet G$ if F has a Gâteaux differential and G has a Fréchet differential. Fréchet differentiability of F at x_0 implies continuity of F at x_0.

Consider an operator $F: \mathcal{V} \to \mathcal{W}$, where $\mathcal{V}$ is an open subset of the product space $\mathcal{P} = \mathcal{E}_1 \times \ldots \times \mathcal{E}_n$. The *Gâteaux partial differential* at $u \equiv (u_1, \ldots, u_n)$ of F with respect to u_i is the bounded linear operator $D_i F(u_1, \ldots, u_n; h_i): \mathcal{E}_i \to \mathcal{W}$ defined such that the following relation holds:

$$F(u_1, \ldots, u_{i-1}, u_i + h_i, u_{i+1}, \ldots, u_n) - F(u_1, \ldots, u_n) = D_i F(u_1, \ldots, u_n; h_i) \\ + R(u_1, \ldots, u_n; h_i) \quad \text{(I.D.7)}$$

where $\lim_{t \to 0} \frac{R(u_1, \ldots, u_n; th_i)}{t} = 0$.

The operator F is said to be *totally Gâteaux differentiable* at u_0 if F, considered as a mapping on $\mathcal{V} \subset \mathcal{P}$ into $\mathcal{W}$, is Gâteaux differentiable at u_0. This means that

$$F(u_1 + h_1, \ldots, u_n + h_n) - F(u_1 \ldots u_n) = L(u_1, \ldots, u_n; h_1, \ldots, h_n) \\ + R(u_1, \ldots, u_n; h_1, \ldots, h_n), \quad \text{(I.D.8)}$$

where the total Gâteaux differential L is a continuous linear operator in $h = (h_1 \ldots h_n)$ and where

$$\lim_{t \to 0} (t^{-1}) R(u_1, \ldots, u_n; th_1, \ldots, th_n) = 0.$$

The *Fréchet partial differential* at $(u_1, \ldots, u_n)$ of F with respect to u_i is the bounded linear operator $d_i F(u_1, \ldots, u_n; h_i)$ defined such that, for all $h_i \in \mathscr{E}_i$, with $(u_1, \ldots, u_{i-1}, u_i + h_i, u_{i+1}, \ldots, u_n) \in \mathscr{V}$, the following relation holds:

$$F(u_1, \ldots, u_{i-1}, u_i + h_i, \ldots, u_n) - F(u_i, \ldots, u_n) = d_i F(u_1, \ldots, u_n; h_i) + R(u_1, \ldots, u_n; h_i), \quad \text{(I.D.9)}$$

where $\dfrac{\| R(u_1, \ldots, u_n; h_i) \|}{\| h_i \|} \to 0$ as $h_i \to 0$.

The *total Fréchet differential* of F is denoted by $dF(u_1, \ldots, u_n; h_1, \ldots, h_n)$ and is defined as the linear mapping on $\mathscr{V} \subset \mathscr{E}_1 \times \ldots \times \mathscr{E}_n$ into $\mathscr{W}$, which is continuous in $h = (h_1, \ldots, h_n)$, such that the following relation holds:

$$\lim_{h \to 0} \frac{\| F(u_1 + h_1, \ldots, u_n + h_n) - F(u_1 \ldots u_n) - dF(u_1, \ldots, u_n; h_1, \ldots, h_n) \|}{\| h_1 \| + \ldots + \| h_n \|} = 0.$$

(I.D.10)

An operator $F: \mathscr{V} \subset \mathscr{P} \to \mathscr{W}$ that is totally differentiable at $(u_1, \ldots, u_n)$, is partially differentiable with respect to each variable, and its total differential is the sum of the differentials with respect to each of the variables. If F is totally differentiable at each point of $\mathscr{V}$, then a necessary and sufficient condition for $F': \mathscr{V} \to L(\mathscr{E}_1 \times \ldots \times \mathscr{E}_n; \mathscr{W})$ to be continuous is that the partial derivatives $F'_i: \mathscr{V} \to L(\mathscr{E}_i; \mathscr{W})$, $(i = 1, \ldots, n)$, be continuous.

Higher order partial derivatives are defined by induction. Note, in particular, that if $F: \mathscr{V} \to \mathscr{W}$ is twice Fréchet (totally) differentiable at x_0, then the second order partial derivatives $\dfrac{\partial^2 F(x_0)}{\partial x_i \partial x_j} \in L(\mathscr{E}_i, \mathscr{E}_j, \mathscr{W}), (i, j = 1, \ldots, n)$ exist, and

$$d^2 F(x_0; k_1, \ldots, k_n; h_1, \ldots, h_n) = \sum_{i,j=1}^{n} \frac{\partial^2 F(x_0)}{\partial x_i \partial x_j} k_i h_j. \quad \text{(I.D.11)}$$

Thus the second order derivative $F''(x_0)$ may be represented by the array $\{\partial^2 F(x_0)/\partial x_i \partial x_j ; (i,j = 1,\ldots,n)\}$. Note that $d^2 F(x_0;h,k)$ is symmetric in h and k; consequently, the mixed partial derivatives are also symmetric, i.e.,

$$\frac{\partial^2 F(x_0)}{\partial x_i \partial x_j} = \frac{\partial^2 F(x_0)}{\partial x_j \partial x_i}, \quad (i,j = 1,\ldots,n). \tag{I.D.12}$$

If the operator $F: \mathcal{D} \to \mathcal{W}$ has a nth-variation on $\mathcal{D}$, $\delta^n F(x_0 + th; h)$, which is continuous in t on $[0,1]$, then the following Taylor-expansion with integral remainder holds for $x_0, x_0 + h \in \mathcal{D}$:

$$F(x_0 + h) = F(x_0) + \delta F(x_0; h) + \frac{1}{2}\delta^2 F(x_0; h) + \ldots + \frac{1}{(n-1)!}\delta^{n-1} F(x_0; h)$$

$$+ \int_0^1 \frac{(1-t)^{n-1}}{(n-1)!} \delta^n F(x_0 + th; h) dt. \tag{I.D.13}$$

Note that the integral in Eq.(I.D.13) exists and is a Banach space-valued integral in the Riemann sense; note also that $\delta^k F(x_0; h) = \delta^k F(x_0; h_1, \ldots, h_k)$ for $h_1 = \ldots = h_k = h$. Furthermore, if an operator $F: \mathcal{D} \to \mathcal{W}$ has a nth-order Fréchet differential, and if the map $\varphi: [0,1] \to F^{(n)}(x_0 + th)$ is bounded and its set of discontinuities is of measure zero, then the Taylor expansion in Eq. (I.D.13) becomes

$$F(x_0 + h) = F(x_0) + F'(x_0)h + \frac{1}{2}F''(x_0)h^2 + \ldots + \frac{1}{(n-1)!}F^{(n-1)}(x_0)h^{n-1}$$

$$+ \int_0^1 \frac{(1-t)^{n-1}}{(n-1)!} F^{(n)}(x_0 + th)h^n\, dt. \tag{I.D.14}$$

In the above expansion, $F^{(k)}(x_0)h^k$ denotes the value of the k-linear operator $F^{(k)}$ at $(h,\ldots,h)$. The relation between the various differentiation concepts for a nonlinear operator F is shown schematically below, where "uniform in h" indicates the validity of the relation $\lim\limits_{h \to 0} \dfrac{\|F(x+h) - F(x) - \delta F(x; h)\|}{\|h\|} = 0$:

$$\text{G-Differential} \xrightarrow{\text{uniform in } h} \text{F-Differential} \xrightarrow{\text{linear in } h} \text{F-Derivative},$$

$$\text{G-Differential} \xrightarrow{\text{linear in } h} \text{G-Derivative} \xrightarrow{\text{uniform in } h} \text{F-Derivative}.$$

Chapter II

CONCEPTS OF PROBABILITY THEORY FOR SENSITIVITY AND UNCERTAINTY ANALYSIS

This chapter is intended to provide a summary of the main definitions and features of the concepts of probability theory and statistics that underlie sensitivity and uncertainty analysis of data and models. The main interpretations of probability commonly encountered in data and model analysis are that of *relative frequency* (which is used, in particular, for assigning statistical errors to measurements) and that of *subjective probability* (which is used, in particular, to quantify systematic uncertainties). From a mathematical point of view, however, the concepts of probability theory are optimally introduced by using Kolmogorov's axiomatic approach, in which probability is postulated in terms of abstract functions operating on well-defined event spaces. This axiomatic approach avoids both the mathematical ambiguities inherent to the concept of relative frequencies and the pitfalls of inadvertently misusing the concept of inductive reasoning. Nevertheless, all three interpretations of probability will be employed for the purposes of sensitivity and uncertainty analysis of models and data, in order to take advantage of their respective strengths.

Since the exact form of mathematical models and/or exact values of data are rarely, if ever, available in practice, their mathematical form must be *estimated*. The use of observations to estimate the underlying features of models forms the objective of *statistics*, which embodies both inductive and deductive reasoning, encompassing procedures for estimating parameters from incomplete knowledge and for refining prior knowledge by consistently incorporating additional information. Thus, the solution to practical problems requires a synergetic use of the various interpretations of probability, including the axiomatic, frequency, and Bayesian interpretations and methodologies.

The concepts of probability theory and statistics covered in this chapter are intended to serve as a self-contained basis for sensitivity and uncertainty analysis, and are provided for easy reference. Note, though, that the material presented in this chapter does not contain detailed illustrative examples, and is not intended to serve as a replacement for the many excellent textbooks on probability theory and statistics. A list of suggested reference mathematical textbooks is provided in the Reference Section of this book.

II.A. INTRODUCTION AND TERMINOLOGY

Probability theory is a branch of mathematical sciences that provides a model for describing the process of observation. The need for probability theory arises from the fact that most observations of natural phenomena do not lead to uniquely predictable results. Probability theory provides the tools for dealing with actual variations in the outcome of realistic observations and measurements. The challenging pursuit to develop a theory of probability that is mathematically rigorous and also describes many phenomena observable in nature has generated over the years notable disputes over conceptual and logical problems. Modern probability theory is based on postulates constructed from three axioms attributed to A. Kolmogorov, all of which are consistent with the notion of frequency of occurrence of events. The alternative approach, traceable to P. Laplace, is based on the concept that probability is simply a way of providing a numerical scale to quantify our reasonable beliefs about a situation, which we know incompletely. This approach is consistent with Bayes' theorem, conditional probabilities, and inductive reasoning. Either approach to probability theory would completely describe a natural phenomenon if sufficient information were available to determine the underlying probability distribution exactly. In practice, though, such exact knowledge is seldom, if ever, available so the features of the probability distribution underlying the physical phenomenon under consideration must be *estimated*. Such estimations form the study object of *statistics*, which is defined as the branch of mathematical sciences that uses the results of observations and measurements to estimate, in a mathematically well-defined manner, the essential features of probability distributions.

Both statistics and probability theory use certain generic terms for defining the objects or phenomena under study. A *system* is the object or phenomena under study. It represents the largest unit being considered. A system can refer to a nuclear reactor, corporation, chemical process, mechanical device, biological mechanism, society, economy, or any other conceivable object that is under study. The *output or response* of a system is a result that can be measured quantitatively or enumerated. The temperature of a nuclear reactor, the profit of a corporation, yield of a chemical process, torque of a motor, life span of an organism, and the inflation rate of an economy are all examples of system outputs. A *model* is a mathematical idealization that is used as an approximation to represent the output of a system. Models can be quite simple or highly complex; they can be expressed in terms of a single variable, many variables, or sets of nonlinear integro-differential equations. Regardless of its complexity, the model is an idealization of the system, so it cannot be exact: usually, the more complex the system, the less exact the model, particularly since the ability to solve exactly mathematically highly complex expressions diminishes with increasing complexity. The dilemma facing the analyst is: the more the model is simplified, the easier it is to analyze but the less precise the results.

A *statistical model* comprises mathematical formulations that express the various outputs of a system in terms of probabilities. Usually, a statistical model is used when the system's output cannot be expressed as a fixed function of the input variables. Statistical models are particularly useful for representing the behavior of a system based on a limited number of measurements, and for summarizing and/or analyzing a set of data obtained experimentally or numerically. For example, there exist families of probability distributions, covering a wide range of shapes, which can be used for representing experimentally obtained data sets. The resulting statistical model can then be used to perform extrapolations or interpolations, for determining the probability of occurrence of some event in some specific interval of values, etc. Statistical models are constructed based on the physical properties of the system. Of course, when the physical properties and, hence, the principles of operation of the system are well understood, the model will be derived from these underlying principles. Most often, though, the basic physical properties of the system are incompletely understood. In simple cases and when sufficient data are available, it is possible to select a general yet flexible family of statistical models and fit one member of this family to the observations. The validity of such fits is then verified by performing a sensitivity analysis to determine the sensitivity of the system's output to the selection of a specific model to represent the system. In situations when the physical properties of the system under investigation are ill understood, but measurements are nevertheless available, the model for the system is selected on a trial-and-error basis, usually based not only on objective criteria but also on subjective expert opinion. Finally, when no data are available, it is possible to study the conceptual model on a computer to generate numerically synthetic observations. The previously mentioned statistical methods can then be applied to analyze the synthetic observations generated numerically, just as if they had been physical observations.

The group study to be measured or counted, or the conceptual entity for which predictions are to be made, is called *population*. A *parameter of a model* is a quantity that expresses a characteristic of the system; a parameter could be constant or variable. A *sample* is a subset of the population selected for study. An *experiment* is a sequence of a limited (or, occasionally, unlimited) number of trials. An *event* is an outcome of an experiment. The set of all events (i.e., the set that represents all outcomes of an experiment) is called the *event space* or *sample space*, of the experiment; the respective events are also referred to as *sample points*.

In order to consider multiple events, the following basic ideas from set theory need to be recalled: if $(\mathcal{E}_1, \mathcal{E}_2, \ldots, \mathcal{E}_k)$ denote any k events (sets) in an event space $\mathcal{S}$, then:

The event consisting of all sample points belonging to either $\mathcal{E}_1$ or $\mathcal{E}_2, \ldots$, or $\mathcal{E}_k$ is called the *union* of the events $(\mathcal{E}_1, \mathcal{E}_2, \ldots, \mathcal{E}_k)$ and is denoted as $\bigcup_{i=1}^{k} \mathcal{E}_i$.

The event consisting of all sample points belonging to $\mathcal{E}_1$ and $\mathcal{E}_2, \ldots$, and $\mathcal{E}_k$ is called the *intersection* of the events $(\mathcal{E}_1, \mathcal{E}_2, \ldots, \mathcal{E}_k)$ and is denoted as $\bigcap_{i=1}^{k} \mathcal{E}_i$.

The *null event* is the event Ø that contains no sample points; Ø is also termed the *vacuous event*, the *empty event*, i.e., the event that never occurs.

The *universal event* is the same as *the entire sample space* $\mathcal{S}$, i.e., it is the set of all sample points in the sample space.

If $\mathcal{E}$ is any event, then the *complement* of $\mathcal{E}$, denoted as $\overline{\mathcal{E}}$, is the event consisting of all sample points in the sample space $\mathcal{S}$ that do not belong to $\mathcal{E}$.

The event $\mathcal{E}_1$ is said to be a *subevent* of $\mathcal{E}_2$, denoted as $\mathcal{E}_1 \subset \mathcal{E}_2$, if every sample point of $\mathcal{E}_1$ is also a sample point of $\mathcal{E}_2$. If the relations $\mathcal{E}_2 \subset \mathcal{E}_1$ and $\mathcal{E}_1 \subset \mathcal{E}_2$ hold simultaneously, then $\mathcal{E}_2$ and $\mathcal{E}_1$ contain the same sample points, i.e., $\mathcal{E}_1 = \mathcal{E}_2$.

The event $\mathcal{E}_1 - \mathcal{E}_2$ is the set of all sample points that belong to $\mathcal{E}_1$ but not to $\mathcal{E}_2$. Equivalently, $\mathcal{E}_1 - \mathcal{E}_2 = \mathcal{E}_1 \cap \overline{\mathcal{E}}_2$.

Two events $\mathcal{E}_1$ and $\mathcal{E}_2$ are said to be *mutually exclusive* if their intersection is the null event, i.e., if $\mathcal{E}_1 \cap \mathcal{E}_2 = \emptyset$.

The outcome of experiments and/or observations is described in terms of compound events, consisting of combinations of elementary events. In turn, the combinations of elementary events are analyzed in terms of k-tuples; a k-*tuple* is a collection of k quantities generated from elementary events in a space $\mathcal{S}$. If the position of every event in the k-tuple is unimportant, then the respective k-tuple is called *unordered*; otherwise, the respective k-tuple is called *ordered* (or an *arrangement*). Two k-tuples containing the same elementary events are called *distinguishable* if ordering is considered; otherwise, they are called *undistinguishable*.

A process is said to take place *without replacement* if each of the elementary events in $\mathcal{S}$ can be used only once in forming a k-tuple. If no restriction is placed, then the k-tuples are formed *with replacement*. Each distinguishable way of forming an ordered k-tuple from an unordered one is called a *permutation* of the unordered k-tuple. Recall that the number of *permutations of n objects taken r at a time*, p_r^n, is defined as $p_r^n = n!/(n-r)!$, where $n \geq r$ and $n! \equiv n(n-1)\ldots 1;\ 0! \equiv 1$. Also, the number of *combinations* of n objects taken r at a time, denoted by $\binom{n}{r}$, or C_r^n, is defined as $p_r^n/r!$, that is,

$$\binom{n}{r} \equiv C_r^n = \frac{n!}{r!(n-r)!}.$$

In statistical, nuclear, or particle physics, it is useful to visualize k-tuples in terms of a collection of unique cells, which may or may not be occupied. Then, a k-tuple is equivalent to a k-fold specification of the *occupancy* of these cells. The k elements occupying these cells may or may not be distinct and multiple occupancy of cells may or may not be allowed. This way, for example, the basis for the Maxwell-Boltzmann law of classical statistical physics is obtained by calculating the number, Z_{MB}, of ordered k-tuples that can be formed by selecting events (atoms or molecules) with replacements (i.e., the atoms or molecules are "distinguishable"!) from an event set (the classical gas) $\mathcal{S}$ of finite size n; this yields $Z_{MB} = n^k$. On the other hand, nonclassical particles such as *fermions* behave like *undistinguishable* particles that are forbidden (by Pauli's exclusion principle) to multiple-occupy a cell. The corresponding *Fermi-Dirac statistics*, obtained by calculating the number, Z_{FD}, of unordered k-tuples in which multiple occupancy of available states is forbidden, is given by $Z_{FD} = C_k^n = n!/[k!(n-k)!]$. The *bosons* are also nonclassical particles (i.e., they, too, are "indistinguishable") but, in contradistinction to fermions, bosons favor multiple occupancies of available energy states. Counting, in this case, the total number, Z_{BE}, of unordered k-tuples with replacement and multiple occupancy of cells gives $Z_{BE} = C_k^{n+k-1}$.

When a process operates on an event space such that the individual outcomes of the observations cannot be controlled, the respective process is called a *random process* and the respective events are called *random events*. Although the exact outcome of any single *random trial* is unpredictable, a sample of random trials of reasonable size is expected to yield a *pattern of outcomes*. In other words, *randomness implies lack of deterministic regularity but existence of statistical regularity*. Thus, random phenomena must be distinguished from *totally unpredictable* phenomena, for which no pattern can be construed, even for exceedingly large samples. Observations of physical phenomena tend to fall somewhere between total unpredictability, at the one extreme, and statistically well-behaved processes, at the other extreme. For this reason, experimentalists must invest a considerable effort to identify and eliminate, as much as possible, statistically unpredictable effects (e.g., malfunctioning equipment), in order to perform experiments under controlled conditions conducive to generating statistically meaningful results.

The quantitative analysis of statistical models relies on concepts of probability theory. The basic concepts of probability theory can be introduced in several ways, ranging from the intuitive notion of *frequency of occurrence* to the *axiomatic development* initiated by A. Kolmogorov in 1933, and including the subjective *inductive reasoning* ideas based on "degree of belief" as originally formulated by Laplace and Bayes. For the purposes of sensitivity and uncertainty analysis of models and data, all three interpretations of probability will be employed in order to take advantage of their respective strengths. From a mathematical point of view, however, the concepts of probability theory are optimally introduced by using Kolmogorov's axiomatic approach, in which probability is postulated in terms of abstract functions operating on well-defined event spaces. This axiomatic approach avoids both the mathematical ambiguities inherent to the concept of relative frequencies and the pitfalls of inadvertently misusing the concept of inductive reasoning.

Thus, consider that $\mathcal{S}$ is the sample space consisting of a certain number of events, the interpretation of which is left open for the moment. Assigned to each subset $\mathcal{A}$ of $\mathcal{S}$, there exists a real number $P(\mathcal{A})$, called *a probability*, defined by the following three axioms (after A. Kolmogorov, 1933):

AXIOM I (EXISTENCE): For every subset $\mathcal{A}$ in $\mathcal{S}$, the respective probability exists and is nonnegative, i.e., $P(\mathcal{A}) \geq 0$.

AXIOM II (ADDITIVITY): For any two subsets $\mathcal{A}$ and $\mathcal{B}$ that are disjoint (i.e., $\mathcal{A} \cap \mathcal{B} = \emptyset$), the probability assigned to the union of $\mathcal{A}$ and $\mathcal{B}$ is the sum of the two corresponding probabilities, i.e., $P(\mathcal{A} \cup \mathcal{B}) = P(\mathcal{A}) + P(\mathcal{B})$.

AXIOM III (NORMALIZATION): The probability assigned to the entire sample space is one, i.e., $P(\mathcal{S}) = 1$; in other words, the certain event has unit probability.

Note that the statements of the three axioms mentioned above are not entirely rigorous from the standpoint of pure mathematics, but have been deliberately simplified somewhat, in order to suit the scope of sensitivity and uncertainty analysis of models and data. A mathematically more precise definition of probability requires that the set of subsets to which probabilities are assigned constitute a so-called σ-field. Several useful properties of probabilities can be readily derived from the axiomatic definition introduced above:

$$\begin{aligned}
&P(\overline{\mathcal{A}}) = 1 - P(\mathcal{A}), \text{ where } \overline{\mathcal{A}} \text{ is the complement of } \mathcal{A}; \\
&P(\mathcal{A} \cup \overline{\mathcal{A}}) = 1; \\
&0 \leq P(\mathcal{A}) \leq 1; \\
&P(\emptyset) = 0, \text{ but } P(\mathcal{A}) = 0 \text{ does NOT mean that } \mathcal{A} = \emptyset; \\
&\text{if } \mathcal{A} \subset \mathcal{B}, \text{ then } P(\mathcal{A}) \leq P(\mathcal{B}); \\
&P(\mathcal{A} \cup \mathcal{B}) = P(\mathcal{A}) + P(\mathcal{B}) - P(\mathcal{A} \cap \mathcal{B}).
\end{aligned} \qquad (\text{II.A.1})$$

The last relation above is known as Poincaré's theorem of probability addition. This theorem can be extended to any *finite sequence* of events $(\mathcal{E}_1, \mathcal{E}_2, \ldots, \mathcal{E}_k)$ in a sample space $\mathcal{S}$, in which case it becomes

$$P\left(\bigcup_{i=1}^{n} \mathcal{E}_i\right) = \sum_{i=1}^{n} P(\mathcal{E}_i) - \sum_{i \neq j} P(\mathcal{E}_i \cap \mathcal{E}_j) + \sum_{i \neq j \neq k} P(\mathcal{E}_i \cap \mathcal{E}_j \cap \mathcal{E}_k) \\ + \ldots + (-1)^{n+1} P\left(\bigcap_{i=1}^{n} \mathcal{E}_i\right). \qquad (\text{II.A.2})$$

When the events $(\mathcal{E}_1, \mathcal{E}_2, \ldots, \mathcal{E}_n)$ form a *finite sequence of mutually exclusive events*, Eq. (II.A.2) reduces to

$$P\left(\bigcup_{i=1}^{n} \mathcal{E}_i\right) = \sum_{i=1}^{n} P(\mathcal{E}_i). \tag{II.A.3}$$

When extended to the infinite case $n \to \infty$, Eq. (II.A.3) actually expresses one of the defining properties of the *probability measure*, which makes it possible to introduce the concept of *probability function* defined on an event space $\mathcal{S}$. Specifically, a function P is a probability function defined on $\mathcal{S}$ iff it possesses the following properties:

$P(\emptyset) = 0$, where $\emptyset$ is the null set.
$P(\mathcal{S}) = 1$.
$0 \le P(\mathcal{E}) \le 1$, where $\mathcal{E}$ is any event in $\mathcal{S}$.
$P\left(\bigcup_{i=1}^{\infty} \mathcal{E}_i\right) = \sum_{i=1}^{\infty} P(\mathcal{E}_i)$, where $(\mathcal{E}_1, \mathcal{E}_2, \ldots)$ is any sequence of mutually exclusive events in $\mathcal{S}$.

The concept of probability defined thus far cannot address *conditions* (implied or explicit) such as "What is the probability that event $\mathcal{E}_2$ occurs if it is known that event $\mathcal{E}_1$ has actually occurred?" In order to address such conditions, an additional concept, namely the concept of conditional probability, needs to be introduced. For this purpose, consider that $\mathcal{E}_1$ and $\mathcal{E}_2$ are any two events in an event space $\mathcal{S}$, as depicted in Fig. II.1, with $P(\mathcal{E}_1) > 0$. Then, the *conditional probability* $P(\mathcal{E}_2|\mathcal{E}_1)$ that event $\mathcal{E}_2$ occurs given the occurrence of $\mathcal{E}_1$ is defined as

$$P(\mathcal{E}_2|\mathcal{E}_1) \equiv \frac{P(\mathcal{E}_1 \cap \mathcal{E}_2)}{P(\mathcal{E}_1)}. \tag{II.A.4}$$

From the definition introduced in Eq. (II.A.4), it follows that

$$P(\mathcal{E}_1 \cap \mathcal{E}_2) = P(\mathcal{E}_2|\mathcal{E}_1) P(\mathcal{E}_1). \tag{II.A.5}$$

Figure II.1: Relationship between the sets $\mathcal{E}_2$, $\mathcal{E}_1$, and $\mathcal{S}$ in the definition of conditional probability

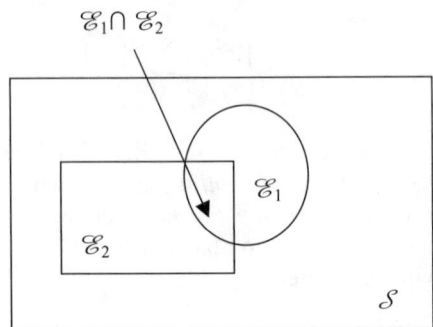

Conditional probabilities also satisfy the axioms of probability, as can be seen from the definition introduced in Eq. (II.A.4). In particular, this definition implies that the usual probability $P(\mathcal{E}_1)$ can itself be regarded as a conditional probability, since $P(\mathcal{E}_1) = P(\mathcal{E}_1|\mathcal{S})$ [i.e., $P(\mathcal{E}_1)$ is the conditional probability for $\mathcal{E}_1$ given $\mathcal{S}$].

The definition of conditional probability can be readily extended to more than two events by partitioning the sample space $\mathcal{S}$ into k mutually exclusive events $(\mathcal{E}_1, \mathcal{E}_2, \ldots, \mathcal{E}_k)$, and by considering that $\mathcal{E} \subset \mathcal{S}$ is an arbitrary event in $\mathcal{S}$ with $P(\mathcal{E}) > 0$. Then the probability that event $\mathcal{E}_i$ occurs given the occurrence of $\mathcal{E}$, $P(\mathcal{E}_i|\mathcal{E})$, is expressed as

$$P(\mathcal{E}_i|\mathcal{E}) = \frac{P(\mathcal{E}|\mathcal{E}_i)P(\mathcal{E}_i)}{\sum_{j=1}^{k} P(\mathcal{E}|\mathcal{E}_j)P(\mathcal{E}_j)}, \quad (i = 1, 2, \ldots, k), \tag{II.A.6}$$

where $P(\mathcal{E}|\mathcal{E}_i)$ denotes the probability that event $\mathcal{E}$ occurs, given the occurrence of $\mathcal{E}_i$. Equation (II.A.6) is known as *Bayes' theorem*, and is of fundamental importance to practical applications of probability theory.

Two events, $\mathcal{E}_1$ and $\mathcal{E}_2$, are said to be *statistically independent* if $P(\mathcal{E}_2|\mathcal{E}_1) = P(\mathcal{E}_2)$. This means that the occurrence (or nonoccurrence) of $\mathcal{E}_1$ does not affect the occurrence of $\mathcal{E}_2$. Note that if $\mathcal{E}_1$ and $\mathcal{E}_2$ are statistically independent, then $P(\mathcal{E}_1\mathcal{E}_2) = P(\mathcal{E}_1)P(\mathcal{E}_2)$, $P(\mathcal{E}_1|\mathcal{E}_2) = P(\mathcal{E}_1)$, and also conversely. Actually, the following statements are equivalent:

$\mathscr{E}_1$ and $\mathscr{E}_2$ are statistically independent;

$$P(\mathscr{E}_1\mathscr{E}_2) = P(\mathscr{E}_1)P(\mathscr{E}_2); \tag{II.A.7}$$
$$P(\mathscr{E}_1|\mathscr{E}_2) = P(\mathscr{E}_1).$$

The numerical representation of the elementary events $\mathscr{E}$ in a set $\mathscr{S}$ is accomplished by introducing a function, say X, which operates on all events $\mathscr{E} \subset \mathscr{S}$ in such a way as to establish a particular correspondence between $\mathscr{E}$ and the real number $x = X(\mathscr{E})$. The function X defined in this way is called a *random variable*. It is very important to note that X itself is the random variable, rather than the individual values $x = X(\mathscr{E})$, which X generates by operating on $\mathscr{E}$. Thus, although X has the characteristics of a "function" rather than those of a "variable," the usual convention in probability theory and statistics is to call X a "random variable" rather than a "random function." As will be seen shortly, there also exist "functions of random variables," which may, or may not, be random themselves.

Random events $\mathscr{E} \subset \mathscr{S}$ that can be completely characterized by a *single-dimensional random variable* X are called *single-variable events*. The qualifier "single-dimensional" is omitted when it is apparent from the respective context. The concept of single-dimensional random variable, introduced to represent numerically a single-variable event, can be extended to a *multi-variable event* $\mathscr{E} \subset \mathscr{S}$. Thus, a n-*variable event* $\mathscr{E} \subset \mathscr{S}$ is represented numerically by establishing the correspondence $\boldsymbol{x} = \boldsymbol{X}(\mathscr{E})$, where $\boldsymbol{x} = (x_1, x_2, ... x_n)$ is a vector of real numbers and $\boldsymbol{X} = (X_1, X_2, ... X_n)$ is *a multi-dimensional random variable* (or *random vector,* for short) that takes on the specific values $\boldsymbol{x}$.

The dimensionality of a random number is defined by the dimensionality of the particular set of values $x = X(\mathscr{E})$, generated by the function X when operating on the events $\mathscr{E}$. Although the cardinalities (i.e., dimensions) of event sets and corresponding random variables are usually equal, it is important to note that the explicit dimensionality of $\mathscr{E}$ is irrelevant in this context. By definition, *finite random variables* are those for which the set of values x obtainable from X operating on $\mathscr{E}$ is in one to one correspondence with a finite set of integers. *Infinite but discrete random variables* are those for which the set of values x obtainable from X operating on $\mathscr{E}$ is in one-to-one correspondence with the infinite set of all integers. *Unaccountable or nondenumerable random variables* are those for which the set of values x obtainable from X operating on $\mathscr{E}$ is in one-to-one correspondence with the infinite set of all real numbers.

As discussed above, random variables are actually well-behaved functions that operate on event spaces to yield numerical values that characterize, in turn, the respective events. Random variables can also serve as arguments of other functions, whenever these functions are well behaved in the sense that such functions are bounded and are devoid of singularities except, perhaps, for a finite number of jump-discontinuities. For the purposes of uncertainty analysis, four such types of functions are of particular interest, namely: transformations and statistics; discrete and continuous density functions; cumulative distribution functions; and indicator functions.

II.B. INTERPRETATION OF PROBABILITY

The main interpretations of probability commonly encountered in data and model analysis are that of *relative frequency* (which is used, in particular, for assigning statistical errors to measurements) and that of *subjective probability* (which is used, in particular, to quantify systematic uncertainties). These two interpretations are described in more detail below.

II.B.1. Probability as a Relative Frequency

In data analysis, probability is most commonly interpreted as a *limiting relative frequency*. In this interpretation, the elements of the set $\mathcal{S}$ correspond to the possible outcomes of a measurement, assumed to be (at least hypothetically) repeatable. A subset $\mathcal{E}$ of $\mathcal{S}$ corresponds to the occurrence of any of the outcomes in the subset. Such a subset is called an *event*, which is said to occur if the outcome of a measurement is in the subset. A subset of $\mathcal{S}$ consisting of only one element denotes an *elementary outcome*.

When interpreting probability as a limiting relative frequency, the probability of an elementary outcome $\mathcal{E}$ is defined as the fraction of times that $\mathcal{E}$ occurs in the limit when the measurement is repeated infinitely many of times, namely:

$$P(\mathcal{E}) = \lim_{n \to \infty} \frac{\text{number of occurences of outcome } \mathcal{E} \text{ in } n \text{ measurements}}{n}.$$

(II.B.1)

The probability for the occurrence of any one of several outcomes (i.e., for a nonelementary subset $\mathcal{E}$) is determined from the probabilities for individual elementary outcomes by using the addition rule provided by the axioms of probability. These individual probabilities correspond, in turn, to relative frequencies of occurrence. The relative frequency interpretation is consistent with the axioms of probability, since the fraction of occurrence is always greater than or equal to zero, the frequency of any outcome is the sum of the individual frequencies of the individual outcomes (as long as the set of individual outcomes is disjoint), and the measurement must, by definition, eventually yield some outcome [i.e., $P(\mathcal{S}) = 1$]. Correspondingly, the conditional probability $P(\mathcal{E}_2|\mathcal{E}_1)$ represents the number of cases where both $\mathcal{E}_2$ and $\mathcal{E}_1$ occur divided by the number of cases in which $\mathcal{E}_1$ occurs, regardless of whether $\mathcal{E}_2$ occurs. In other words, $P(\mathcal{E}_2|\mathcal{E}_1)$ gives the frequency of $\mathcal{E}_2$ with the subset $\mathcal{E}_1$ taken as the sample space.

The interpretation of probability as a relative frequency is straightforward when studying physical laws, since such laws are assumed to act the same way in repeated experiments, implying that the validity of the assigned probability values can be tested experimentally. This point of view is appropriate, for example, in nuclear and particle physics, where repeated collisions of particles constitute repetitions of an experiment. Note, though, that the probabilities based on such an interpretation can never be determined experimentally with perfect precision. Hence, the fundamental tasks of classical statistics are to interpret the experimental data by (a) estimating the probabilities (assumed to have some definite but unknown values) of occurrence of events of interest, given a finite amount of experimental data, and (b) testing the extent to which a particular model or theory that predicts the probabilities estimated in (a) is compatible with the observed data.

The concept of probability as a relative frequency becomes questionable when attempting to assign probabilities for very rare (or even uniquely occurring) phenomena such as a core meltdown in a nuclear reactor or the big bang. For such rare events, the frequency interpretation of probabilities might perhaps be rescued by imagining a large number of similar universes, in some fraction of which the rare event under consideration would occur. However, such a scenario is pure utopia, even in principle; therefore the frequency interpretation of probability must be abandoned in practice when discussing extremely rare events. In such cases, probability must be considered as a mental construct to assist us in expressing a degree of belief about the single universe in which we live; this mental construct provides the premises of the Bayesian interpretation of probability, which will be discussed next.

II.B.2. Subjective Probability

Complementary to the frequency interpretation of probability is the so-called *subjective* (also called *Bayesian*) *probability*. In this interpretation, the elements of the sample space are considered to correspond to *hypotheses* or *propositions*, i.e., statements that are either true or false; the sample space is often called the *hypothesis space*. Then, the probability associated with a cause or hypothesis $\mathcal{A}$ is interpreted as a measure of degree of belief, namely:

$$P(\mathcal{A}) \equiv a\ priori \text{ measure of the rational degree of belief} \qquad \text{(II.B.2)}$$
$$\text{that } \mathcal{A} \text{ is the correct cause or hypothesis.}$$

The sample space $\mathcal{S}$ must be constructed so that the elementary hypotheses are mutually exclusive, i.e., only one of them is true. A subset consisting of more than one hypothesis is true if any of the hypotheses in the subset is true. This means that the union of sets corresponds to the Boolean *OR* operation, while the intersection of sets corresponds to the Boolean *AND* operation. One of the hypotheses must necessarily be true, implying that $P(\mathscr{S}) = 1$.

Since the statement "a measurement will yield a given outcome for a certain fraction of the time" can be regarded as a hypothesis, it follows that the framework of subjective probability includes the relative frequency interpretation. Furthermore, subjective probability can be associated with (for example) the value of an unknown constant; this association reflects one's confidence that the value of the respective probability is contained within a certain fixed interval. This is in contrast with the frequency interpretation of probability, where the "probability for an unknown constant" is not meaningful, since if we repeat an experiment depending on a physical parameter whose exact value is not certain, then its value is either never or always in a given fixed interval. Thus, in the frequency interpretation, the "probability for an unknown constant" would be either zero or one, but we do not know which. For example, the mass of a physical quantity (e.g., neutron) may not be exactly known, but there is considerable evidence, in practice, that it lays between some upper and lower limits of a given interval. In the frequency interpretation, the statement "the probability that the mass of the neutron lies within a given interval" is meaningless. By contrast, though, a subjective probability of 90% that the neutron mass is contained within the given interval is a meaningful reflection of one's state of knowledge.

The use of subjective probability is closely related to Bayes' theorem and forms the basis of *Bayesian* (as opposed to classical) *statistics,* as can be readily illustrated by considering Eq. (II.A.6), for simplicity, for the particular case of two subsets. Thus, in Bayesian statistics, the subset $\mathscr{E}_2$ appearing in the definition of conditional probability is interpreted as the hypothesis that "a certain theory is true," while the subset $\mathscr{E}_1$ designates the hypothesis that "an experiment will yield a particular result (i.e., data)." In this interpretation, Bayes' theorem takes on the form

$$P(theory|data) \propto P(data|theory) \cdot P(theory). \qquad \text{(II.B.3)}$$

In the above expression of proportionality, $P(theory)$ represents the *prior probability* that the theory is true, while the *likelihood* $P(data|theory)$ expresses the probability of observing the data that were actually obtained under the assumption that the theory is true. The *posterior probability,* that the theory is correct after seeing the result of the experiment, is given by $P(theory|data)$. Note that the prior probability for the data, $P(data)$, does not appear explicitly, so the above relation expresses a proportionality rather than an equality. Furthermore, Bayesian statistics provides no fundamental rule for assigning the prior probability to a theory. However, once a prior probability has been assigned, Bayesian statistics indicates how one's degree of belief should change after obtaining additional information (e.g., experimental data).

The choice of the "most appropriate" prior distribution lies at the heart of applying Bayes' theorem to practical problems and has caused considerable debates over the years. Thus, *when prior information related to the problem under consideration is available and can be expressed in the form of a probability distribution, this information should certainly be used*; in such cases, the repeated application of Bayes' theorem will serve to refine the knowledge about the respective problem. *When only scant information is available, the maximum entropy principle* (as described in statistical mechanics and information theory) *is the recommended choice for constructing a prior distribution*. Finally, *in the extreme case when no information is available, the general recommendation is to use a continuous uniform distribution as the prior*. In any case, the proper repeated use of Bayes' theorem ensures that the impact of the choice of priors on the final result diminishes as additional information (e.g., measurements) containing consistent data is successively incorporated.

II.C. MULTIVARIATE PROBABILITY DISTRIBUTIONS

Consider that $\mathcal{S}$ is a sample space in k-dimensions. As has been discussed in the Section II.A, if each random variable X_i $(i=1,2,\ldots,k)$ is a real-valued function defined on a domain $\mathcal{N}_i$ (representing the ith dimension of $\mathcal{S}$), then $(X_1,\ldots,X_k)$ is a *multivariate random variable* or *random vector*. Furthermore, consider that each domain $\mathcal{N}_i$ $(i=1,2,\ldots,k)$ is a *discrete* set, either finite or denumerably infinite ($\mathcal{N}_i$ is usually the set of nonnegative integers or a subset thereof). Then, *a probability function*, $p(x_1,\ldots,x_k)$, of the discrete random vector $(X_1,\ldots,X_k)$ is defined by requiring $p(x_1,\ldots,x_k)$ to satisfy, for each value x_i taken on by X_i $(i=1,2,\ldots,k)$, the following properties:

$$\text{(i)} \quad p(x_1, x_2, \ldots, x_k) = P\{X_1 = x_1, X_2 = x_2, \ldots, X_k = x_k\} \tag{II.C.1}$$

and

$$\text{(ii)} \quad P\{\mathcal{A}\} = \sum_{(x_1,\ldots,x_k) \in \mathcal{A}} p(x_1,\ldots,x_k) \tag{II.C.2}$$

for any subset $\mathcal{A}$ of $\mathcal{N}$, where $\mathcal{N}$ is the k-dimensional set whose i^{th} component is $\mathcal{N}_i$ $(i=1,2,\ldots,k)$ with $P\{\mathcal{N}\}=1$.

Consider that $\mathcal{A}$ is the set of all random vectors $(X_1,\ldots,X_k)$ such that $X_i \leq x_i$ $(i = 1,2,\ldots,k)$. Then

$$P\{\mathcal{A}\} = P\{X_1 \leq x_1, X_2 \leq x_2, \ldots, X_k \leq x_k\} \qquad \text{(II.C.3)}$$

is called the *cumulative distribution function* (*CDF*) of $(X_1,\ldots,X_k)$. The usual notation for the *CDF* of $(X_1,\ldots,X_k)$ is $F(x_1,\ldots,x_k)$. Note that the cumulative distribution function is not a random variable; rather, it is a real numerical-valued function whose arguments represent compound events.

To define the probability density function of a *continuous random vector* $(X_1,\ldots,X_k)$, consider that $(X_1,\ldots,X_k)$ is a random vector whose i^{th} component, X_i, is defined on the real line $(-\infty,\infty)$ or on a subset thereof. Suppose $p(x_1,\ldots,x_k) > 0$ is a function such that for all $x_i \in [a_i, b_i](i = 1,2,\ldots,k)$, the following properties hold:

$$\text{(i) } P\{a_1 < X_1 < b_1, \ldots, a_k < X_k < b_k\} = \int_{a_k}^{b_k} \ldots \int_{a_1}^{b_1} p(x_1,\ldots,x_k)dx_1 \ldots dx_k \qquad \text{(II.C.4)}$$

and if $\mathcal{A}$ is any subset of k-dimensional intervals,

$$\text{(ii) } P\{\mathcal{A}\} = \int \ldots \int_{(x_1,\ldots,x_k) \in \mathcal{A}} p(x_1, x_2,\ldots,x_k)dx_1 dx_2 \ldots dx_k \qquad \text{(II.C.5)}$$

then $p(x_1,\ldots,x_k)$ is said to be a *joint probability density function* (*PDF*) of the continuous random vector $(X_1,\ldots,X_k)$, if it is normalized to unity over its domain.

Consider that the set $(x_1,\ldots,x_k)$ represents a collection of random variables with a multivariate joint probability density $p(x_1,\ldots,x_k) > 0$; then, the *marginal probability density* of x_i, denoted by $p_i(x_i)$, is defined as

$$p_i(x_i) = \int_{-\infty}^{\infty} dx_1 \ldots \int_{-\infty}^{\infty} dx_{i-1} \int_{-\infty}^{\infty} dx_{i+1} \ldots \int_{-\infty}^{\infty} p(x_1, x_2,\ldots,x_k)dx_k. \qquad \text{(II.C.6)}$$

Also, the conditional probability density function (*PDF*) $p(x_1,\ldots,x_{i-1},x_{i+1},\ldots,x_k | x_i)$ can be defined whenever $p_i(x_i) \neq 0$ by means of the expression

Concepts of Probability Theory

$$p(x_1,\ldots,x_{i-1},x_{i+1},\ldots,x_k \mid x_i) = p(x_1,\ldots,x_k)/p_i(x_i). \tag{II.C.7}$$

The meaning of marginal and conditional probability can be illustrated by considering bivariate distributions. Thus, if (X,Y) is a discrete random vector whose joint probability function is $p(x,y)$, where X is defined over $\mathcal{N}_1$ and Y is defined over $\mathcal{N}_2$, then the *marginal probability distribution function (PDF)* of X, $p_x(x)$, is defined as

$$p_x(x) = \sum_{y \in \mathcal{N}_2} p(x,y). \tag{II.C.8}$$

On the other hand, if (X,Y) is a continuous random vector whose joint *PDF* is $p(x,y)$, where X and Y are each defined over the domain $(-\infty,\infty)$, then the marginal *PDF* of X, $p_x(x)$, is defined by the integral

$$p_x(x) = \int_{-\infty}^{\infty} p(x,y)\,dy. \tag{II.C.9}$$

Consider that (X,Y) is a random vector (continuous or discrete) whose joint *PDF* is $p(x,y)$. The *conditional PDF of y given $X = x$ (fixed)*, denoted by $h(y|x)$, is defined as

$$h(y|x) = \frac{p(x,y)}{p_x(x)}, \tag{II.C.10}$$

where the domain of y may depend on x, and where $p_x(x)$ is the marginal *PDF* of X, with $p_x(x) > 0$.

Similarly, the conditional *PDF*, say $g(x|y)$, for x given y is

$$g(x|y) = \frac{p(x,y)}{p_y(y)} = \frac{p(x,y)}{\int p(x',y)\,dx'}. \tag{II.C.11}$$

Combining equations (II.C.10) and (II.C.11) gives the following relationship between $g(x|y)$ and $h(y|x)$,

$$g(x|y) = \frac{h(y|x)p_x(x)}{p_y(y)}, \qquad (\text{II.C.12})$$

which expresses Bayes' theorem for the case of continuous variables.

Consider that (X,Y) is a random vector whose joint *PDF* is $p(x,y)$. The random variables X and Y are called *stochastically independent* iff

$$p(x,y) = p_x(x)p_y(y) \qquad (\text{II.C.13})$$

over the entire domain of (X,Y) (i.e., for all x and y). From this definition and from Eq. (II.C.11), it follows that X and Y are independent iff $g(x|y) = p_x(x)$ over the entire domain of (X,Y). Of course, this definition of stochastic independence can be generalized to random vectors.

For certain cases, the *PDF* of a function of random variables can be found by using integral transforms. In particular, Fourier transforms are well-suited for dealing with the *PDF* of sums of random variables, while Mellin transforms are well suited for dealing with *PDF* of products of random variables. For example, the *PDF* $f(z)$, of the sum $z = x + y$, where x and y are two independent random variables distributed according to $g(x)$ and $h(y)$, respectively, is obtained as

$$f(z) = \int_{-\infty}^{\infty} g(x)h(z-x)dx = \int_{-\infty}^{\infty} g(z-y)h(y)dy. \qquad (\text{II.C.14})$$

The function f in the above equation is actually the *Fourier convolution* of g and h, and is often written in the form $f = g \otimes h$.

On the other hand, the *PDF*, $f(z)$, of the product $z = xy$, where x and y are two random variables distributed according to $g(x)$ and $h(y)$, is given by

$$f(z) = \int_{-\infty}^{\infty} g(x)h(z/x)\frac{dx}{|x|} = \int_{-\infty}^{\infty} g(z/y)h(y)\frac{dy}{|y|}, \qquad (\text{II.C.15})$$

where the second equivalent expression is obtained by reversing the order of integration. The function f defined by Eq. (II.C.15) is actually the *Mellin convolution* of g and h, often written (also) in the convolution form $f = g \otimes h$. Taking the Mellin transform of Eq. (II.C.15), or taking the Fourier transform of Eq.(II.C.14), respectively, converts the respective convolution equations $f = g \otimes h$ into the respective products $\tilde{f} = \tilde{g} \cdot \tilde{h}$ of the transformed density functions. The actual *PDF* is subsequently obtained by finding the inverse transform of the respective $\tilde{f}$'s.

Consider that $x = (x_1,\ldots,x_n)$ and $y = (y_1,\ldots,y_n)$ are two distinct vector random variables that describe the same events. Consider, further, that the respective multivariate probability densities $p_x(x)$ and $p_y(y)$ are such that the mappings $y_i = y_i(x_1,\ldots,x_n)$, $(i = 1,2,\ldots,n)$, are continuous, one-to-one, and all partial derivatives $\partial y_i / \partial x_j$, $(i,j = 1,\ldots,n)$, exist. Then, the transformation from one *PDF* to the other is given by the relationship

$$p_y(y)|dy| = p_x(x)|dx|, \text{ or } p_x(x) = |J|\, p_y(y), \qquad \text{(II.C.16)}$$

where $|J| \equiv \det|\partial y_i / \partial x_j|, (i,j = 1,\ldots,n)$, is the Jacobian of the respective transformation.

II.D. FUNDAMENTAL PARAMETERS OF PROBABILITY FUNCTIONS

Probabilities cannot be measured directly; they can be inferred from the results of observations or they can be postulated and (partially) verified through accumulated experience. In practice, though, certain random vectors tend to be more probable, so that most probability functions of practical interest tend to be localized. Therefore, the essential features regarding probability distributions of practical interest are measures of *location* and of *dispersion*. These measures are provided by the *expectation* and *moments* of the respective probability function. If the probability function is known, then these moments can be calculated directly, through a process called *statistical deduction.* Otherwise, if the probability function is not known, then the respective moments must be estimated from experiments, through a process called *statistical inference.*

II.D.1. Expectations and Moments

Consider that $x = (x_1,\ldots,x_n)$ is a collection of random variables that represent the events in a space $\mathscr{E}$, and consider that $\mathscr{S}_x$ represents the n-dimensional space formed by all possible values of x. The space $\mathscr{S}_x$ may encompass the entire range of real numbers (i.e., $-\infty < x_i < \infty, i = 1,\ldots,n$) or a subset thereof. Furthermore, consider a real-valued function, $g(x)$, and a probability density,

$p(x)$, both defined on $\mathcal{S}_x$. Then, *the expectation of* $g(x)$, denoted as $E(g(x))$, is defined as:

$$E(g(x)) \equiv \int_{\mathcal{S}_x} g(x)p(x)dx \qquad \text{(II.D.1)}$$

if the condition of *absolute convergence*, namely

$$E(|g|) = \int_{\mathcal{S}_x} |g(x)| p(x)dx < \infty \qquad \text{(II.D.2)}$$

is satisfied. When x is discrete with the domain $\mathcal{N}$, the expectation of g is defined as

$$E(g(x)) = \sum_{x \in \mathcal{N}} g(x)p(x) \qquad \text{(II.D.3)}$$

provided that $E(|g(x)|) < \infty$.

In particular, *the moment of order k about a point c* is defined for a univariate probability function as

$$E((x-c)^k) \equiv \int_{\mathcal{S}_x} (x-c)^k p(x)dx, \qquad \text{(II.D.4)}$$

where $\mathcal{S}_x$ denotes the set of values of x for which $p(x)$ is defined, and the integral above is absolutely convergent.

For a multivariate probability function, given a collection of n random-variables $(x_1,\ldots,x_n)$ and a set of constants $(c_1,\ldots,c_n)$, the *mixed moment of order k* is defined as

$$E\left((x_1-c_1)^{k_1},\ldots,(x_n-c_n)^{k_n}\right) \equiv \int_{\mathcal{S}_{x_1}} dx_1 \ldots \int_{\mathcal{S}_{x_n}} dx_n (x_1-c_1)^{k_1} \ldots (x_n-c_n)^{k_n} p(x_1 \ldots x_n).$$

$$\text{(II.D.5)}$$

The *zeroth-order moment* is obtained by setting $k = 0$ (for univariate probability) in Eq. (II.D.4) or by setting $k_1 = \ldots = k_n = 0$ (for multivariate probability) in Eq. (II.D.5), respectively. *Since probability functions are required to be normalized, the zeroth moment is always equal to unity.*

Concepts of Probability Theory   57

In particular, when $c = 0$ (for univariate probability) or when $c_1 = \ldots = c_n = 0$ (for multivariate probability), the quantities defined as $v_k \equiv E(x^k)$, and, respectively, $v_{k_1 \ldots k_n} \equiv E(x_1^{k_1} \ldots x_n^{k_n})$ are called the *moments about the origin* (often also called *raw* or *crude moments*). If $\sum_{i=1,n} k_i = k$, then the moments of the form $v_{k_1 \ldots k_n}$ are called the *mixed raw moments of order* k. For $k = 1$, these moments are called *mean values*, and are denoted as $m_o \equiv v_1 \equiv E(x)$ for univariate probability, and $m_{oi} \equiv v_{o \ldots 1 \ldots o} \equiv E(x_i)$, $(i = 1, \ldots, n)$, for multivariate probability, respectively. Note that a "0" in the jth subscript position signifies that $k_j = 0$ for the particular raw moment in question, while a "1" in the ith subscript position indicates that $k_i = 1$ for the respective moment.

The moments about the mean or central moments are defined as

$$\mu_k \equiv E\big((x - m_o)^k\big), \text{ for univariate probability,} \qquad (II.D.6)$$

and

$$\mu_{k_1 \ldots k_n} \equiv E\big((x_1 - m_{o1})^{k_1} \ldots (x_n - m_{on})^{k_n}\big), \text{ for multivariate probability.} \qquad (II.D.7)$$

Furthermore, if $\sum_{i=1,n} k_i = k$, then the above moments are called the *mixed central moments of order* k. Note that the central moments vanish whenever one particular $k_i = 1$ and all other $k_j = 0$, i.e., $\mu_{o \ldots 1 \ldots o} = 0$. Note also that all even-power central moments of univariate probability functions (i.e., μ_k, for $k = even$) are nonnegative.

The central moments for $k = 2$ play very important roles in statistical theory, and are therefore assigned special names. Thus, for univariate probability, the second moment, $\mu_2 \equiv E\big((x - m_o)^2\big)$, is called *variance*, and is usually denoted as $\mathrm{var}(x)$ or σ^2. The positive square root of the variance is called the *standard deviation*, denoted as σ, and defined as

$$\sigma \equiv [\mathrm{var}(x)]^{1/2} \equiv \mu_2^{1/2} \equiv \big[E\big((x - m_o)^2\big)\big]^{1/2}. \qquad (II.D.8)$$

The terminology and notation used for univariate probability are also used for multivariate probability. Thus, for example, the standard deviation of the ith component is defined as:

$$\mu_{o \ldots 2 \ldots o} \equiv \mathrm{var}(x_i) \equiv \sigma_i^2 \equiv E\big((x_i - m_o)^2\big). \qquad (II.D.9)$$

To simplify the notation, the subscripts accompanying the moments v and μ for multivariate probability functions are usually dropped and alternative notation is used. For example, μ_{ii} is often employed to denote $\text{var}(x_i)$, and μ_{ij} signifies $E((x_i - m_{oi})(x_j - m_{oj}))$, $(i, j = 1, \ldots, n)$.

The raw and central moments are related to each other through the important relationship

$$\mu_k = \sum_{i=0}^{k} C_i^k (-1)^i v_{k-i} v_1^i \qquad (k \geq 1), \qquad \text{(II.D.10)}$$

where $C_i^k = k!/[(k-i)!i!]$ is *the binomial coefficient*. This formula is very useful for estimating central moments from sampling data, since, in practice, it is more convenient to estimate the raw moments directly from the data, and then derive the central moments by using the above equation.

II.D.2. Variance, Standard Deviation, Covariance, and Correlation

Since measurements rarely yield true values, it is necessary to introduce surrogate parameters to measure location and dispersion for the observed results. Practice indicates that *location is best described by the mean value*, while *dispersion* of observed results appears to be *best described by the variance, or standard deviation*. In particular, the mean value can be interpreted as a locator of the center of gravity, while the variance is analogous to the moment of inertia (which linearly relates applied torque to induced angular acceleration in mechanics). Also very useful for the study of errors is the *Minimum Variance Theorem*, which states that: *if c is a real constant and x is a random variable, then* $\text{var}(x) \leq E((x-c)^2)$.

Henceforth, when we speak of *errors in physical observations,* they *are to be interpreted as standard deviations*. In short, errors are simply the measures of dispersion in the underlying probability functions that govern observational processes. The *fractional relative error* or *coefficient of variation*, f_x, is defined by $f_x = \sigma/|E(x)|$, when $E(x) \neq 0$. The reciprocal, $(1/f_x)$, is commonly called (particularly in engineering applications) the *signal-to-noise ratio*. Finally, the term *percent error* refers to the quantity $100 f_x$.

When the probability function is known and the respective mean and variance (or standard deviation) exist, they can be computed directly from their definitions. However, when the actual distribution is not known, it is considerably more difficult to interpret the knowledge of the mean and standard deviation in terms of confidence that they are representative of the distribution of measurements. The difficulty can be illustrated by considering a *confidence indicator associated with the probability function* p, $C_p(k\sigma)$, defined by means of the integral

$$C_p(k\sigma) \equiv \int_{m_o-k\sigma}^{m_o+k\sigma} p(x)dx, \qquad \text{(II.D.11)}$$

where σ is the standard deviation and $k \geq 1$ is an integer. Since the probability density integrated over the entire underlying domain is normalized to unity, it follows that $C_p(k\sigma) < 1$ for all k. However, $C_p(k\sigma) \approx 1$ whenever $k \gg 1$. Thus, $C_p(k\sigma)$ can vary substantially in magnitude for different types of probability functions p, even for fixed values of σ and k. This result indicates that although the variance or standard deviation are useful parameters for measuring dispersion (error), knowledge of them alone does not provide an unambiguous measure of confidence in a result, unless the probability family to which the distribution in question belongs is *a priori* known. Consequently, when an experiment involves several observational processes, each governed by a distinct law of probability, it is difficult to interpret overall errors (which consist of several components) in terms of confidence. In practice, though, the consequences are mitigated by a very important theorem of statistics, called the *Central Limit Theorem*, which will be presented in Section II.G.

For multivariate probability, the second-order central moments comprise not only the variances $\mu_{ii} = \text{var}(x_i) = E((x_i - m_{oi})^2)$, $(i = 1,\ldots,n)$, but also the moments $\mu_{ij} = E((x_i - m_{oi})(x_j - m_{oj}))$, $(i,j = 1,\ldots,n)$. These moments are called *covariances*, and the notation $\text{cov}(x_i, x_j) \equiv \mu_{ij}$ is often used. The collection of all second-order moments of a multivariate probability function involving n random variables forms an $n \times n$ matrix, denoted as V_x, and called the *variance-covariance matrix*, or, simply, the *covariance matrix*. Since $\mu_{ij} = \mu_{ji}$ for all i and j, covariance matrices are *symmetric*.

When $\mu_{ii} > 0$, $(i = 1,\ldots,n)$, it is often convenient to use the quantities ρ_{ij} defined by the relationship

$$\rho_{ij} \equiv \mu_{ij}/(\mu_{ii}\mu_{jj})^{1/2}, \qquad (i,j = 1,\ldots,n), \qquad \text{(II.D.12)}$$

and called *correlation parameters* or, simply, *correlations*. The matrix obtained by using the correlations, ρ_{ij}, is called the *correlation matrix*, and will be denoted as C_x.

Using the *Cauchy-Schwartz inequality*, it can be shown that the elements of V_x always satisfy the relationship

$$|\mu_{ij}| \le (\mu_{ii}\mu_{jj})^{1/2} \quad (i,j = 1,n), \qquad (\text{II.D.13})$$

while the elements ρ_{ij} of the correlation matrix C_x satisfy the relationship

$$-1 \le \rho_{ij} \le 1. \qquad (\text{II.D.14})$$

In the context of covariance matrices, the *Cauchy-Schwartz inequality* provides an *indicator of data consistency* that is very useful to verify practical procedures for processing experimental information. Occasionally, practical procedures may generate covariance matrices with negative eigenvalues (thus violating the condition of positive-definiteness), or with coefficients that would violate the Cauchy-Schwartz inequality; such matrices would, of course, be unsuitable for representing physical uncertainty. Although the mathematical definition of the variance only indicates that it must be nonnegative, the variance for physical quantities should in practice be positive, because it provides a mathematical basis for the representation of physical uncertainty. Since zero variance means no error, probability functions for which some of the random variables have zero variance are not realistic choices for the representation of physical phenomena, since such probability functions would indicate that some parameters were without error, which is never the case in practice. Furthermore, a situation where $\mu_{ii} < 0$ would imply an imaginary standard deviation (since $\sigma_i = \mu_{ii}^{1/2}$), a manifestly absurd situation! The reason for mentioning these points here is because, in practice, the elements of covariance matrices are very seldom obtained from direct evaluation of expectations, but are obtained by a variety of other methods, many of them ad hoc. Practical considerations also lead to the requirement that $|\rho_{ij}| < 1$, for $i \ne j$, but a presentation of the arguments underlying this requirement is beyond the purpose of this book. These and other constraints on covariance and correlation matrices lead to the conclusion that *matrices which properly represent physical uncertainties are positive definite*.

Since covariance matrices are symmetric, an $n \times n$ matrix contains no more than $n + [n(n-1)/2]$ distinct elements, namely the off-diagonal covariances and the n variances along the diagonal. Often, therefore, only the diagonal and upper or lower triangular part of covariance and correlation matrices are listed in the literature. A formula often used in practical computations of covariances is obtained by rewriting the respective definition in the form

$$\operatorname{cov}(x_i, x_j) = E(x_i x_j) - m_{oi} m_{oj}. \tag{II.D.15}$$

Note that if any two random variables, x_i and x_j, in a collection of n random variables are independent, then $\operatorname{cov}(x_i, x_j) = 0$. Because the converse of this statement is false, it is important to state emphatically here that $\operatorname{cov}(x_i, x_j) = 0$ *does not necessarily imply that x_i and x_j are independent.*

A very useful tool for practical applications is the so-called *scaling and translation theorem*, which states that if x_i and x_j are any two members of a collection of n random variables, then the following relations hold for the random variables $y_i = a_i x_i + b_i$ and $y_j = a_j x_j + b_j$:

$$E(y_i) = a_i E(x_i) + b_i, \quad \operatorname{var}(y_i) = a_i^2 \operatorname{var}(x_i), \quad (i = 1, \ldots, n);$$

$$\operatorname{cov}(y_i, y_j) = a_i a_j \operatorname{cov}(x_i, x_j), \quad (i, j = 1, \ldots, n, \, i \neq j).$$

The constants a_i and a_j are called *scaling parameters*, while the constants b_i and b_j are called *translation parameters*. The above relationships show that mean values are affected by both scaling and translation, while the variances and covariances are only affected by scaling. In particular, the above relationships can be used to establish the following theorem regarding the relationship between ordinary random variables x_i and their standard random variable counterparts $u_i = (x_i - m_{oi})/\sigma_i$, $(i = 1, \ldots, n)$: *The covariance matrix for the standard random variables $u_i = (x_i - m_{oi})/\sigma_i$ is the same as the correlation matrix for the random variables x_i.*

The *determinant*, $\det(V_x)$, of the variance matrix is often referred to as the *generalized variance*, since it degenerates to a simple variance for univariate distributions. The probability distribution is called *nondegenerate* when $\det(V_x) \neq 0$; when $\det(V_x) = 0$, however, the distribution is called *degenerate*. Degeneracy is an indication that the information content of the set x of random variable is less than rank n, or that the probability function is confined to a hyperspace of dimension lower than n. Of course, the determinant of a covariance matrix vanishes iff there exist (one or more) linear relationships among the random variables of the set x.

Since V_x must be positive definite in order to provide a meaningful representation of uncertainty, it follows that $\det(V_x) > 0$. Due to the properties of V_x mentioned above, it also follows that

$$\det(\underline{V}_x) \le \prod_{i=1,n} \mathrm{var}(x_i) = \prod_{i=1,n} \sigma_i^2 . \qquad (\text{II.D.16})$$

The equality in the above relation is reached only when V_x is diagonal, i.e., when $\mathrm{cov}(x_i, x_j) = 0$, $(i, j = 1, n, i \ne j)$; in this case, $\det(V_x)$ attains its maximum value, equal to the product of the respective variances. The determinant $\det(V_x)$ is related to the determinant of the correlation matrix, $\det(C_x)$, by the relationship

$$\det(V_x) = \det(C_x) \prod_{i=1,n} \sigma_i^2 . \qquad (\text{II.D.17})$$

From Eqs. (II.D.16) and (II.D.17), it follows that $\det(C_x) \le 1$. It further follows that $\det(C_x)$ attains its maximum value of unity only when $\mathrm{cov}(x_i, x_j) = 0$, $(i, j = 1, n, i \ne j)$. In practice, $\det(C_x)$ is used as a *measure of degeneracy* of the multivariate probability function. In particular, *the quantity* $[\det(C_x)]^{1/2}$ is called the *scatter coefficient for the probability function*. Note that $\det(C_x) = 0$ when $\rho_{ij} = \rho_{ji} = 1$ for at least one pair (x_i, x_j), with $i \ne j$.

Two random variables, x_i and x_j, with $i \ne j$, are called *fully-correlated if* $\mathrm{cor}(x_i, x_j) = 1$; this situation arises iff the corresponding standard random variables u_i and u_j are identical, i.e., $u_i = u_j$. On the other hand, *if* $\mathrm{cor}(x_i, x_j) = -1$, *then* x_i *and* x_j *are fully anti-correlated*, which can happen iff $u_i = -u_j$. Therefore, the statistical properties of fully correlated or fully anti-correlated random variables are identical, so that only one of them needs to be considered, a fact reflected by the practice of referring to such random variables as being *redundant*.

In addition to covariance matrices, V_x, and their corresponding correlation matrices, C_x, a third matrix, called the *relative covariance matrix* or *fractional error matrix*, can also be defined when the elements of the covariance matrix satisfy the condition that $m_{oi} \ne 0$, $(i = 1, \ldots, n)$. This matrix is usually denoted as R_x, and its elements $(R_x)_{ij} = \eta_{ij}$ are defined as

$$\eta_{ij} = \mu_{ij} / (m_{oi} m_{oj}) . \qquad (\text{II.D.18})$$

The matrix R_x is used for representing uncertainty in evaluated nuclear data files, such as the U.S. Evaluated Nuclear Data File (ENDF) system. Evaluated data files are intended to provide suggested mean values and (if available) associated uncertainty information. The most compact and least redundant way to accomplish this is by representing the uncertainties with a relative covariance matrix, R_x.

Since variances (or standard deviations) provide a measure for the dispersion of the probability distribution with respect to its mean, they can be visualized as providing a measure of the region of (n-dimensional) random-variable space where most of the probability is concentrated. The intuitive understanding of the meaning of correlation is less straightforward; perhaps the simplest way to appreciate intuitively the meaning of correlation is to consider a bivariate distribution for the random variables x_1 and x_2. For simplicity, suppose that $-\infty \leq x_1 \leq \infty$ and $-\infty \leq x_2 \leq \infty$, and that p is a probability density for x_1 and x_2. Considering, without loss of generality, that x_1 and x_2 correspond to orthogonal coordinates of a two-dimensional Cartesian plane, the surface $x_3 = p(x_1, x_2)$ will appear as a "hill" in the third (Cartesian) coordinate x_3; note that x_3 is *not* a random variable. The surface $x_3 = p(x_1, x_2)$ is "centered," in the (x_1, x_2)-plane, on the point (m_{o1}, m_{o2}), while the lateral extents of this surface are measured by the standard deviations σ_1 and σ_2; furthermore, the surface $x_3 = p(x_1, x_2)$ would be symmetric when $\sigma_1 = \sigma_2$. Since p is normalized and since $p(x_1, x_2) \geq 0$ for all (x_1, x_2), the surface $p(x_1, x_2)$ will have at least one maximum in the direction x_3, say $(x_3)_{max}$. Slicing through the surface $x_3 = p(x_1, x_2)$ with a horizontal plane $x_3 = h$, where $0 < h < (x_3)_{max}$, and projecting the resulting planar figure onto the (x_1, x_2)-plane yields elliptical shapes that can be inscribed in the rectangle $(x_1 = 2l\sigma_1, x_2 = 2l\sigma_2)$, with $0 < l < \infty$. The covariance $cov(x_i, x_j)$, or equivalently, the *correlation parameter* $\rho_{12} = \rho$ indicates the orientation of the probability distribution in the random variable space (x_1, x_2). This is particularly clear when $\sigma_1 \neq \sigma_2$, in which case the surface $x_3 = p(x_1, x_2)$ is not symmetric. Thus, a nonzero correlation implies that the surface $x_3 = p(x_1, x_2)$ is *tilted* relative to the x_1 and x_2 axes. On the other hand, a zero correlation implies that the surface $x_3 = p(x_1, x_2)$ is somehow aligned with respect to these axes. Since the intrinsic shape of the surface $x_3 = p(x_1, x_2)$ is governed by the underlying probability but not by the choice of variables, it is possible to perform an orthogonal transformation such as to align the surface relative to the new, transformed coordinate system. The transformed random variables generated by an orthogonal transformation are called *orthogonal random variables*, and they are independent from one another.

Moments of first and second order (i.e., means and covariance matrices) provide information only regarding the location and dispersion of probability distributions. Additional information on the nature of probability distributions is carried by the higher-order moments, although moments beyond fourth-order are seldom examined in practice. The nature of such information can be intuitively understood by considering the third- and fourth-order moments of univariate probability functions. For this purpose, it is easiest to consider the respective *reduced central moments*, α_k, defined in terms of central moments and the standard deviation by the relationship $\alpha_k \equiv \mu_k/\sigma^k$. The reduced central moment α_3 is called the *skewness* of the probability distribution, because it measures quantitatively the departure of the probability distribution from symmetry (a symmetric distribution is characterized by the value $\alpha_3 = 0$). Thus, if $\alpha_3 < 0$, the distribution is skewed toward the left (i.e., it favors lower values of x relative to the mean), while $\alpha_3 > 0$ indicates a distribution skewed toward the right (i.e., it favors higher values of x relative to the mean). The reduced central moment α_4 measures the degree of sharpness in the peaking of a probability distribution and it is called *kurtosis*. Kurtosis is always nonnegative. The standard for comparison of kurtosis is the normal distribution (Gaussian) for which $\alpha_4 = 3$. This distribution is discussed, along with other distributions of practical interest, in Section II.F. Distributions with $\alpha_4 < 3$ are called *platykurtic distributions*. Those with $\alpha_4 = 3$ are called *mesokurtic distributions*. Finally, distributions with $\alpha_4 > 3$ are called *leptokurtic distributions*.

Very often in practice, the details of the distribution are unknown, and only the mean and standard deviations can be estimated from the limited amount of information available. Even under such circumstances, it is still possible to make statements regarding confidence by relying on *Chebyshev's theorem*, which can be stated as follows: *Consider that m_o and $\sigma > 0$ denote the mean value and standard deviation, respectively, of an otherwise unknown multivariate probability density p involving the random variable x. Furthermore, consider that P represents cumulative probability, C_p represents confidence, and $k \geq 1$ is a real constant (not necessarily an integer). Then, Chebyshev's theorem states that the following relationship holds:*

$$C_p(k\sigma) = P(|x - m_o| \leq k\sigma) \geq 1 - (1/k^2). \qquad \text{(II.D.19)}$$

Chebyshev's theorem is a weak law of statistics in that it *provides an upper bound on the probability of a particular deviation* ε. The actual probability of such a deviation (if the probability function were known in detail so that it could be precisely calculated) would always be smaller (implying greater confidence) than Chebyshev's limit. This important point is illustrated in the table below, which presents probabilities for observing particular deviations, ε, from the mean when sampling from a normal distribution, and the corresponding bounds predicted by Chebyshev's theorem:

Probability of Occurrence

Deviation ε	Normal distribution	Chebyshev's limit
$>\sigma$	<0.3173	<1.0
$>2\sigma$	<0.0455	<0.25
$>3\sigma$	<0.00270	<0.1111
$>4\sigma$	<0.0000634	<0.0625
$>5\sigma$	<$5.73*10^{-7}$	<0.04
$>6\sigma$	<$2.0*10^{-9}$	<0.02778

The above table clearly underscores the fact that normally distributed random variables are much more sharply localized with respect to the mean than indicated by Chebyshev's theorem.

II.E. CHARACTERISTIC AND MOMENT GENERATING FUNCTIONS

The *characteristic function* $\phi_x(k)$ for a random variable x with PDF $p(x)$ is defined as the expectation value of e^{ikx}, namely

$$\phi_x(k) \equiv E(e^{ikx}) = \int_{-\infty}^{\infty} e^{ikx} p(x) dx. \qquad \text{(II.E.1)}$$

The above definition is essentially the Fourier transform of the probability density function; there is a one-to-one correspondence between the PDF and the characteristic function, so that knowledge of one is equivalent to knowledge of the other. The PDF, $p(x)$, is obtained from the inverse Fourier transform of the characteristic function $\phi_x(k)$, namely

$$p(x) = \frac{1}{2\pi} \int_{-\infty}^{\infty} \phi_x(k) e^{-ikx} dk. \qquad \text{(II.E.2)}$$

The moments of $p(x)$ can be obtained by differentiating the characteristic function $\phi_x(k)$ m times and evaluating the resulting expression at $k = 0$; this yields the relation

$$\left\{\frac{d^m}{dk^m}\phi_x(k)\right\}_{k=0} = \left\{\frac{d^m}{dk^m}\int e^{ikx}p(x)dx\right\}_{k=0} = i^m \int x^m p(x)dx = i^m v_m \quad \text{(II.E.3)}$$

where $v_m = E(x^m)$ is the mth raw moment (about the origin) of x.

For easy reference, the table below lists characteristic functions for some commonly used probability functions.

Distribution	PDF	$\phi(k)$
Binomial	$p(x) = \dfrac{n!}{x!(n-x)!}s^x(1-s)^{n-x}$	$\left[s(e^{ik}-1)+1\right]^n$
Poisson	$p(x) = \dfrac{v^x}{x!}e^{-v}$	$\exp\left[v(e^{ik}-1)\right]$
Uniform	$p(x) = \begin{cases} \dfrac{1}{\beta-\alpha}, & \alpha \leq x \leq \beta \\ 0, & \text{otherwise} \end{cases}$	$\dfrac{e^{i\beta k} - e^{i\alpha k}}{(\beta-\alpha)ik}$
Exponential	$p(x) = \dfrac{1}{\xi}e^{-x/\xi}$	$\dfrac{1}{1-ik\xi}$
Gaussian	$p(x) = \dfrac{1}{\sqrt{2\pi\sigma^2}}\exp\left(\dfrac{-(x-v)^2}{2\sigma^2}\right)$	$\exp\left(ivk - \dfrac{1}{2}\sigma^2 k^2\right)$
Chi-square	$p(x) = \dfrac{1}{2^{n/2}\Gamma(n/2)}x^{n/2-1}e^{-z/2}$	$(1-2ik)^{-n/2}$
Cauchy	$f(x) = \dfrac{1}{\pi}\dfrac{1}{1+x^2}$	$e^{-|k|}$

Related to the characteristic function is the *moment generating function* (*MGF*), usually denoted by $M_x(t)$, and defined as the expectation of e^{tx}, namely:

$$M_x(t) = E(e^{tx}) = \int_{-\infty}^{\infty} e^{tx} p(x) dx, \quad \text{(II.E.4)}$$

where x is a random variable and t is a real number, $t \in \mathbb{R}$. Occasionally, t

may be restricted to an interval subspace of the real line, say, $t \in (a,b) \subset \mathbb{R}$, but the interval (a,b) must necessarily contain the value $t=0$. For convenience, one occasionally refers to t as a *transform variable*. Denoting the kth-derivative of $M_x(t)$ with respect to t by $M_x^{(k)}(t)$, and differentiating Eq. (II.E.4) repeatedly with respect to t leads to the following expressions:

$$M_x^{(1)}(t) = \int_{-\infty}^{\infty} x e^{tx} p(x) dx,$$

$$M_x^{(2)}(t) = \int_{-\infty}^{\infty} x^2 e^{tx} p(x) dx,$$

$$M_x^{(3)}(t) = \int_{-\infty}^{\infty} x^3 e^{tx} p(x) dx,$$

$$M_x^{(k)}(t) = \int_{-\infty}^{\infty} x^k e^{tx} p(x) dx.$$

(II.E.5)

Setting $t=0$ in Eq. (II.E.5) and recalling the definition of the raw moment, v_k, of order k of x (about the origin), it becomes apparent that $M_x^{(k)}(0) = E(x^k) = v_k$. Thus, if the expression of the *MGF*, $M_x(t)$, is known for a probability distribution of x, then all the raw moments of the respective distribution can be calculated by taking derivatives of the *MGF* with respect to t and evaluating them at $t=0$.

The most important feature of *MGF*'s, namely uniqueness, is established by the following theorem: two random variables x and y are identically distributed iff their corresponding *MGF*'s, $M_x(t)$ and $M_y(t)$, are equal for every real number t in the range $-t_o < t < t_o$ $(t_o > 0)$. This theorem implies that the information contents of a probability function and of its corresponding *MGF* are completely equivalent. Knowledge of this fact is very important for practical applications since, in statistical problems, it is often easier to deal with equivalent *MGF*'s rather than with the underlying probability functions.

The *MGF* function, $M_x(t)$, generates directly the raw moments v_k of a probability function, including the mean value $m_o = v_1$. In addition, the *MGF* can also be used in conjunction with scaling and translations of random variables to obtain, perhaps more easily, the central moments μ_k. The procedures most often used for this purpose are embodied in the following four theorems, which deal with random-variable translation and scaling:

If the MGF, $M_x(t)$, of the PDF of a random variable x exists, and if c is a constant, then $M_{x+c}(t) = e^{ct} M_x(t)$.

If the conditions of the preceding theorem apply, and if $m_o = v_1$ is the mean value, then the MGF, $M_{x-m_o}(t)$, is the generator of the central moments of the underlying probability function.

If the MGF, $M_x(t)$, of the PDF of a random variable x exists, and if c is a constant, then $M_{cx}(t) = M_x(ct)$.

The MGF for the PDF of a random variable x is related to the MGF of the PDF for the standard random variable $u = (x - m_o)/\sigma$ by the formula $M_u(t) = \exp(-m_o t/\sigma) M_x(t/\sigma)$.

The development of *MGF*'s for probability functions of several random variables (random vectors) is analogous to the development for the univariate case. Thus, consider that $x = (x_1, x_2, \ldots, x_n)$ denotes a random vector with n components, where each $x_i \in \mathbb{R}$, $(i = 1, \ldots, n)$. Furthermore, consider that $\mathcal{S}_x$ represents the corresponding space of all such random vectors x, and that $p(x)$ denotes the associated probability density. Then the multivariate *MGF* is defined as

$$M_x(t) = E\left(\exp\left(\sum_{i=1}^n x_i t_i \right) \right) = \int_{\mathcal{S}_x} \exp\left(\sum_{i=1}^n x_i t_i \right) p(x) dx, \qquad (\text{II.E.6})$$

where the vector $(t_1, t_2, \ldots, t_n)$ has components t_i defined symmetrically around the origin $(0, 0, \ldots, 0)$, namely $-t_{oi} < t_i < t_{oi}$, with $t_{oi} > 0$, $(i = 1, \ldots, n)$. Just as for the univariate case, the *MGF* defined in Eq. (II.E.6) is unique, and there is a complete equivalence between it and the underlying probability function. Note that *MGF*'s for discrete multivariate distributions are defined similarly as above, except that the multiple integrals in Eq. (II.E.6) are replaced by corresponding multiple sums.

Mutual stochastic independence of the random variables $(x_1, x_2, \ldots, x_n)$ leads to factorization of the corresponding *MGF*, just as it does for the underlying probability function: *iff the random variables* $(x_1, x_2, \ldots, x_n)$ *are mutually independent, then* $M_x(t) = \prod_{i=1,n} M_{x_i}(t_i)$.

The raw moments of multivariate probability functions can be generated from the partial derivatives of $M_x(t)$ with respect to the parameters t_i, $(i = 1, \ldots, n)$, evaluated at $t = 0$. For example,

$$m_{oi} = v_i = E(x_i) = [\partial M_x(t)/\partial t_i]_{t=o} \quad (i = 1,\ldots,n),$$
$$E(x_i x_j) = [\partial^2 M_x(t)/\partial t_i \partial t_j]_{t=o} \quad (i, j = 1,\ldots,n),$$
(II.E.7)

and so on. In particular, the elements of the covariance matrix V_x, $(i, j = 1,\ldots,n)$, can be readily generated by using Eq. (II.E.7) in conjunction with the relation $\text{cov}(x_i, x_j) = E[(x_i - m_{oi})(x_j - m_{oj})] = \mu_{ij} = E(x_i x_j) - m_{oi} m_{oj}$, and using the convention $\text{var}(x_i) = \text{cov}(x_i, x_i)$, $(i = 1,\ldots,n)$.

II.F. COMMONLY ENCOUNTERED PROBABILITY DISTRIBUTIONS

This Section presents, mostly without proofs, the most important features of the probability distributions and density functions most commonly used for uncertainty analysis of data and models.

Degenerate Distribution: Consider that x is a random variable which can assume only a single value, namely, $x = c$, where c is a real number. The distribution for x is called the *degenerate distribution*, and the corresponding probability density is given by

$$p(x) = \delta(x - c),$$
(II.F.1)

where $-\infty \leq x \leq \infty$, and δ denotes the Dirac delta functional. The *MGF* for this distribution is

$$M_x(t) = e^{tc}, \quad (t \in \mathbb{R}),$$
(II.F.2)

while the mean value is $m_o = c$, and the variance is $\mu_2 = 0$.

Discrete Uniform Distribution: Consider that x is a random variable, which can assume only the integer values $x = 1,\ldots,n$. Each of these values carries equal probability. The distribution for x is called the *discrete uniform distribution*, and the corresponding probability is given by

$$p(x) = \begin{cases} 1/n, & (x = 1, 2, \ldots, n) \\ 0, & \text{otherwise.} \end{cases}$$
(II.F.3)

Using the Dirac delta functional, Eq. (II.F.3) can also be written in the alternative form $p(x) = (1/n) \sum_{i=1,n} \delta(x - i)$. The *MGF* for this distribution is

$$M_x(t) = \left[e^t\left(1-e^{nt}\right)\right]/\left[n\left(1-e^t\right)\right], \qquad (t \in \mathbb{R}). \qquad \text{(II.F.4)}$$

Using either the above *MGF* or by direct calculations, the mean value of x is obtained as $m_o = (n+1)/2$, while the variance of x is obtained as $\mu_2 = (n^2 - 1)/12$.

Continuous Uniform Distribution: Consider that x is a random variable, which can assume any real value in the nondegenerate (i.e., $a < b$) interval $I(a,b)$. The distribution for x is called the *continuous uniform distribution*, and the corresponding probability density function is given by

$$p(x) = \begin{cases} 1/(b-a), & a \leq x < b;\ a,b \in \mathbb{R}, \\ 0, & \text{otherwise}. \end{cases} \qquad \text{(II.F.5)}$$

The *MGF* for this distribution is

$$M_x(t) = \begin{cases} \left[\left(e^{tb} - e^{ta}\right)\right]/[(b-a)t], & t \in \mathbb{R},\ t \neq 0, \\ 1, & t = 0. \end{cases} \qquad \text{(II.F.6)}$$

From the above *MGF*, the mean value is obtained as $m_o = (a+b)/2$, while the variance is obtained as $\mu_2 = (b-a)^2/12$. This distribution is employed wherever the range of a finite random variable is bounded and there is no *a priori* reason for favoring one value over another within that range. In practical applications, the continuous uniform distribution is often used in Monte Carlo analysis or computational methods, and as a prior for applying Bayes' theorem in the extreme situation when no information is available prior to performing an experiment.

Bernoulli Distribution: Bernoulli trials are defined as random trials in which the outcomes can be represented by a random variable having only two values, say $x = 0$ and $x = 1$. Such a model can be applied to a random trial whose outcomes (events) are described by "yes or no," "on or off," "black or white," "success or failure," etc. Suppose that only one Bernoulli trial is performed, for which the probability of "success" $(x = 1)$ is denoted by s, for $0 < s < 1$, while the probability of "failure" $(x = 0)$ is $(1-s)$. The probability distribution that describes this trial is called the *Bernoulli distribution*, and has the form

Concepts of Probability Theory

$$P(x) = \begin{cases} s^x(1-s)^{1-x}, & x = 0,1 \\ 0, & \text{otherwise} \end{cases} \quad \text{(II.F.7)}$$

The *MGF* for this distribution is given by

$$M_x(t) = se^t + 1 - s, \quad (t \in \mathbb{R}). \quad \text{(II.F.8)}$$

From the above *MGF*, the mean value is obtained as $m_o = s$, while the variance is obtained as $\mu_2 = s(1-s)$. The Bernoulli distribution provides a basis for the binomial and related distributions.

Binomial and multinomial distributions: Consider a series of n independent trials or observations, each having two possible outcomes, usually referred to as "success" and "failure," respectively. Consider further that the probability for success takes on a constant value, s, and consider that the quantity of interest is the accumulated result of n such trials (as opposed to the outcome of just one trial). The set of n trials can thus be regarded as a single measurement characterized by a *discrete* random variable x, *defined to be the total number of successes*. Thus, the sample space is defined to be the set of possible values of x successes given n observations. If the measurement were repeated many times with n trials each time, the resulting values of x would occur with relative frequencies given by the *binomial distribution*, which is defined as

$$P(x) = \frac{n!}{x!(n-x)!} s^x (1-s)^{n-x}, \quad (x = 0,1,\ldots,n), \quad \text{(II.F.9)}$$

where x is the random variable, while n and s are parameters characterizing the binomial distribution. Note that the binomial distribution is symmetric for $s = 1/2$. The *MGF* is obtained as

$$M_x(t) = (se^t + 1 - s)^n, \quad (t \in \mathbb{R}). \quad \text{(II.F.10)}$$

The mean value of x is given by

$$m_o = \sum_{n=0}^{\infty} x \frac{n!}{x!(n-x)!} s^x (1-s)^{n-x} = ns, \quad \text{(II.F.11)}$$

while the variance is given by $\mu_2 = ns(1-s)$.

If the space $\mathcal{E}$ of all possible simple events is partitioned in $m+1$ (instead of just two, as was the case for the binomial distribution) compound events $A_i, (i = 1,\ldots,m+1)$, the binomial distribution can be generalized to the *multinomial distribution* by considering a set of nonnegative integer variables $(x_1, x_2,\ldots, x_{m+1})$, satisfying the conditions $\sum_{i=1}^{m+1} x_i = n$, $\sum_{i=1}^{m+1} A_i = \mathcal{E}$, and $\sum_{i=1}^{m+1} s_i = 1$ (since one of the outcomes must ultimately be realized).

Then, the multinomial distribution for the joint probability for x_1 outcomes of type 1, x_2 of type 2, etc. is given by

$$P(x_1,\ldots,x_{m+1}) = \frac{n!}{x_1!\ldots x_{m+1}!} s_1^{x_1}\ldots s_{m+1}^{x_{m+1}}. \qquad \text{(II.F.12)}$$

The *MGF* for this distribution is

$$M_{x_1\ldots x_{m+1}}(t_1,\ldots,t_m) = \left(\sum_{i=1}^{m} s_i e^{t_i} + s_{m+1}\right)^n, \quad (t_i \in \mathbb{R}, i = 1,\ldots,m). \qquad \text{(II.F.13)}$$

The variances and covariances for this distribution are, respectively:

$$\mu_{ii} = \sigma_i^2 = n s_i(1-s_i), \text{ for } i = j, \text{ and } \mu_{ij} = -n s_i s_j, \text{ for } i \neq j \; (i, j = 1,\ldots,m+1). \qquad \text{(II.F.14)}$$

Since μ_{ij} is negative (anti-correlated variables), it follows that, if in n trials, bin i contains a larger than average number of entries ($x_i > n s_i$), then the probability is increased that bin j will contain a smaller than average number of entries.

Geometric Distribution: The geometric distribution is also based on the concept of a Bernoulli trial. Consider that s, $0 < s < 1$, is the probability that a particular Bernoulli trial is a success, while $1-s$ is the corresponding probability of failure. Also, consider that x is a random variable that can assume the infinite set of integer values $(1, 2,\ldots)$. The *geometric distribution gives the probability that the first* $x-1$ *trials will be failures, while the* x^{th} *trial is a success*. Therefore, it is the distribution of the "waiting time" for a success. Thus, the probability function characterizing the geometric distribution is

$$P(x) = s(1-s)^{x-1}, \quad (x = 1, 2,\ldots). \qquad \text{(II.F.15)}$$

The *MGF* for this distribution is

$$M_x(t) = se^t / \left[1 - (1-s)e^t\right], \quad (t < -\ln(1-s)). \tag{II.F.16}$$

From the above *MGF*, the mean value is obtained as $m_o = (1/s)$, while the variance is obtained as $\mu_2 = (1-s)/s^2$.

Negative Binomial (Pascal) Distribution: The negative binomial (Pascal) distribution also employs the concept of a Bernoulli trial. Thus, consider that s, $0 < s < 1$, is the probability of success in any single trial and $1-s$ is the corresponding probability of failure. This time, though, the result of interest is the *number of trials that are required in order for r successes to occur*, $(r = 1,2,...)$. Note that at least r trials are needed in order to have r successes. Consider, therefore, that x is a random variable that represents the number of additional trials required (beyond r) before obtaining r successes, so that $(x = 0,1,2...)$. Then, the form of the Pascal probability distribution is found to be

$$P(x) = C_{mx} s^r (1-s)^x, \quad (x = 0,1,2,...), \tag{II.F.17}$$

where $m = x + r - 1$ and C_{mx} is the binomial coefficient. The *MGF* for the binomial distribution is found from Eq. (II.F.17) to be

$$M_n(t) = \left[s / \left(1 - (1-s)e^t\right)\right]^r, \quad (t < -\ln(1-s)). \tag{II.F.18}$$

It follows from Eq. (II.F.18) that the mean value for the Pascal distribution is $m_o = [r(1-s)/s]$, while the variance is $\mu_2 = [r(1-s)/s^2]$.

Poisson distribution: In the limit of many trials, as n becomes very large and the probability of success s becomes very small, but such that the product ns (i.e., the expectation value of the number of successes) remains equal to some finite value v, the binomial distribution takes on the form

$$P(x) = \frac{v^x}{x!} e^{-v}, \quad (x = 0,1,...), \tag{II.F.19}$$

which is called the *Poisson distribution* for the integer random variable x. The corresponding *MGF* is given by the expression

$$M_x(t) = \exp[\nu(e^t - 1)], \quad (t \in \mathbb{R}). \tag{II.F.20}$$

From the above *MGF*, or from a direct calculation, the expectation value of the Poisson random variable x is obtained as

$$m_o = E(x) = \sum_{x=0}^{\infty} x \frac{\nu^x}{x!} e^{-\nu} = \nu, \tag{II.F.21}$$

and the variance is obtained as

$$\mu_2 = V(x) = \sum_{x=0}^{\infty} (x - \nu)^2 \frac{\nu^x}{x!} = \nu. \tag{II.F.22}$$

Although the Poisson variable x is discrete, it can be treated as a continuous variable if it is integrated over a range $\Delta x \gg 1$. An example of a Poisson random variable is the number of decays of a certain amount of radioactive material in a fixed time period, in the limit that the total number of possible decays (i.e., the total number of radioactive atoms) is very large and the probability for an individual decay within the time period is very small.

Exponential distribution: The exponential *PDF*, $p(x)$, of the continuous variable x ($0 \le x \le \infty$) is defined by

$$p(x) = \frac{1}{\xi} e^{-x/\xi}, \tag{II.F.23}$$

where ξ a is real-valued parameter. The expectation value of x is given by

$$m_o = E(x) = \frac{1}{\xi} \int_0^\infty x e^{-x/\xi} dx = \xi, \tag{II.F.24}$$

and the variance of x is given by

$$\mu_2 = V(x) = \frac{1}{\xi} \int_0^\infty (x - \xi)^2 e^{-x/\xi} dx = \xi^2. \tag{II.F.25}$$

The *MGF* is given by the expression

$$M_x(t) = \lambda/(\lambda - t); \quad \lambda \equiv 1/\xi, \quad (t \in \mathbb{R}, t < \lambda). \tag{II.F.26}$$

The exponential distribution is widely used in radioactivity applications and in equipment *failure rate* analysis. The failure rate is defined as the reciprocal of the mean time to failure, i.e., $\xi = 1/\lambda$. The quantity $R(t_o) = e^{-\lambda t_o}$ is usually denoted as the reliability (of the equipment) at time $t_o > 0$. It should not be surprising that the reliability of a piece of equipment should decline with age, and statistical experience has shown that this decline is well represented by the exponential distribution.

Gaussian distribution: Perhaps the single most important distribution in theoretical as well as in applied statistics is the *Gaussian (or normal) PDF* of the continuous random variable x (with $-\infty < x < \infty$), defined as

$$p(x) = \frac{1}{\sqrt{2\pi\sigma^2}} \exp\left(\frac{-(x-v)^2}{2\sigma^2}\right). \tag{II.F.27}$$

Note that the Gaussian distribution has two parameters, v and σ^2, which are, by design, the mean and variance of x, i.e.,

$$E(x) = \int_{-\infty}^{\infty} x \frac{1}{\sqrt{2\pi\sigma^2}} \exp\left(\frac{-(x-v)^2}{2\sigma^2}\right) dx = v, \tag{II.F.28}$$

$$V(x) = \int_{-\infty}^{\infty} (x-\mu)^2 \frac{1}{\sqrt{2\pi\sigma^2}} \exp\left(\frac{-(x-v)^2}{2\sigma^2}\right) dx = \sigma^2. \tag{II.F.29}$$

The *MGF* for the Gaussian distribution is

$$M_x(t) = \exp\left[vt + \left(\sigma^2 t^2/2\right)\right], \quad (t \in \mathbb{R}). \tag{II.F.30}$$

The *standard normal distribution* is obtained by setting $v = 0$ and $\sigma = 1$ in Eq. (II.F.27), to obtain

$$p(x) = \exp\left(-x^2/2\right)/(2\pi)^{1/2}, \tag{II.F.31}$$

with *MGF*

$$M_x(t) = \exp\left(t^2/2\right), \quad (t \in \mathbb{R}). \tag{II.F.32}$$

The function

$$P(x) = \int_{-x}^{x} p(z)dz, \quad (x > 0), \tag{II.F.33}$$

with $p(z)$ given by Eq. (II.F.31), represents the integrated probability of an event with $-x \leq z < x$, for the standard normal distribution.

The first four derivatives of the *MGF*, $M_x(t)$, of the standard normal distribution are obtained as:

$$M_x^{(1)}(t) = t\exp(t^2/2), \quad M_x^{(2)}(t) = (1+t^2)\exp(t^2/2),$$
$$M_x^{(3)}(t) = (3t+t^3)\exp(t^2/2), \quad M_x^{(4)}(t) = (3+6t^2+t^4)\exp(t^2/2).$$

Evaluating the above expressions at $t=0$ yields:

$$M_x^{(1)}(0) = v_1 = m_o = 0,$$
$$M_x^{(2)}(0) = v_2 = \mu_2 = \sigma^2 = 1,$$
$$M_x^{(3)}(0) = v_3 = \mu_3 = \alpha_3 = 0,$$
$$M_x^{(4)}(0) = v_4 = \mu_4 = \alpha_4 = 3.$$

Thus, a normal distribution is symmetric (since it has skewness $\alpha_3 = 0$) and is mesokurtic (since it has kurtosis $\alpha_4 = 3$). Furthermore, since $m_o = 0$ for all standard distributions, the respective raw and central moments are equal to each other. In particular, it can be shown that $\mu_{2k-1} = 0$, $(k = 1,2,...)$, which indicates that all of the odd central moments of the normal distribution vanish. Furthermore, it can be shown that the even central moments are given by $\mu_{2k} = (1)(3)...(2k-1)\sigma^{2k}$, $(k = 1,2,...)$. These results highlight the very important feature of the normal distribution that all its nonzero higher-order central moments can be expressed in terms of a single parameter, namely the standard deviation. This is one of several reasons why the Gaussian distribution is arguably the single most important *PDF* in statistics.

Another prominent practical role played by the Gaussian distribution is as a replacement for either the binomial distribution or the Poisson distribution. The circumstances under which such a replacement is possible are given by *DeMoivre-Laplace theorem* (given below without proof), which can be stated as follows: *Consider the binomial distribution, denoted as $p_b(k)$, $(k = 1,...,n)$ (n and s denote the usual parameters of the binomial distribution, while the subscript "b" denotes "binomial"), and consider that a and c are two nonnegative integers satisfying $a < c < n$. Furthermore, consider the standard* *normal distribution, denoted as* $p_{sn}(x)$ *(the subscript "sn" denotes "standard normal"). Finally, define the quantities* α *and* β *as*

$$\alpha \equiv (a - ns - 0.5)/[ns(1-s)]^{1/2} \text{ and } \beta \equiv (c - ns + 0.5)/[ns(1-s)]^{1/2}.$$

Then, for large n, the following relation holds

$$\sum_{k=a}^{c} p_b(k) \approx \int_{\alpha}^{\beta} p_{sn}(x)dx. \qquad (II.F.34)$$

Thus, the DeMoivre-Laplace theorem indicates that the sum of the areas of contiguous histogram segments, representing discrete binomial probabilities, approximately equals the area under the corresponding continuous Gaussian curve spanning the same region. The binomial becomes increasingly symmetrical as $s \to 0.5$, so the approximation provided by the DeMoivre-Laplace theorem is accurate even for relatively small n, since the Gaussian is intrinsically symmetric. When $s \ll 1$, and n is so large that $\sigma^2 = ns(1-s) = v(1 - v/n) \approx v$, even though $v \gg 1$, then the normal distribution can be shown to be a reasonably good approximation to both the corresponding Poisson distribution and to the binomial distribution (particularly when $x \approx v$).

Another extremely important role played by the Gaussian distribution is highlighted by the *Central Limit Theorem*, which essentially states that *the sum of n independent continuous random variables* x_i *with means* μ_i *and variances* σ_i^2 *becomes a Gaussian random variable with mean* $\mu = \sum_{i=1}^{n} \mu_i$ *and variance* $\sigma^2 = \sum_{i=1}^{n} \sigma_i^2$ *in the limit that n approaches infinity*. This statement holds under fairly general conditions, regardless of the form of the individual *PDF*'s of the respective random variables x_i. The central limit theorem provides the formal justification for treating measurement errors as Gaussian random variables, as long as the total error is the sum of a large number of small contributions.

The behavior of certain distributions for limiting cases of their parameters can be investigated more readily by using characteristic or moment generating functions, rather than their *PDF*'s. For example, taking the limit $s \to 0, n \to \infty$, with $v = sn$ (constant), in the characteristic function for the binomial distribution yields the characteristic function of the Poisson distribution:

$$\phi(k) = \left[s\left(e^{ik}-1\right)+1\right]^n = \left(\frac{v}{n}\left(e^{ik}-1\right)+1\right)^n \to \exp\left[v\left(e^{ik}-1\right)\right].$$

Note also that a Poisson variable x with mean v becomes a Gaussian variable in the limit as $v \to \infty$. This fact can be shown as follows: although the Poisson variable x is discrete, when it becomes large, it can be treated as a continuous variable as long as it is integrated over an interval that is large compared to unity. Next, the Poisson variable x is transformed to the variable

$$z = \frac{x-v}{\sqrt{v}}.$$

The characteristic function of z is

$$\phi_z(k) = E\left(e^{ikz}\right) = E\left(e^{ikx/\sqrt{v}}e^{-ik\sqrt{v}}\right) = \phi_x\left(\frac{k}{\sqrt{v}}\right)e^{-ik\sqrt{v}},$$

where ϕ_x is the characteristic function of the Poisson distribution. Expanding the exponential term and taking the limit as $v \to \infty$ yields

$$\phi_z(k) = \exp\left[v\left(e^{ik/\sqrt{v}}-1\right)-ik\sqrt{v}\right] \to \exp\left(-\frac{1}{2}k^2\right).$$

The last term on the right side of the above expression is the characteristic function for a Gaussian with zero mean and unit variance. Transforming back to the original Poisson variable x, one finds, therefore, that for large v, x follows a Gaussian distribution with mean and variance both equal to v.

Multivariate Normal Distribution: Consider that x is a (column) vector of dimension n, with components $x_i \in \mathbb{R}$, $(i=1,\ldots,n)$. Consider further that $\boldsymbol{m}_o$ is a (column) vector of constants with components denoted as $m_{oi} \in \mathbb{R}$, $(i=1,\ldots,n)$. Finally, consider that V is a real, symmetric, and positive definite $n \times n$ matrix with elements denoted by μ_{ij}, $(i,j=1,\ldots,n)$. Then, the *multivariate normal distribution for n variables* is defined by the probability density function

$$p(\boldsymbol{x}) = (2\pi)^{-n/2}[\det(V)]^{-1/2}\exp\left[-(1/2)(\boldsymbol{x}-\boldsymbol{m}_o)^+ V^{-1}(\boldsymbol{x}-\boldsymbol{m}_o)\right] \qquad \text{(II.F.35)}$$

where the superscript "+" denotes matrix transposition, and where V^{-1} is the inverse of V. The quantity

$$Q \equiv (x - m_o)^+ V^{-1}(x - m_o) \tag{II.F.36}$$

is called a *multivariate quadratic form*. The *MGF* corresponding to Eq. (II.F.35) is given by

$$M_x(t) = \exp\left[t^+ m_o + (1/2) t^+ V^{-1} t\right], \quad (t_i \in \mathbb{R}; \, i = 1, \ldots, n). \tag{II.F.37}$$

Note that m_o is the *mean vector*, while V is actually the covariance matrix for this distribution; note also that

$$E(x_i) = \left[\partial M_x(t)/\partial t_i\right]_{t=0} = m_{oi}, \quad E(x_i x_j) = \left[\partial^2 M_x(t)/\partial t_i \partial t_j\right]_{t=0} = m_{oi} m_{oj} + \mu_{ij}.$$

In particular, the Gaussian *PDF* for two random variables x_1 and x_2 becomes

$$p(x_1, x_2; m_{o1}, m_{o2}, \sigma_1, \sigma_2, \rho) = \frac{1}{2\pi \sigma_1 \sigma_2 \sqrt{1-\rho^2}}$$

$$\times \exp\left\{-\frac{1}{2(1-\rho^2)}\left[\left(\frac{x_1 - m_{o1}}{\sigma_1}\right)^2 + \left(\frac{x_2 - m_{o2}}{\sigma_2}\right)^2 - 2\rho\left(\frac{x_1 - m_{o1}}{\sigma_1}\right)\left(\frac{x_2 - m_{o2}}{\sigma_2}\right)\right]\right\}$$

where $\rho = \mathrm{cov}(x_1, x_2)/(\sigma_1 \sigma_2)$ is the correlation coefficient.

The following properties of the multivariate normal distribution are often used in applications to data evaluation and analysis:

1. *If x is a normally distributed random vector of dimension n with corresponding mean vector m_o and positive definite covariance matrix V, and if A is an $m \times n$ matrix of rank m, with $m \leq n$, then the m-dimensional random vector $y = Ax$ is also normally distributed, with a mean (vector) equal to Am_o and a positive definite covariance matrix equal to AVA^+*, where the superscript "+" denotes "transposition".

2. *If x is a normally distributed random vector of dimension n, with corresponding mean vector m_o and positive definite covariance matrix V, and if A is a nonsingular $n \times n$ matrix such that $A^+ V^{-1} A = I$ (the diagonal unit matrix), then $y = A^{-1}(x - m_o)$ is a random vector with a zero mean vector, 0,*

and unit covariance matrix, I. This linear transformation is the multivariate equivalent of the standard-variable transformation for the univariate Gaussian distribution.

3. If x is a normally distributed random vector of dimension n with corresponding mean vector m_o and positive definite covariance matrix V, then the quadratic form

$$Q = (x - m_o)^{\dagger} V^{-1}(x - m_o) \tag{II.F.38}$$

is a statistic distributed according to the chi-square distribution with n degrees of freedom.

4. If x is a normally distributed vector with mean vector m_o and covariance matrix V, then the individual variables $x_1, x_2, \ldots, x_n$ are mutually independent iff all the elements μ_{ij} in V are equal to zero for $i \neq j$.

The geometrical features of multivariate normal distributions can be most readily visualized by considering the bivariate distribution, for which the quadratic form of Eq. (II.F.38) becomes

$$Q = \frac{\{[(x_1 - m_{o1})^2/\sigma_1^2] - [2\rho(x_1 - m_{o1})(x_2 - m_{o2})/(\sigma_1\sigma_2)] + [(x_2 - m_{o2})^2/\sigma_2^2]\}}{(1 - \rho^2)},$$

(II.F.39)

where σ_1 and σ_2 are the standard deviations for x_1 and x_2, respectively, and ρ is the correlation coefficient (i.e., $\mu_{11} = \sigma_1^2$, $\mu_{22} = \sigma_2^2$, and $\mu_{12} = \mu_{21} = \rho\sigma_1\sigma_2$ are the elements of the covariance matrix V). Considering, in three dimensions, that x_1 and x_2 are two rectangular coordinates in a plane while x_3 is the coordinate of elevation above this plane, a plot of the bivariate Gaussian surface $x_3 = p(x_1, x_2)$ forms a smooth "hill" centered on the position $x_1 = m_{o1}$ and $x_2 = m_{o2}$. Regardless of their azimuthal orientation, planes $x_1 = const.$ or $x_2 = const.$ (parallel to x_3) that pass through the central point (m_{o1}, m_{o2}) will cut the surface $x_3 = p(x_1, x_2)$ into univariate Gaussian profiles. Planes perpendicular to x_3 (i.e., $x_3 = c = const.$) will cut the surface $x_3 = p(x_1, x_2)$ into ellipses, provided that $0 < c < [2\pi\sigma_1\sigma_2(1 - \rho^2)]^{-1/2}$; note that $Q = -2\ln\left[2\pi\sigma_1\sigma_2 c(1 - \rho^2)^{1/2}\right]$ when $x_3 = p(x_1, x_2) = c$. The projections of these ellipses onto the $x_3 = 0$ plane will be imbedded in rectangles centered at (m_{o1}, m_{o2}), with sides $a_1 = 2\beta\sigma_1$ (along the x_1-coordinate axis) and $a_2 = 2\beta\sigma_2$

(along the x_2-coordinate axis). The correlation coefficient ρ determines the shape and orientation of the surface. Thus, when $\rho = \pm 1$, the surface degenerates and lies entirely in a plane perpendicular to the $x_3 = 0$ plane. More generally, when $|\rho| < 1$, the ellipses take on various orientations that depend upon the magnitude and sign of ρ. In addition, the shapes of the profiles produced by horizontal and vertical planes cutting through the surface $x_3 = p(x_1, x_2)$ also depend upon the relative magnitudes of σ_1 and σ_2.

Log-normal distribution: The random variable $x = e^y$, where y is a Gaussian random variable with mean μ and variance σ^2, is distributed according to the log-normal distribution

$$p(x) = \frac{1}{\sqrt{2\pi\sigma^2}} \frac{1}{x} \exp\left(-\frac{(\log x - \mu)^2}{2\sigma^2}\right). \qquad \text{(II.F.40)}$$

Since the log-normal and Gaussian distributions are closely connected to one another, they share many common properties. For example, the moments of the log-normal distribution can be obtained directly from those of the normal distribution, by noting that

$$E(x^k) = E(e^{ky}) = M_y(t)\Big|_{t=k}. \qquad \text{(II.F.41)}$$

Thus, the expectation value and variance, respectively, for the log-normal distribution are

$$E(x) = \exp\left(\mu + \frac{1}{2}\sigma^2\right), \quad V(x) = \exp(2\mu + \sigma^2)\left[\exp(\sigma^2) - 1\right].$$

Applying the central limit theorem to the variable $x = e^y$ shows that a variable x that stems from the product of n factors (i.e., $x = x_1 x_2 \ldots x_n$) will follow a log-normal distribution in the limit as $n \to \infty$. Therefore, the log-normal distribution is often used to model random errors which change a result by a multiplicative factor. Since the log-normal function is distributed over the range of positive real numbers, and has only two parameters, it is particularly useful for modeling nonnegative phenomena, such as analysis of incomes, classroom sizes, masses or sizes of biological organisms, evaluation of neutron cross sections, scattering of subatomic particles, etc.

Cauchy Distribution: The Cauchy distribution is defined by the probability density function

$$p(x) = (\lambda/\pi)\left[\lambda^2 + (x-v)^2\right]^{-1}, \qquad (\text{II.F.42})$$

for $x \in \mathbb{R}$, $v \in \mathbb{R}$, and $\lambda \in \mathbb{R}$ ($\lambda > 0$). Although this distribution is normalized (i.e., the zeroth raw moment exists), the expectations that define the higher-order raw moments are divergent. Mathematical difficulties can be alleviated, however, by confining the analysis to the vicinity of $x = v$; if needed, this distribution can be arbitrarily truncated.

Gamma distribution: Consider that x is a nonnegative real variable ($x \in \mathbb{R}$, $x > 0$), and α and β are positive real constants ($\alpha \in \mathbb{R}, \alpha > 0$; $\beta \in \mathbb{R}, \beta > 0$). Then the probability density function for the *gamma distribution* is defined as

$$p(x) = x^{\alpha-1} e^{-(x/\beta)} / \left[\beta^\alpha \Gamma(\alpha)\right], \qquad (\text{II.F.43})$$

where the well-known gamma function is defined as

$$\Gamma(\alpha) \equiv \int_0^\infty t^{\alpha-1} e^{-t} dt. \qquad (\text{II.F.44})$$

Recall that, for all $\alpha > 0$, $\Gamma(\alpha+1) = \alpha \Gamma(\alpha)$, $\Gamma(1) = 1$, $\Gamma(1/2) = \pi^{1/2}$, and

$$\Gamma[n + (1/2)] = [(2n-1)(2n-2)\ldots 3 \cdot 2 \cdot 1] \Gamma(1/2)/2^n. \qquad (\text{II.F.45})$$

The *MGF* for the gamma distribution is

$$M_x(t) = 1/(1-\beta t)^\alpha, \quad (t \in \mathbb{R},\ t < 1/\beta), \qquad (\text{II.F.46})$$

while the mean value and variance are $m_o = \alpha\beta$ and $\mu_2 = \alpha\beta^2$, respectively. Note that when $\alpha = 1$ and $\beta = 1/\lambda$, the gamma distribution reduces to the exponential distribution.

Beta Distribution: Consider that x is a real variable in the range $0 \le x \le 1$, and α and β are positive real parameters ($\alpha \in \mathbb{R}, \alpha > 0$; $\beta \in \mathbb{R}, \beta > 0$). The probability density function for the beta distribution is defined as

$$p(x) = \{\Gamma(\alpha+\beta)/[\Gamma(\alpha)\Gamma(\beta)]\} x^{\alpha-1}(1-x)^{\beta-1}. \qquad (\text{II.F.47})$$

Since the *MGF* for the beta distribution is inconvenient to use, it is easier to derive the mean value and the variance directly from Eq. (II.F.47); this yields $m_o = \alpha/(\alpha+\beta)$ and $\mu_2 = \alpha\beta/[(\alpha+\beta)^2(\alpha+\beta+1)]$. The beta distribution is often used for weighting probabilities along the unit interval.

Student's t-Distribution: The *t-distribution* was discovered by W. Gosset (who published it under the pseudonym "Student"), and arises when considering the quotient of two random variables. The probability density for the *t*-distribution is defined as

$$p(x) = \Gamma[(n+1)/2]\left[1+\left(x^2/n\right)\right]^{-(n+1)/2} / \left[(n\pi)^{1/2}\Gamma(n/2)\right], \ (n=1,2,3,\ldots) \qquad (\text{II.F.48})$$

where x is a real variable ($x \in \mathbb{R}$).

The *t*-distribution function does *not* have an *MGF*. However, certain moments do exist and can be calculated directly from Eq. (II.F.48). Thus, the mean value is $m_o = 0$, for $n > 1$, and the variance is $\mu_2 = n/(n-2)$, for $n > 2$. Note that the conditions $n > 1$ and $n > 2$ for the existence of the mean value and variance, respectively, arise from the fact that the *t*-distribution for $n=1$ is equivalent to the Cauchy distribution for $\nu = 0$ and $\lambda = 1$.

F-Distribution: Just as the *t*-distribution, the *F*-distribution also arises when considering the quotient of two random variables; its probability density is given by

$$p(x) = (n/m)^{n/2}\Gamma[(n+m)/2]x^{(n/2)-1} / \left\{\Gamma(n/2)\Gamma(m/2)[1+(nx/m)]^{(n+m)/2}\right\}, \qquad (\text{II.F.49})$$

where $x \in \mathbb{R}$, $x > 0$, and the parameters m and n are positive integers called *degrees of freedom*. The *MGF* for the *F*-distribution does not exist, but the mean value is $m_o = m/(m-2)$, for $m > 2$, and the variance is

$$\mu_2 = m^2(2m+2n-4)/[n(m-2)^2(m-4)], \ \text{ for } m > 4.$$

Few-Parameter Distribution and Pearson's Equation: K. Pearson made the remarkable discovery that the differential equation

$$(dp/dx) = p(x)(d-x)/(a+bx+cx^2) \qquad (\text{II.F.50})$$

yields several of the univariate probability density functions considered in the foregoing. In particular, Eq. (II.F.50) yields:
  i)   The normal distribution if $a > 0$, $b = c = 0$, while d is arbitrary,
  ii) The exponential distribution if $a = c = d = 0$, $b > 0$,
  iii) The gamma distribution if $a = c = 0$, $b > 0$, and $d > -b$,
  iv) The beta distribution if $a = 0$, $b = -c$, and $d > 1 - b$.

The solutions to Eq. (II.F.50) are known as Pearson's curves, and they underscore the close relationships among the probability distributions mentioned above.

Chi-square χ^2-Distribution: The χ^2 (chi-square) distribution of the continuous variable x ($0 \leq x < \infty$) is defined as

$$p(x) = \frac{1}{2^{n/2}\Gamma(n/2)} x^{n/2-1} e^{-x/2}, \quad (n = 1, 2, \ldots), \tag{II.F.51}$$

where the parameter n is the *number of degrees of freedom*. The mean value and variance of x are given by $m_o = n$ and $\mu_2 = 2n$, respectively.

The χ^2-distribution is related to the sum of squares of normally distributed variables: given N independent Gaussian random variables x_i with known mean μ_i and variance σ_i^2, the random variable

$$x = \sum_{i=1}^{N} \frac{(x_i - \mu_i)^2}{\sigma_i^2} \tag{II.F.52}$$

is distributed according to the χ^2-distribution with N degrees of freedom. More generally, if the random variables x_i are not independent but are described by an N-dimensional Gaussian *PDF*, then the random variable

$$Q = (x - \mu)^+ V^{-1} (x - \mu) \tag{II.F.53}$$

also obeys a χ^2-distribution with N degrees of freedom [cf. Eq. (II.F.38)]. Random variables following the χ^2-distribution play an important role in tests of goodness-of-fit, as highlighted, for example, by the method of least squares.

II.G. CENTRAL LIMIT THEOREM

Although several versions of this theorem can be found in the literature, it is convenient for the purposes of uncertainty analysis to consider the following formulation:

Consider that $(x_1, x_2, \ldots, x_n)$ is a random sample of the parent random variable x, whose moment generating function $M_x(t)$ exists in a range $-t_o < t < t_o$, $t_o > 0$, around $t = 0$. This requirement implies that the mean value m_o and $\text{var}(x) = \sigma^2$ both exist, while the sample average is $\xi_n = \left(\sum_{i=1}^{n} x_i\right)/n$. Note that the MGF of ξ_n is $M_{\xi_n}(t) = [M_x(t/n)]^n$. Furthermore, define $z_n \equiv (\xi_n - m_o)/(\sigma/n^{1/2})$ to be the reduced random-variable equivalent of ξ_n. Then, the central limit theorem states that z_n, ξ_n, and $n\xi_n$ are all asymptotically normal in the limit as $n \to \infty$.

The central limit theorem can be proven as follows: the existence of $M_x(t)$ in the range $-t_o < t < t_o$ ensures the existence of all moments of x. Therefore, z_n is well defined for each n. Hence, the MGF for z_n, $M_{z_n}(t)$, is obtained as

$$M_{z_n}(t) = \exp(-n^{1/2} m_o t/\sigma) \{M_x[t/(n^{1/2}\sigma)]\}^n. \qquad \text{(II.G.1)}$$

Expanding the definition of $M_x(\beta t) = E[\exp(\beta t x)]$, for a nonzero constant β, in a Taylor-expansion around $t = 0$ gives

$$M_x(\beta t) = 1 + E(x)\beta t + E(x^2)(\beta^2 t^2/2) + O(t^3 \beta^3), \qquad \text{(II.G.2)}$$

It follows from Eq. (II.G.2) that the Taylor-expansion of $M_x[t/(n^{1/2}\sigma)]$ around $t = 0$ is

$$M_x[t/(n^{1/2}\sigma)] = 1 + m_o[t/(n^{1/2}\sigma)] + E(x^2)[t^2/(2n\sigma^2)] + O(t^3 n^{-3/2}). \qquad \text{(II.G.3)}$$

On the other hand, taking the natural logarithm of $M_{z_n}(t)$ given in Eq. (II.G.1), and using Eq. (II.G.3), yields

$$\ln M_{z_n}(t) = -nm_o[t/(n^{1/2}\sigma)]$$
$$+ n\ln\{1 + m_o[t/(n^{1/2}\sigma)] + E(x^2)[t^2/(2n\sigma^2)]\} + O(t^3 n^{-3/2}).$$

Applying the series expansion $\ln(1+\alpha) = \alpha - (\alpha^2/2) + (\alpha^3/3) - \ldots$ to the second term on the right side of the above expression leads to

$$\ln M_{z_n}(t) = -nm_o\left[t/(n^{1/2}\sigma)\right] + n\{m_o\left[t/(n^{1/2}\sigma)\right] + E(x^2)\left[t^2/(2n\sigma^2)\right]\} \\ - (n/2)\{m_o\left[t/(n^{1/2}\sigma)\right]\}^2 + O(t^3 n^{-3/2}). \tag{II.G.4}$$

Noting in the above equation that the terms linear in t cancel, and collecting all terms of order less than three yields

$$\ln M_{z_n}(t) = \left(E(x^2) - m_o^2\right)\left[t^2/(2\sigma^2)\right] + O(t^3 n^{-3/2}). \tag{II.G.5}$$

However, the quantity $\left(E(x^2) - m_o^2\right)$ is just the variance σ^2, so that Eq. (II.G.5) reduces to

$$\ln M_{z_n}(t) = (t^2/2) + O(t^3 n^{-3/2}). \tag{II.G.6}$$

Taking the limit as $n \to \infty$ in Eq. (II.G.6) gives

$$\lim_{n\to\infty}\left[\ln M_{z_n}(t)\right] = t^2/2,$$

which implies that

$$\lim_{n\to\infty}\left[M_{z_n}(t)\right] = \exp(t^2/2). \tag{II.G.7}$$

However, Eq.(II.G.7) is just the *MGF* for the standard normal distribution, so the proof of the central limit theorem for the random variable z_n is thus concluded. The proofs of the central limit theorem for the random variables ξ_n and $n\xi_n$, respectively, are carried out using arguments similar to those used in the foregoing for the random variable z_n, as can be inferred from Eq. (II.G.7).

The conditions stated in the preceding theorem are more restrictive than they need to be; in particular, the condition that $(x_1, x_2, \ldots, x_n)$ be equally distributed can be eliminated. The least restrictive necessary and sufficient condition for the validity of the central limit theorem is the *Lindeberg condition*, which states that if the sequence of random variables $(x_1, x_2, \ldots, x_n)$ is uniformly bounded (i.e., if there exists a positive real constant C such that $|x_i| < C$ for each x_i and all possible n) and the sequence is not degenerate, then the central limit theorem holds. *In practice, the Lindeberg condition requires that the mean values and variances exist for each of these variables, and that the overall variance in the sum ξ_n of these random variables be not dominated by just a few of the components.* Application of the central limit theorem to correlated random variables is still an open field of research in mathematical statistics.

Rather than specify the conditions under which the central limit theorem holds exactly in the limit $n \to \infty$, *in practice it is more important to know the extent to which the Gaussian approximation is valid for finite n*. This is difficult to quantify exactly, but the rule of thumb is that the central limit theorem holds as long as the sum is built up of a large number of small contributions. Discrepancies arise if, for example, the distributions of the individual terms have long tails, so that occasional large values make up a large part of the sum. Such contributions lead to "non-Gaussian" tails in the sum, which can significantly alter the probability to find values with large departures from the mean. In such cases, the main assumption underlying the central limit theorem, namely the assumption that the measured value of a quantity is a normally distributed variable centered about the mean value, breaks down. Since this assumption is often used when constructing a confidence interval, such intervals can be significantly underestimated if non-Gaussian tails are present. In particular, the relationship between the confidence level and the size of the interval will differ from the Gaussian prescription (i.e., 68.3% for a "1 σ" interval, 95.4% for "2 σ," etc.) A better understanding of the non-Gaussian tails can often be obtained from a detailed Monte Carlo simulation of the individual variables making up the sum.

An example where the central limit theorem breaks down is provided by the distribution of the total number of electron-ion pairs created when a charged particle traverses a layer of matter. Actually, the number of pairs in a layer of a given thickness can be described by the Landau distribution. If this layer were subdivided into a large number of very thin layers, then the total ionization would be given by the sum of a large number of individual contributions, so the central limit theorem would be expected to apply. But the Landau distribution has a long tail extending to large values, so that relatively rare highly ionizing collisions can make up a significant fraction of the total ionization, a fact that invalidates the application of the central limit theorem to this situation.

Another example where the central limit theorem cannot be used is for calculating the angle by which a charged particle is deflected upon traversing a layer of matter. Although the total angle can be regarded as the sum of a small number of deflections caused by multiple Coulomb scattering collisions with nuclei in the substance being traversed, and although there are many such collisions, the total angle cannot be calculated by using the central limit theorem. This is because the distribution for individual deflections has a long tail extending to large angles, which invalidates the main assumption underlying the central limit theorem.

II.H. BASIC CONCEPTS OF STATISTICAL ESTIMATION

In practice, the exact form of mathematical models and/or exact values of data are rarely, if ever, available. Rather, the available information comes in the form of observations, usually associated with a frequency distribution, which must be used, in turn, to *estimate* the mathematical form and/or the parameters describing the underlying probability distribution function. The use of observations to estimate the underlying features of probability functions forms the objective of a branch of mathematical sciences called *statistics*. Conceptually, therefore, the objective of *statistical estimation* is to estimate the parameters $(\theta_1, \ldots \theta_k)$ that describe a particular statistical model, by using observations, x_n, of a frequency function $f(x; \theta_1, \ldots \theta_k)$. Furthermore, this statistical estimation process must provide reasonable assurance that the model based on these estimates will fit the observed data within acceptable limits. Furthermore, *the statistical estimates obtained from observational data must be consistent, unbiased, and efficient.* Therefore, the science of statistics embodies both inductive and deductive reasoning, encompassing procedures for estimating parameters from incomplete knowledge and for refining prior knowledge by consistently incorporating additional information. Hence, the solution to practical problems requires a synergetic use of the various interpretations of probability, including the axiomatic, frequency, and Bayesian interpretations and methodologies.

Consider that x represents a random variable that describes events in a certain event space. For simplicity, the symbol x will be used in this Section to represent both the random variable and a typical value; a distinction between these two uses will be made only when necessary to avoid confusion. Thus, x is considered to be described by a probability density $p(x)$, with a mean value $E(x) = m_o$ and $\text{var}(x) = \sigma^2$. Without loss of generality, x can be considered to be continuous, taking values in an uncountably infinite space $\mathcal{S}_x$; the statistical formalisms to be developed in the following can be similarly developed for finite or countably infinite random variables.

In both classical and Bayesian statistics, the estimation procedures are applied to samples of data. A *sample*, $\boldsymbol{x}_s$, *of size* n is defined as a collection of n equally distributed random variables $(x_1, x_2, \ldots, x_n)$; each x_i is associated with the same event space and has the same probability density, $p_i(x_i) = p(x_i)$. The random variable x, which each x_i resembles, is usually called the *parent random variable*. Each sampling step corresponds to the selection of a sample; thus, the first step selects sample 1, which corresponds to x_1, while the last step (the nth-step) selects sample n, which corresponds to x_n. The selection of values x_i is called a *sampling process*, and the result of this process is the n-*tuple* of values $\boldsymbol{x}_s \equiv (x_1, x_2, \ldots, x_n)$. If the sampling is random (i.e., the selection of each x_i is unaffected by the selection of all other x_j, $j \neq i$), then the collection of random variables can be treated as a random vector $\mathbf{x}_s \equiv (x_1, x_2, \ldots, x_n)$ distributed according to the multivariate probability density $p(x_1, x_2, \ldots, x_n) = p(x_1)p(x_2)\ldots p(x_n)$. For random sampling, the components x_i are uncorrelated (i.e., the covariance matrix is diagonal), and have mean values $E(x_i) = m_o$ and variances $\text{var}(x_i) = E\left[(x_i - m_o)^2\right] = \sigma^2$, identical to the mean value and standard deviation of the parent distribution x.

A function $\Theta(x_1, \ldots, x_n)$ that acts only on the sample random variables (and, possibly, on well-defined constants) is called a *statistic*. An *estimator*, $T = \Theta(x_1, \ldots, x_n)$, is a statistic specifically employed to provide estimated values for a particular, true yet unknown, constant value T_o of the underlying probability distribution for the parent variable x; the function Θ is called the *estimation rule*. Since this rule is designed to provide specific values of T that are meant to approximate the constant T_o, the estimator T is called a *point estimator*. The process of selecting estimators is not unique, so the criteria used for particular selections are very important, since they determine the properties of the resulting estimated values for the parameters of the chosen model.

An estimator T of a physical quantity T_o is called *consistent* if it approaches the true value T_o of that quantity (i.e., it converges in probability to T_o) as the number of observations x_n of T_o increases:

$$T(x_1, \ldots, x_n) \xrightarrow{n \to \infty} T_o. \qquad (\text{II.H.1})$$

An estimator T of T_o is called *unbiased* if its expectation is equal to the true value of the estimated quantity:

$$E(T) = T_o. \qquad (\text{II.H.2})$$

The bias $B(T, T_o)$ of an estimator is defined as

$$B(T, T_o) \equiv E(T) - T_o. \qquad (\text{II.H.3})$$

If $B(T, T_o) > 0$, then T tends to overestimate T_o; if $B(T, T_o) < 0$, then T tends to underestimate T_o. The quantity $E\left[(T - T_o)^2\right]$ is called the mean-squared error.

If the estimator $T = \Theta(x_1, \ldots, x_n)$ utilizes all the information in the sample that pertains to T_o, then the respective estimator is called a *sufficient* estimator. In practice, the choice of estimators is further limited by considering unbiased estimators, which, among all similar estimators, have the smallest variance. A consistent, unbiased, and minimum variance estimator is called an *efficient estimator*.

Intuitively, it would be expected that the smaller the variance of an unbiased estimator, the closer the estimator is to the respective parameter value. This intuitive expectation is indeed correct. For example, $T(x_1, x_2, \ldots, x_n) = \sum_{i=1}^{n} a_i x_i$, where $(a_1, \ldots, a_n)$ are constants, would be a best linear unbiased estimator (*BLUE*) of a parameter θ, if $T(x_1, \ldots, x_n)$ is a linear unbiased estimator such that $\text{var}\{T(x_1, \ldots, x_n)\} \leq \text{var}\{T'(x_1, \ldots, x_n)\}$, where $T'(x_1, \ldots, x_n)$ is any other linear unbiased estimator of θ.

The sample moment of order k, x_k^S, is the statistic defined as

$$x_k^S \equiv (1/n) \sum_{i=1}^{n} x_i^k, \quad (k = 1, 2, \ldots). \tag{II.H.4}$$

In the above definition, the superscript "S" denotes "sample." In the special case when $k = 1$, Eq. (II.H.4) defines the *sample mean value*

$$x^S \equiv (1/n) \sum_{i=1}^{n} x_i. \tag{II.H.5}$$

Note that

$$E(x^S) = \frac{1}{n} \sum_{i=1}^{n} E(x_i) = \frac{1}{n} nE(x) = m_o, \tag{II.H.6}$$

which indicates that the expectation value, $E(x^S)$, of the sample mean is an unbiased estimator for the distribution mean value $E(x) = m_o$. It can also be shown that the variance of the sample mean, $\text{var}(x^S)$, is related to $\sigma^2 = \text{var}(x)$ by means of the relation

$$\text{var}(x^S) = \sigma^2 / n = \text{var}(x)/n. \tag{II.H.7}$$

The sample central moment statistic of order k is defined as

$$\mu_k^S \equiv (1/n) \sum_{i=1}^{n} (x_i - x^S)^k, \quad (k = 2, 3, \ldots). \tag{II.H.8}$$

In particular, the *sample variance* is obtained by setting $k = 2$ in the above definition, to obtain:

$$\mu_2^S \equiv (1/n)\sum_{i=1}^{n}(x_i - x^S)^2, \qquad (\text{II.H.9})$$

while the *sample standard deviation* is calculated using the formula

$$SD(x^S) = \sqrt{\left(\frac{1}{n-1}\right)\sum_{i=1}^{n}(x_i - x^S)^2}. \qquad (\text{II.H.10})$$

The properties of sampling distributions from a normally distributed parent random variable x are of particular practical importance due to the prominent practical and theoretical role played by the central limit theorem. Thus, consider a normally distributed parent random variable x, with mean m_o and variance σ^2. Furthermore, consider a sample $(x_1, x_2, \ldots, x_n)$ with sample mean value x^S and sample variance μ_2^S, as defined in Eqs. (II.H.5) and (II.H.9), respectively. Then, the following theorems hold and are often used in practice:

(i) The quantities x^S and μ_2^S are independent random variables; note that the converse also holds, namely, if x^S and μ_2^S are independent, then the distribution for x must be normal.

(ii) The random variable $\left(n\mu_2^S/\sigma^2\right)$ is distributed according to a chi-square distribution with $(n-1)$ degrees of freedom.

If y is a random variable distributed according to a χ^2-square distribution with n degrees of freedom, and z is distributed as a standard normal random variable, and y and z are independent, then the ratio random variable $r \equiv z/\sqrt{y/n}$ is distributed according to Student's t-distribution with n degrees of freedom. In particular, this theorem holds when r, y, z are random variables defined as $\left(n\mu_2^S/\sigma^2\right) \equiv y$, $(x^S - m_o)/\sigma^2 \equiv z$, and $r \equiv z/\sqrt{y/n}$; the ratio r is frequently used in practice to measure the scatter of the actual data relative to the scatter that would be expected from the parent distribution with standard deviation σ.

If y is distributed according to a χ^2-square distribution with n degrees of freedom, and w is distributed according to a χ^2-square distribution with m degrees of freedom, and y and w are independent, then the ratio variable

$R \equiv (w/m)/(y/n)$ is distributed according to an F-distribution with degrees of freedom m and n.

It is important to note that the sample mean $x^S \equiv (1/n)\sum_{i=1}^{n} x_i$ defined in Eq. (II.H.5) is the *BLUE* for the mean $E(x) = m_o$ of the parent distribution. This property is demonstrated by first showing that a linear estimator $T(x_1, x_2, \ldots, x_n) \equiv \sum_{i=1}^{n} a_i x_i$ is unbiased iff the constraint $\sum_{i=1}^{n} a_i = 1$ is satisfied. Then, using this constraint, the minimum of

$$\text{var}\{T(x_1, x_2, \ldots, x_n)\} = \left[\sum_{i=1}^{n-1} a_i^2 + \left(1 - \sum_{i=1}^{n-1} a_i\right)^2 \right] \sigma^2$$

is calculated by setting its first-derivative to zero, i.e.,

$$\frac{\partial \text{var}\{T(x_1, x_2, \ldots, x_n)\}}{\partial a_i} = \left[2a_i - 2\left(1 - \sum_{i=1}^{n-1} a_i\right) \right] \sigma^2 = (2a_i - 2a_n)\sigma^2 = 0.$$

The solution of the above equation is $a_i = a_n$, $(i = 1, \ldots, n-1)$. Since $\sum_{i=1}^{n} a_i = 1$, it follows that $a_i = 1/n$, $i = 1, \ldots, n$, which proves that x^S is the *BLUE* for $E(x) = m_o$.

A concept similar to the *BLUE* is the *maximum likelihood estimator* (*MLE*), which can be introduced, for simplicity, by considering that a single parameter, say θ_o, is to be estimated from a random sample $(x_1, x_2, \ldots, x_n)$ of size n. Since each sample is selected independently, the conditional multivariate probability density for the observed sample data set $(x_1, x_2, \ldots, x_n)$ is

$$\prod_{i=1}^{n} p(x_i \mid \theta) \equiv L(x_1, \ldots, x_n \mid \theta), \tag{II.H.11}$$

where $p(x_i|\theta)$ is the conditional probability density that the value x_i will be observed in a single trial. The function $L(x_1, \ldots, x_n|\theta)$ defined in Eq. (II.H.11) is called the *likelihood function*. The maximum-likelihood method for estimating θ_o consists of finding the particular value, say $\tilde{\theta} \equiv \tilde{\theta}(x_1, \ldots, x_n)$, which maximizes $L(x_1, \ldots, x_n|\theta)$ for the observed data set $(x_1, x_2, \ldots, x_n)$. Thus, the *MLE*, $\hat{\theta}$, of θ_o is found as the solution to the equation

Concepts of Probability Theory

$$\frac{d \ln L(x_1,\ldots,x_n \mid \theta)}{d\theta}\bigg|_{\theta=\hat{\theta}} = 0. \qquad \text{(II.H.12)}$$

If Eq. (II.H.12) admits multiple solutions, then the solution that yields the largest likelihood function $L(x_1,\ldots,x_n \mid \theta)$ is defined to be the *MLE*. The maximum likelihood method sketched above can, of course, be extended to more than a single parameter θ_o.

For a normally distributed sample $(x_1,\ldots,x_n)$, the sample mean x^S is the *MLE* for the parent's distribution mean value, m_o, while the sample variance, μ_2^S, is the *MLE* for the variance σ^2 of the parent's distribution. Suppose, therefore, that the parent distribution is normal, so that the probability distribution for observing x_i is

$$p_i(x_i \mid m_o, \sigma^2) = \frac{1}{\sigma\sqrt{2\pi}} e^{-(x_i - m_o)^2/2\sigma^2},$$

where the *unknown parameters* m_o and σ^2 are to be estimated by finding their respective *MLE's*. First, the likelihood function is obtained according to Eq. (II.H.11) as

$$L = \left(\frac{1}{\sigma\sqrt{2\pi}}\right)^n \exp\left(-\frac{1}{2\sigma^2}\sum_{i=1}^n (x_i - m_o)^2\right).$$

The maximum of L can be conveniently calculated by setting the first partial derivatives of $\ln L$ to zero, to obtain:

$$\frac{\partial L}{L\partial(m_o)} = \frac{1}{\sigma^2}\sum_{i=1}^n (x_i - m_o) = 0,$$

$$\frac{\partial L}{L\partial(\sigma^2)} = -\frac{n}{2\sigma^2} + \frac{1}{2\sigma^4}\sum_{i=1}^n (x_i - m_o)^2 = 0. \qquad \text{(II.H.13)}$$

The first of the equations in (II.H.13) yields the estimate

$$\tilde{m}_o = \frac{1}{n}\sum_{i=1}^n x_i \equiv x^S. \qquad \text{(II.H.14)}$$

As has already been discussed in the foregoing, x^S is an unbiased estimate for m_o; furthermore, according to the law of large numbers, it is apparent that $x^S \to m_o$ as $n \to \infty$, which indicates that x^S is also a consistent estimate for m_o.

The second equation in (II.H.13) gives the estimate

$$\tilde{\sigma}^2 = (1/n)\sum_{i=1}^{n}(x_i - m_o)^2. \qquad \text{(II.H.15)}$$

Since m_o is unknown, it is replaced in Eq. (II.H.15) by the estimate $\tilde{m}_o = x^S$, as obtained in Eq. (II.H.14); this replacement leads to the estimate

$$\mu_2^S = \frac{1}{n}\sum_{i=1}^{n}(x_i - x^S)^2. \qquad \text{(II.H.16)}$$

The quantity $E(\mu_2^S)$ is calculated as follows:

$$E(\mu_2^S) = E\left(\frac{1}{n}\sum_{i=1}^{n}(x_i - m_o + m_o - x^S)^2\right) = E\left(\frac{1}{n}\sum_{i=1}^{n}(x_i - m_o)^2 - (x^S - m_o)^2\right)$$

$$= E\left(\frac{1}{n}\sum_{i=1}^{n}(x_i - m_o)^2\right) + E\left((x^S - m_o)^2\right) = \sigma^2 - \sigma^2/n = \sigma^2(n-1)/n.$$

As the above expression indicates, $E(\mu_2^S) \to \sigma^2$ as $n \to \infty$, which means that the *MLE* estimate μ_2^S is consistent; however, the *MLE* estimate μ_2^S is biased. This result underscores the limitations of *MLE*'s, namely that they may be biased.

Multiplying μ_2^S with the correction factor $n/(n-1)$ yields the estimate

$$[n/(n-1)]\mu_2^S = [1/(n-1)]\sum_{i=1}^{n}(x_i - x^S)^2, \qquad \text{(II.H.17)}$$

which deviates from the *MLE* value but, on the other hand, is both consistent and unbiased. In practice, a small deviation from the maximum of the likelihood function is less important than a potential bias in the estimate.

The sample standard deviation is computed using the formula

$$SD(x^S) = \sqrt{\frac{\sum_{i=1}^{n}(x_i - x^S)^2}{n-1}} \; .\tag{II.H.18}$$

Similarly, it can be shown that the quantity

$$\hat{V}_{xy} \equiv \frac{1}{n-1}\sum_{i=1}^{n}(x_i - x^S)(y_i - y^S)\tag{II.H.19}$$

is an unbiased estimator of the covariance V_{xy} of two random variables x and y of unknown mean.

The *PDF of an estimator* can be found by using either the characteristic function or the moment generating function techniques. As an illustrative example, consider n independent observations of a random variable x from an exponential distribution $p(x;\xi) = (1/\xi)\exp(-x/\xi)$. As has been previously shown, the maximum likelihood estimator (*MLE*) $\hat{\xi}$ for ξ is the sample mean of the observed x_i, namely

$$\hat{\xi} = x^S = \frac{1}{n}\sum_{i=1}^{n}x_i \; .\tag{II.H.20}$$

If the experiment were repeated many times, one would obtain values of $\hat{\xi}$ distributed according to a *PDF*, $f(\hat{\xi}\,|\,n,\xi)$, that depends on the number, n, of observations per experiment and the true value of the parameter ξ. To calculate $f(\hat{\xi}\,|\,n,\xi)$, recall that the characteristic function for the exponentially distributed random variable x is

$$\phi_x(k) = \int e^{ikx} p(x;\xi)dx = \frac{1}{1-ik\xi} \; .\tag{II.H.21}$$

On the other hand, the characteristic function for the random variable $\varsigma \equiv \sum_{i=1}^{n}x_i = n\hat{\xi}$ is

$$\phi_\varsigma(k) = \frac{1}{(1-ik\xi)^n} \; .\tag{II.H.22}$$

The *PDF*, $f_\varsigma(\varsigma)$, for $\varsigma \equiv n\hat{\xi}$ is the inverse Fourier transform of $\phi_\varsigma(k)$, namely

$$f_\varsigma(\varsigma) = \frac{1}{2\pi} \int_{-\infty}^{\infty} \frac{e^{-ik\varsigma}}{(1-ik\xi)^n} dk. \qquad \text{(II.H.23)}$$

The integrand in Eq. (II.H.23) has a pole of order n at $-i/\xi$ in the complex k-plane and can be evaluated using the residue theorem to obtain

$$f_\varsigma(\varsigma) = \frac{1}{(n-1)!} \frac{\varsigma^{n-1}}{\xi^n} e^{-\varsigma/\xi}. \qquad \text{(II.H.24)}$$

Transforming variables $\varsigma \to \hat{\xi}$ in Eq. (II.H.24) yields the *PDF* for the *MLE* $\hat{\xi}$ as

$$f(\hat{\xi} \mid n, \xi) = \frac{n^n}{(n-1)!} \frac{\hat{\xi}^{n-1}}{\xi^n} e^{-n\hat{\xi}/\xi}. \qquad \text{(II.H.25)}$$

Note that Eq. (II.H.25) is a special case of the *gamma distribution*. In particular, Eq. (II.H.25) can be used to compute the expectation values and *PDF*'s for mean lifetimes and decay constants in nuclear radioactive decay processes. For example, consider n decay-time measurements $(t_1, \ldots, t_n)$, for which $\hat{\tau} = (1/n) \sum_{i=1}^{n} t_i$ provides an estimate of the mean lifetime τ of the respective particle. Then, the expectation of the estimated mean lifetime $\hat{\tau}$ is obtained, using Eq. (II.H.25), as

$$E(\hat{\tau}) = \int_0^\infty \hat{\tau} f(\hat{\tau} \mid n, \tau) d\hat{\tau} = \tau. \qquad \text{(II.H.26)}$$

The *PDF*, say $p(\hat{\lambda} \mid n, \lambda)$, for the *MLE* $\hat{\lambda} = 1/\hat{\tau}$ of the decay constant $\lambda = 1/\tau$ can be computed from Eq. (II.H.25) by changing the variables $\hat{\tau} \to 1/\hat{\lambda}$, $\tau \to 1/\lambda$, as follows:

$$p(\hat{\lambda} \mid n, \lambda) = f(\hat{\tau} \mid n, \tau) \left| d\hat{\tau}/d\hat{\lambda} \right| = \frac{n^n}{(n-1)!} \frac{\lambda^n}{\hat{\lambda}^{n+1}} e^{-n\lambda/\hat{\lambda}}. \qquad \text{(II.H.27)}$$

Using the above expression, the expectation value of the MLE $\hat{\lambda}$ is obtained as

$$E(\hat{\lambda}) = \int_0^\infty \hat{\lambda}\, p(\hat{\lambda} \mid n, \lambda)\, d\hat{\lambda} = \frac{n}{n-1}\lambda. \qquad (\text{II.H.28})$$

Note that, even though the MLE $\hat{\tau} = (1/n)\sum_{i=1}^n t_i$ is an unbiased estimator for τ, the estimator $\hat{\lambda} = 1/\hat{\tau}$ is not an unbiased estimator for $\lambda = 1/\tau$. The bias, however, vanishes in the limit as n goes to infinity, since $E(\hat{\lambda}) \to \lambda$ as $n \to \infty$.

In many practical instances when estimating a set of parameters $(\theta_1, \ldots \theta_k)$, a certain amount of relevant knowledge is already available prior to performing the experiment. Such *a priori* knowledge is usually available in the form of a regression (or fitting) model $y = f(x; \theta_1, \ldots \theta_k)$, *a priori* parameter values $\theta_a = (\theta_1^a, \ldots \theta_k^a)$, and corresponding *a priori* covariance matrix V_a (the letter "a" indicates *a priori*). In such instances, it is important to combine the *a priori* information *consistently* with the new information derived from the experimental data set $(\eta_1, \ldots \eta_n)$, of observations of y, with a covariance V. The most popular method to accomplish such a consistent combination of *a priori* with newly obtained information is the *generalized least-squares method*. For example, when the *a priori* information is uncorrelated with the current information, the least squares method yields the array of values $(\theta_1, \ldots \theta_k)$ obtained by satisfying the condition

$$\chi^2 = (\theta - \theta_a)^+ V_a^{-1}(\theta - \theta_a) + (\eta - y)^+ V^{-1}(\eta - y) = \min. \qquad (\text{II.H.29})$$

The above expression can be generalized to include also the case when the prior information is correlated with the newly obtained information. The statistic χ^2 to be minimized follows a χ^2-square distribution, and provides a very valuable index for testing the consistency of data and numerical solutions (if applicable) of the regression model $y = f(x; \theta_1, \ldots \theta_k)$.

The estimators of the form $T = \Theta(x_1, \ldots, x_n)$ that have been discussed in the foregoing are specifically employed to provide estimated values for a particular, true yet unknown, value $T_o = \Theta(x_1^o, \ldots, x_n^o)$ of the underlying probability distribution. Since the estimation rule Θ is designed to provide specific values of T for approximating T_o, the estimator T is called a point estimator. Once the estimator T has been obtained, it becomes of interest to determine by how much the estimator can change when measurements are repeatedly performed under the same conditions. This question is addressed by constructing the so-called confidence interval for the true value $T_o = \Theta(x_1^o,...,x_n^o)$ of the measured quantity. The *confidence interval* is an interval that contains, with a prescribed probability called the *confidence probability*, the true value of the measured quantity. This concept can be illustrated by considering that $(x_1,...,x_n)$ is a set of random variables defining the sample of data under study, and θ_p is a fundamental parameter of the underlying distribution that produced the data. If it were now possible to introduce two statistics, say $\theta_1 = \Theta_1(x_1,...,x_n)$ and $\theta_2 = \Theta_2(x_1,...,x_n)$ which would guarantee that $P\{\theta_1 < \theta_p < \theta_2\} = \alpha$, then the interval $I(\theta_1,\theta_2)$ would be called the $100\alpha\%$ *confidence interval*. The procedure employed to determine an estimate of the confidence interval is called *interval estimation*. Note that a single experiment involving n samples would produce a single sample of data, say $(x_1,...,x_n)$; in turn, this sample would yield a single interval. Additional similar experiments would generate different data and, consequently, different intervals. However, if $I(\theta_1,\theta_2)$ is the $100\alpha\%$ *confidence interval*, then $100\alpha\%$ of all intervals that might be generated this way would contain the true value, θ_p, of the parameter in question.

There are several methods for constructing confidence intervals. Perhaps the simplest and the most general method to construct confidence intervals is based on Chebyshev's theorem, as indicated in the Section II.D. As has also been indicated there, however, the intervals obtained by using Chebyshev's theorem are too large for practical purposes. When the distribution of the sample observations can be assumed to follow a normal distribution with sample mean x^S and *known* standard deviation σ, then the confidence interval which would contain the true value, θ_p, of the parameter in question is constructed based on the expression

$$P\left\{\left|x^S - \theta_p\right| \leq z_\alpha \frac{\sigma}{\sqrt{n}}\right\} = \alpha, \qquad (\text{II.H.30})$$

where z_α is the *quantile of the normalized Gaussian distribution, corresponding to the selected confidence probability* α. Standard tables of tabulated values for the Gaussian function are customarily used in conjunction with Eq. (II.H.30).

In practice, however, the sample standard deviation is rarely known; only its estimate, $SD(x^S)$, as given in Eq. (II.H.10) can be calculated. In such case, the confidence intervals are constructed based on Student's t-distribution, which is the distribution of the random quantity

Concepts of Probability Theory

$$t \equiv \frac{x^S - \theta_p}{SD(x^S)}, \qquad (II.H.31)$$

where $SD(x^S)$ is the sample standard deviation as calculated from Eq. (II.H.10). The confidence interval $[x^S - t_q SD(x^S), \; x^S + t_q SD(x^S)]$ corresponds to the probability

$$P\{|x^S - \theta_p| \le t_q SD(x^S)\} = \alpha, \qquad (II.H.32)$$

where t_q is the *q-percent point of Student's t-distribution with* $(n-1)$ *degrees of freedom and significance level* $q \equiv (1-\alpha)$. The significance level, q, should be consistent with the significance level adopted for verifying the normality of the sample. Although it is possible to verify the admissibility of the hypothesis that the observations are described by a normal distribution (and therefore verify the hypothesis that Student's *t*-distribution is admissible), in practice, however, confidence intervals are directly constructed based on Student's *t*-distribution without verifying its admissibility. The observation that this procedure works in practice indirectly confirms the tacit assumption that the truncated distributions usually obeyed by experimental data are often even narrower than normal distributions. In practice, the confidence probability is usually set equal to 0.95.

Confidence intervals for the standard deviation can also be constructed by using the χ^2-distribution. The confidence interval thus constructed has the limits $(\sqrt{n-1}/\chi_L)SD(x^S)$ and $(\sqrt{n-1}/\chi_U)SD(x^S)$ for the probability

$$P\left\{\left(\frac{\sqrt{n-1}}{\chi_L}\right)SD(x^S) < \sigma < \left(\frac{\sqrt{n-1}}{\chi_U}\right)SD(x^S)\right\} = \alpha, \qquad (II.H.33)$$

where χ_U^2 and χ_L^2 are found from tables, with χ_U^2 corresponding to $(1+\alpha)/2$, and χ_L^2 corresponding to $(1-\alpha)/2$. Note that the purpose of a confidence interval is to obtain a set of values which, despite sampling variations, yields a reasonable range of values for the estimated parameter, based on the data available. Thus, a confidence interval is an *interval estimator* of the parameter under investigation.

A confidence interval should not be confused with a *statistical tolerance interval*, which is defined as the interval that contains, with prescribed probability α, not less than a prescribed fraction p_o of the entire collection of values of the random quantity. Thus, the statistical tolerance interval is an interval for a random quantity, and this distinguishes it from the confidence interval, which is constructed in order to cover the values of a nonrandom quantity. For example, a group of instruments can be measured to find the interval with limits l_1 and l_2 within which not less than the fraction p_o of the entire batch of instruments will fail, with prescribed probability α. The interval between l_1 and l_2 is the statistical tolerance interval. Care must be exercised to avoid confusing the limits of statistical tolerance and/or confidence intervals with the *tolerance range* for the size of some parameter. The tolerance (or the limits of the tolerance range) is (are) determined prior to the fabrication of a manufactured object, so that the objects for which the value of the parameter of interest falls outside the tolerance range are unacceptable and are discarded. Thus, the limits of the tolerance range are strict limits, which are not associated with any probabilistic relations.

Another context in which inferences are made about parameters is the test of hypotheses. A *statistical hypothesis* is an assumption about the distribution (frequency function) of a random variable. If the frequency function involves parameters, then the hypothesis can be an assumption concerning the respective parameters. A *test of hypothesis* is a rule by which the sample space is divided into two regions: the region in which the hypothesis is accepted and the region in which the hypothesis is rejected. A statistical hypothesis that is being tested is termed the *null hypothesis*, and is usually denoted by H_0. In general, the *alternative hypothesis* is the complement of the null hypothesis. Rejection of the null hypothesis H_0 when it is true is called *type I error*. The size of type I error is the probability of rejecting H_0 when it is true. Acceptance of H_0 when it is false is called *type II error*. The size of type II error is the probability of accepting H_0 when it is false. The sizes of type I and type II errors are usually termed α and β, respectively. By definition, *the power of a test* is $1-\beta$. The concept of power of a test is used to choose between (among) two or more procedures in testing a given null hypothesis, in that the test with the greater power, i.e., the smaller value of β (if one exists), is chosen. Occasionally, the power of a test depends on the value of the specific alternative when H_0 is not true; in such cases, a "most powerful" test does not exist. Tests that are "most powerful" against all alternatives are called *uniformly most powerful tests*.

II.I. ERROR AND UNCERTAINTY

The term *uncertainty* is customarily used to express the inaccuracy of measurement results when the numerical value of the respective inaccuracy is accompanied by a corresponding confidence probability. In this respect, we also note that the second edition of the *International Vocabulary of Basic and*

General Terms in Metrology (2nd edition, ISO 1993) defines the term **uncertainty** as "**an interval having a stated level of confidence**" (3.9, Note 1).

The term *error* is customarily used for all components of uncertainty, while the term *limits of error* is used in cases in which the measurement inaccuracy is caused by the intrinsic error of the measuring instrument, and when a corresponding level of confidence cannot be stated.

Measurements must be reproducible, since otherwise they lose their objective character and become meaningless. The *limits of measurement error or uncertainty* estimated by the experimenter provide a measure of the *non-reproducibility of a measurement permitted by the experimenter*. The validity of the uncertainty calculated for every measurement is based on the validity of the estimates of errors underlying the respective measurement. A correctly estimated measurement uncertainty permits comparisons of the respective result with results obtained by other experimenters.

II.J. PRESENTATION OF RESULTS OF MEASUREMENTS: RULES FOR ROUNDING OFF

When Q_m is the result of a measurement and Δ_U and Δ_L are the upper and lower limits of the error in the measurement, then *the result of a measurement and the respective measurement error* can be written in the form

$$(Q_m, \Delta_U, \Delta_L), \text{ or } (Q_m \pm \Delta) \tag{II.J.1}$$

when $|\Delta_U| = |\Delta_L| = \Delta$. *When the inaccuracy of a measurement is expressed as uncertainty, then the corresponding confidence probability must also be given*, usually in parentheses following the value of the uncertainty. For example, if a measurement yields the value of a current I as $1.56A$, and the uncertainty for this result, say $\pm 0.01A$, was calculated for a confidence probability of 0.95, then the result is written in the form

$$I(0.95) = (1.56 \pm 0.01)A. \tag{II.J.2}$$

If the confidence probability is not indicated in the measurement result, e.g., the result is written as

$$I = (1.56 \pm 0.01)A, \tag{II.J.3}$$

then the inaccuracy must be assumed to have been estimated without the use of probability methods. Although an error estimate obtained without the use of probability methods can be very reliable, it cannot be associated with a probability of one or some other value. Since a probabilistic model was not employed, the probability cannot be estimated and, therefore, should not be indicated.

In many cases, it is of interest to know not only the limiting values of the total measurement error but also the characteristics of the random and systematic error components separately, in order to analyze discrepancies between results of measurements of the same quantity performed under different conditions. Knowing the error components separately is particularly important when the result of a measurement is to be used for calculations together with other data that are not absolutely precise. Furthermore, the main sources of errors together with estimates of their contributions to the total measurement uncertainty should also be described. For a random error, for example, it is of interest to indicate the form and parameters of the distribution function underlying the observations, the method employed for testing the hypothesis regarding the form of the distribution function, the significance level used in the testing, and so on.

The number of significant figures retained in the number expressing the result of a measurement must correspond to the accuracy of the measurement. This means that the uncertainty of a measurement can be equal to 1 or 2 units (and should not exceed 5 units) in the last figure of the number expressing the result of the measurement. Since measurement uncertainty determines only the vagueness of the results, it need not be known precisely. For this reason, the uncertainty is customarily expressed in its final form by a number with one or two significant figures. Two figures are retained for the most precise measurements, and also if the most significant digit of the number expressing the uncertainty is equal to or less than 3. Note, though, that at least two additional significant digits should be retained during intermediate computations, in order to keep the round-off error below the value of the final error. The numerical value of the result of a measurement must be represented such that the last decimal digit is of the same rank as its uncertainty. Including a larger number of digits will not reduce the uncertainty of the result; however, using a smaller number of digits (by further rounding off the number) would increase the uncertainty and would make the result less accurate, thereby offsetting the care and effort invested in the measurement.

The rules for rounding off and for recording the results of measurements have been established by convention, and are listed below:

If the decimal fraction in the numerical value of the result of a measurement terminates in 0's, then the 0's are dropped only up to the digit that corresponds to the rank of the numerical value of the error.

If the digit being discarded is equal to 5 and the digits to its right are unknown or are equal to 0, then the last retained digit is not changed if it is even and it is increased by 1 if it is odd.

Example: If three significant digits are retained, the number 100.5 is rounded off to 100.0 and the number 101.5 is rounded off to 102.0

The last digit retained is not changed if the adjacent digit being discarded is less than 5. Extra digits in integers are replaced by 0's, while extra digits in decimal fractions are dropped.

Example: The numerical value of the result of a measurement 1.5333 with an error in the limits ±0.04 should be rounded off to 1.53. If the error limits are ±0.001, the same number should be rounded off to 1.533.

The last digit retained is increased by 1 if the adjacent digit being discarded is greater than 5, or if it is 5 and there are digits other than 0 to its right.

Example: If three significant digits are retained, the number 2.6351 is rounded off to 2.64.

Chapter III

MEASUREMENT ERRORS AND UNCERTAINTIES: BASIC CONCEPTS

To begin with, this chapter provides a brief description of selected definitions and considerations underlying the theory and practice of measurements and the errors associated with them. After reviewing the main sources and features of errors, the current procedures for dealing with errors and uncertainties are presented for direct and for indirect measurements, to set the stage for a fundamental concept used for assessing the magnitude and effects of errors both in complex measurements and computations. The practical consequences of this fundamental concept are embodied in the *"propagation of errors (moments)" equations*. As will be shown in this chapter, the propagation of errors equations provides a systematic way of obtaining the uncertainties in results of measurements and computations, arising not only from uncertainties in the parameters that enter the respective computational model but also from numerical approximations. The "propagation of errors" equations combine systematically and consistently the parameter errors with the sensitivities of responses (i.e., results of measurements and/or computations) to the respective parameters, thus providing the symbiotic linchpin between the objectives of uncertainty analysis and those of sensitivity analysis. The efficient computation of sensitivities and, subsequently, uncertainties in results produced by various models (algebraic, differential, integrals, etc.) will then form the objectives of subsequent chapters in this book.

III.A. MEASUREMENTS: BASIC CONCEPTS AND TERMINOLOGY

The theory of measurement errors is a branch of metrology - the science of measurements. A measurable quantity is a property of phenomena, bodies, or substances that can be defined qualitatively and expressed quantitatively. Measurable quantities are also called *physical* quantities. The term *quantity* is used both in a general sense, when referring to the general properties of objects, (for example, length, mass, temperature, electric resistance, etc.), and in a particular sense, when referring to the properties of a specific object (for example, the length of a given rod, the electric resistance of a given segment of wire, etc.).

Measurement is the process of finding the value of a physical quantity experimentally with the help of special devices called *measuring instruments*. The *result of a measurement* is a numerical value, together with a corresponding unit, for a physical quantity. Note that a measurement has three features:

(1) The result of a measurement must always be a number expressed in sanctioned units of measurements. The purpose of a measurement is to represent a property of an object by a number.
(2) A measurement is always performed with the help of some measuring instrument; measurement is impossible without measuring instruments.
(3) A measurement is always an experimental procedure.

The true value of a measurable quantity is the value of the measured physical quantity, which, if it were known, would ideally reflect, both qualitatively and quantitatively, the corresponding property of the object. The theory of measurement relies on the following postulates:
    (a)   The true value of the measurable quantity exists;
    (b)   The true value of the measurable quantity is constant (relative to the conditions of the measurement); and
    (c)   The true value cannot be found.

Since measuring instruments are imperfect, and since every measurement is an experimental procedure, the results of measurements cannot be absolutely accurate. This unavoidable imperfection of measurements is generally expressed as measurement inaccuracy, and is quantitatively characterized by measurement errors. Thus, the result of any measurement always contains an error, which is reflected by the deviation of the result of measurement from the true value of the measurable quantity. The measurement error can be expressed in absolute or relative form.

The *absolute measurement error*, δ, of an indication instrument is defined as the difference between the true value of the measured quantity, Q_t, and the measured value indicated by the instrument, Q_m, i.e.,

$$\delta \equiv Q_t - Q_m. \qquad (\text{III.A.1})$$

Note that the absolute error is a physical quantity that has the same units as the measurable quantity. Furthermore, the absolute error may be positive, negative, or be expressed as an interval that contains the measured value. The absolute error should not be confused with the absolute value of the (absolute) error; while the former may be positive or negative, the latter is always positive.

Since absolute errors have units and depend, in general, on the value of the measured quantity, they are awkward to use as a quantitative characteristic of measurement accuracy. In practice, therefore, the error is usually expressed as a fraction (usually percent or per thousand) of the true value of the measurable quantity, by using the *relative measurement error*, ε, defined as:

$$\varepsilon \equiv \frac{\delta}{Q_t}. \qquad (\text{III.A.2})$$

Knowledge of measurement errors would allow statements about measurement accuracy, which is the most important quality of a measurement: the smaller the underlying measurement errors, the more accurate the respective measurement. The accuracy of a measurement can be characterized quantitatively by the inverse of the absolute value of relative error; for example, if the relative error of a measurement is ±0.2%, then the accuracy of this measurement is 500. For the numerous measurements performed around the world, the material base that provides the support for accuracy comprises reference standards. Note, however, that the accuracy of any particular measurement is determined not only by the accuracy of the measuring instruments employed, but also by the method of measurement employed and by the skill of the experimenter. However, since the true value of a measurable quantity is always unknown, the errors of measurements must be estimated theoretically, by computations, using a variety of methods, each with its own degree of accuracy.

Occasionally, measurement errors can be estimated before performing the actual measurement. Such measurements are called measurements with ante-measurement or *a priori* estimation of errors. On the other hand, measurements whose errors are estimated after the measurement are called measurements with post-measurement or *a posteriori* estimation of errors.

Measurements performed during the preliminary study of a phenomenon are called *preliminary measurements*. The purpose of preliminary measurements is to determine the conditions under which some indicator of the phenomenon can be observed repeatedly in order to study, subsequently, its relations (especially the regular ones) with other properties of the object or with an external medium. Since the object of scientific investigation is to establish and study regular relations between objects and phenomena, preliminary measurements are very important to determine the conditions under which a given phenomenon can be observed repeatedly in other laboratories, for the purpose of verifying and confirming it. Preliminary measurements are also required in order to construct a model of the physical quantity under investigation.

When a physical quantity, Q_t, is measured repeatedly, the measuring instrument will not yield identical indications; rather, the indications will differ from one another, in a random manner. This random component of instrument error is referred to as the repeatability error of a measuring instrument. For example, instruments with moving parts have repeatability errors caused by friction in the supports of the movable parts and by hysteresis phenomena. The length of the range of possible values of the random component of instrument error is called the dead band. In other words, the dead band is the maximum interval through which a stimulus may be changed in either direction without producing a change in the response of the measuring instrument. Since instruments are constructed in order to introduce regular relations and certainty into the phenomena under consideration, it is important to reduce the instruments' random errors to levels that are either negligibly small compared with other errors or are within prescribed limits of admissible errors for the respective type of measuring devices.

The properties of measuring instruments will change in time, particularly because of component aging and environmental influences. Such time-dependent changes in any property of the measuring instrument are characterized by the instrument's instability and drift. Instrument instability can be standardized either by prescribing the value of the limits of permissible variations of the error over a definite period of time or by prescribing different error limits to different "lifetimes" of the instrument, after it is calibrated. On the other hand, the drift of an instrument is defined as the change (always in the same direction) that may occur in the output signal over a period of time that is significantly longer than the measurement time. The drift and instability of an instrument do not depend on the input signal but they can depend on external conditions. For this reason, the drift is usually determined in the absence of an input signal.

Measuring instruments are also characterized by their sensitivity, discrimination threshold, and resolution. Thus, the sensitivity of an instrument is defined as the ratio of the variation in the measured quantity at the output of the measuring instrument to variation of the input value of the quantity that causes the output value to change. The discrimination threshold is defined as the minimum variation in the input signal that causes an appreciable variation in the output signal. Finally, the resolution of an instrument is the smallest interval between two distinguishable neighboring discrete values of the output signal.

Measurements are customarily categorized as direct, indirect, and combined measurements. In direct measurements, the quantity to be measured interacts directly with the measuring instrument, and the value of the measured quantity is read from the instrument's indications. In indirect measurements, the value of the physical quantity is found by using a known dependence between the quantity and its basic properties which are themselves measured by means of direct, indirect, or combined measurements. For example, the density of a homogeneous solid body is inferred from an indirect measurement, which would consist of three steps: the first two steps would involve measuring directly the body's mass and volume, respectively, while the third step would involve taking the ratio of the measurements obtained in the first two steps.

Measurements performed with single observations are called single measurements, while measurements performed with repeated observations are called multiple measurements. Customarily, an indirect measurement is regarded as a single measurement if the value of each of the components of the indirectly measured quantity is found as a result of a single measurement. Thus, the indirect measurement of the density of a solid body (performed as mentioned in the previous example) would be considered a single measurement. Combined measurements can also be regarded as single measurements, if the number of measurements is equal to the number of unknowns, so that each unknown is determined uniquely from the system of equations obtained after performing the respective measurements. For example, when measuring the angles of a planar triangle, each angle can be measured in a single direct measurement. Since the sum of all three angles is equal to 180°, it is possible, in principle, to measure two angles only, using single direct measurements, and then deduce the size of the third angle by subtracting the two measured ones from 180°. If, however, all three angles are measured independently and repeatedly, then it is possible to obtain additional and very useful information by using the relation between them (namely their sum being 180°). For example, correlations between the estimated errors in the respective (multiple) measurements could be obtained by using the method of least squares. Since the number of unknowns (two) is less than the number of measurements ultimately performed, the procedure just described would be classified as a combined multiple measurement.

The regime of a measuring instrument in which the output signal can be regarded as constant is called the static regime. For example, for an indicating instrument, the signal is constant for a time sufficient to read the instrument; during this time, therefore, the indicating instrument is in a static regime. Measurements for which the measuring instruments are employed in the static regime are called static measurements. On the other hand, there are situations in which the output signal changes in time so that it is not possible to obtain a result, or to estimate the accuracy of a result, without taking the time-dependent change of the signal into account. In such cases, the measuring instruments are employed in the dynamic regime, and the respective measurements are called dynamic measurements.

III.B. CLASSIFICATION OF MEASUREMENT ERRORS

As has been previously mentioned, the absolute measurement error δ is defined as the difference between the true value, Q_t, of the quantity being measured and the measured value, Q_m, indicated by the instrument [see Eq. (III.A.1)]. This equation cannot be used directly to find the error of a measurement, however, since the true value of the measurable quantity is always unknown. (If the true value were known, then there would be no need for a measurement, of course.) Therefore, measurement errors must be estimated by using indirect data. For this purpose, measurement errors are traditionally classified according to their sources and their properties.

The basic sources of measurement errors are: errors arising from the method of measurement, errors due to the measuring instrument, and personal errors committed by the person performing the experiment. These errors are considered to be additive, so that the general form for the absolute measurement error δ is

$$\delta = \delta_m + \delta_i + \delta_p, \tag{III.B.1}$$

where δ_m, δ_i, and δ_p represent, respectively, the methodological error, the instrumental error, and the personal error. *Methodological errors* are caused by unavoidable discrepancies between the actual quantity to be measured and its model used in the measurement. Most commonly, such discrepancies arise from inadequate theoretical knowledge of the phenomena on which the measurement is based, and also from inaccurate and/or incomplete relations employed to find an estimate of the measurable quantity. *Instrumental measurement errors* are caused by imperfections of measuring instruments. Such errors arise both under reference conditions regarded as normal for the respective measurement (in this case, the respective errors are referred to as *intrinsic errors of measuring instruments*) and also under conditions that cause the quantities that influence the measurement to deviate from their reference values (in such cases, the respective errors are called *additional instrumental errors*). Finally, the individual characteristics of the person performing the measurement may give rise to *personal errors*.

If the results of separate measurements of the same quantity differ from one another, and the respective differences cannot be predicted individually, then the error owing to this scatter of the results is called *random error*. Random errors can be identified by repeatedly measuring the same quantity under the same conditions. On the other hand, a measurement error is called *systematic* if it remains constant or changes in a regular fashion when the measurements of that quantity are repeated. Systematic errors can be discovered experimentally either by using a more accurate measuring instrument or by comparing a given result with a measurement of the same quantity, but performed by a different method. In addition, systematic errors are estimated by theoretical analysis of the measurement conditions, based on the known properties of the measuring instruments and the quantity being measured. Although the estimated systematic error can be reduced by introducing corrections, it is impossible to eliminate completely systematic errors from experiments. Ultimately, a residual error will always remain, and this residual error will then constitute the systematic component of the measurement error.

The quality of measurements that reflects the closeness of the results of measurements of the same quantity performed under the same conditions is called the *repeatability of measurements*. Good repeatability indicates that the random errors are small. On the other hand, the quality of measurements that reflects the closeness of the results of *measurements* of the same quantity performed *under different conditions*, i.e., in different laboratories, at different locations, and/or using different equipment, is called the *reproducibility of measurements*. Good reproducibility indicates that both the random and systematic errors are small.

It is also customary to distinguish between gross or outlying errors and blunders. An error is *gross* or *outlying*, if it significantly exceeds the error justifiable by the conditions of the measurements, the properties of the measuring instrument employed, the method of measurement, and the qualification of the experimenter. For example, if the grid voltage affects the measurements under consideration, then a sharp brief change in the grid voltage can produce an outlying error in a single experiment. Outlying errors in multiple measurements can be discovered by statistical methods and are usually eliminated from the final error analysis. On the other hand, *blunders* occur as a result of errors made by the experimenter. Examples of blunders are a slip of the pen when writing up the results of observations, an incorrect reading of the indications of an instrument, etc. Blunders must be discovered by nonstatistical methods, and they must always be eliminated from the final results.

Measurement errors are also categorized as *static errors* or *dynamic errors*. The types of errors mentioned above fall into the category of static errors. By contrast, dynamic errors are caused by the inertial properties of measuring instruments. If a varying quantity is recorded with the help of a recording device, then the difference between the record obtained by the device and the actual change of the recorded quantity in time is the dynamic error of the respective dynamic measurement. In this case, the dynamic error is also a function of time, and the instantaneous dynamic error can be determined for each moment in time. Note, though, that a dynamic measurement need not necessarily be accompanied by dynamic errors. For example, if the measured quantity varies sufficiently slowly within the time needed for a single measurement, and the single measurements are taken with sufficient frequency, then the collection of measurements thus obtained can be combined into a multiple dynamic measurement that has static but no dynamic errors.

As has already been mentioned, a measurement error cannot be found directly by using its definition as an algorithm, since the true value of the measured quantity is unknown. The measurement error must be found by identifying its underlying sources and reasons, and by performing calculations based on the estimates of all components of the respective measurement inaccuracy.

The smallest of the measurement errors are customarily referred to as *elementary errors (of a measurement)*, and are defined as those components of the overall measurement error that are associated with a single source of inaccuracy for the respective measurement. The total measurement error is calculated, in turn, by using the estimates of the component elementary errors. Even though it is sometimes possible to correct, partially, certain elementary errors (e.g., systematic ones), no amount or combination of corrections can produce an absolutely accurate measurement result; there always remains a residual error. In particular, the corrections themselves cannot be absolutely accurate, and, even after they are effected, there remain residuals of the corresponding errors which cannot be eliminated and which later assume the role of elementary errors.

Since a measurement error can only be calculated indirectly, based on models and experimental data, it is important to identify and classify the underlying elementary errors. This identification and classification is subsequently used to develop mathematical models for the respective elementary errors. Finally, the resulting (overall) measurement error is obtained by synthesizing the mathematical models of the underlying elementary errors.

In the course of developing mathematical models for elementary errors, it has become customary to distinguish four types of elementary errors, namely absolutely constant errors, conditionally constant errors, purely random errors, and quasi-random errors. Thus, *absolutely constant errors* are defined as elementary errors that remain the same (i.e., are constant) in repeated measurements performed under the same conditions, for all measuring instruments of the same type. Absolutely constant errors have definite limits but these limits are unknown. For example, an absolutely constant error arises from inaccuracies in the formula used to determine the quantity being measured, once the limits of the respective inaccuracies have been established. Typical situations of this kind arise in indirect measurements of quantities determined by linearized or truncated simplifications of nonlinear formulas (e.g., analog/digital instruments where the effects of electro-motive forces are linearized). Based on their properties, absolutely constant elementary errors are purely systematic errors, since each such error has a constant value in every measurement, but this constant is nevertheless unknown. Only the limits of these errors are known. Therefore, *absolutely constant errors are modeled mathematically by a determinate (as opposed to random) quantity whose magnitude lies within an interval of known limits.*

Conditionally constant errors are, by definition, elementary errors that have definite limits (just like the absolutely constant errors) but (as opposed to the absolutely constant errors) such errors can vary within their limits due both to the nonrepeatability and the nonreproducibility of the results. A typical example of such an error is the measurement error due to the intrinsic error of the measuring instrument, which can vary randomly between fixed limits. Usually, *the conditionally constant error is mathematically modeled by a random quantity with a uniform probability distribution within prescribed limits.* This mathematical model is chosen because the uniform distribution has the highest uncertainty (in the sense of information theory) among distributions with fixed limits. Note, in this regard, that the round-off error also has known limits, and this error has traditionally been regarded in mathematics as a random quantity with a uniform probability distribution.

Purely random errors appear in measurements due to noise or other random errors produced by the measuring device. The form of the distribution function for random errors can, in principle, be found using data from each multiple measurement. In practice, however, the number of measurements performed in each experiment is insufficient for determining the actual form of the distribution function. Therefore, *a purely random error is usually modeled mathematically by using a normal distribution characterized by a standard deviation that is computed from the experimental data.*

Quasi-random errors occur when measuring a quantity defined as the average of nonrandom quantities that differ from one another such that their aggregate behavior can be regarded as a collection of random quantities. In contrast to the case of purely random errors, though, the parameters of the probability distribution for quasi-random errors cannot be unequivocally determined from experimental data. Therefore, *a quasi-random error is modeled by a probability distribution with parameters (e.g., standard deviation) determined by expert opinion.*

Measurements must be reproducible; otherwise they lose their objective character and become useless. As has been discussed in the foregoing, the limits imposed by errors represent a measure of the irreproducibility of the respective measurement. It is therefore very important to strive to minimize measurement errors, which implies that they should be estimated as accurately as possible. Two considerations are fundamental for estimating measurement errors:

(1) The smaller the relative error, ε, of a given experiment, the more accurate (and, in this sense, the better) the measurement. The definition of ε given in Eq. (III.A.2) is used to obtain the following approximate but practically useful expression for ε:

$$\varepsilon \equiv \frac{\delta}{Q_t} \approx \frac{\Delta}{Q_e}, \qquad (\text{III.B.2})$$

where Q_e is an estimate of the true value, Q_t, of the measured quantity, while Δ represents an estimate of the limits of measurement error δ. Recall also that the value of relative error, ε, does not depend on the value of the measured quantity, but the absolute error δ does.

(2) The estimate, Δ, of the measurement error limits must satisfy the inequality

$$|\delta| \leq |\Delta|. \qquad (\text{III.B.3})$$

The inequality (III.B.3) is obtained from the following considerations: although one strives to eliminate all errors in a measurement, this is not possible in practice; hence $Q_e \neq Q_t$, which implies that $\delta \neq 0$. Furthermore, the error can be negative or positive. The primary estimates of δ aim at establishing lower and upper limits Δ_1 and Δ_2 for δ, such that $\Delta_1 \leq \delta \leq \Delta_2$. In calculations, the larger of $|\Delta_1|$ or $|\Delta_2|$, denoted by $|\Delta|$, is usually used instead of the actual error δ. Therefore, if a measurement error is primarily random, its limits $|\Delta|$ should be estimated such that inequality (III.B.3) is always satisfied, in order to err on the "safe" (or conservative) side. For this reason, inequality (III.B.3), which expresses the requirement that $|\Delta|$ be an upper estimate of the measurement error

δ, is regarded as a fundamental principle of error estimation. Since the measurement error characterizes the spread and nonreproducibility of the measurement, the estimate of error, Δ, cannot be made extremely precise; moreover, this estimate should be weighted toward overestimation rather than underestimation. In practice, therefore, one strives to determine Δ as accurately as possible, since overestimating the error does reduce the measurement's quality; on the other hand, though, one should be aware that underestimating the error could render the entire measurement worthless.

Note that all the information obtained in an experiment, including the experimental data and corrections to the indications of instruments, etc., is employed to find the result of the respective measurement. Thus, the same data cannot be used (in addition to obtaining the measurement result) to verify the correctness of Δ, which must therefore be estimated by using additional information about the properties of the measuring instruments, the conditions of the measurements, and the theory. Thus, if a special experiment aimed at verifying or estimating the measurement error were performed, a second, more accurate, measurement of the same measurable quantity would need to be performed in parallel with the original measurement. In such a situation, though, the original measurement would become meaningless, because its result would be replaced by the result of the more accurate measurement. Hence, the problem of estimating the error of the original measurement would be replaced by the problem of estimating the error of the more accurate measurement, which means, in turn, that the basic problem would remain unresolved.

The correctness of error estimates can nonetheless be verified, albeit indirectly, by the successful use of the measurement result for the purpose intended and by ensuring that the measurement agrees with the results obtained in other experiments measuring the same physical quantity. Furthermore, the correctness of error estimates will inevitably be verified in time, as improvements in measuring instruments and methods lead to increased measurement accuracy. For example, improvements in the reading and regulating mechanisms of modern measuring instruments (e.g., for digital instruments) have all but eliminated personal errors in reading the indications of such instruments.

III.C. NATURAL PRECISION LIMITS OF MACROSCOPIC MEASUREMENTS

For all scientific measurements, it is of fundamental interest to estimate their limits as determined by physical (i.e., natural) laws. Comparing these limits to those allowed by the actual instruments stimulates improvements in their construction. Measurements always involve an interaction between the measuring instruments and the measured quantity. Moreover, both the instrument and the measured quantity interact with the surrounding medium.

Hence, the measuring instrument and the measured quantity must be analyzed together, as a system that interacts, in turn, with the surrounding medium.

Consider, for example, the interaction between the medium and an instrument that has an inertial moving system and elastic elements that keep it in a position of equilibrium. The moving parts of such instruments are continuously interacting with molecules of air or molecules of liquid (if the oscillations are damped by a liquid). Taking a statistical average, the effects of these interactions are the same on all sides of the system, so the moving parts of the instrument are in statistical equilibrium. But, at any given moment in time, the effect of interactions with molecules on a given side of the moving part(s) can be greater than those from the other side, while at the next instant in time the situation can be reversed. Consequently, the movable part of the instrument fluctuates statistically around the position of equilibrium. Such fluctuations limit inherently the possibilities (e.g., precision) of instruments, because the random error caused by such statistical fluctuations cannot be reduced below the level of *thermal noise*. The detailed effects of thermal noise are estimated by using methods of statistical physics in conjunction with the specific characteristics of the measuring instrument and measured quantity. Although the details of the effects of thermal noise do depend on the various types of measuring instruments, surrounding environment, etc., thermal noise has a common behavior for all measuring instruments, because it is always present, and its effects increase with increasing temperature.

The precision of measuring instruments is also limited by *shot noise* (also called *Schottky noise* or *generation-recombination noise*). Shot noise arises in the presence of an electrical current, when the flow of electrons passing per second through the transverse cross section of a conductor becomes sufficiently small so that individual electrons can be counted. In such situations, the number of electrons passing through the respective cross section fluctuates randomly. The randomness of the time instants at which electrons appear in the electric circuit causes random fluctuation of the current strength, which, in turn, produces the shot noise.

Note that shot noise is observed only when current flows along the circuit. By contrast, thermal noise also occurs in the absence of current. Shot and thermal noise are independent of one another. For this reason, when both types of noise occur, the variance of the fluctuations of the measuring instrument can be calculated by summing the variances of the shot and thermal noise components. Furthermore, when shot and thermal noise appear in the input of a measuring instrument, the model of the measured quantity cannot possibly describe the respective quantity exactly, thus adding another factor that must be taken into account when investigating the theoretical measurement limit of the respective instrument.

III.D. DIRECT MEASUREMENTS

When measuring a quantity *directly*, it is useful to distinguish between single and multiple measurements of the respective quantity. The *single measurement* is regarded as the basic form of measurement while *multiple measurements* can be regarded as being derived from single measurements. As has been already discussed, systematic errors cannot be completely eliminated in practice. Some leftover residuals will always remain, and must be taken into account in order to estimate the limits of the leftover systematic error in the result. Henceforth, the systematic error in a measurement should be understood as the leftover residual of the systematic error, if it is too large to be neglected.

III.D.1. Errors and Uncertainties in Single Direct Measurements

A priori, the error for a single measurement can be both systematic and random; however, after the measurement has been performed, the measurement error becomes a systematic error for that measurement. This is because the result of a measurement has a definite numerical value, and the difference between the measurement result and the true value of the measured quantity is a constant. Even if the entire error in a measurement were random, the error seemingly freezes after the measurement result is obtained; thus, the error loses its random character and becomes systematic. Each of the three components of the measurement, namely the method of measurement, the measuring instrument, and the experimenter can be sources of systematic errors. Correspondingly, methodological, instrumental, and personal systematic errors are customarily distinguished from each other.

Methodological systematic errors arise from imperfections of the method of measurement and the limited accuracy of the formulas used to model the phenomena on which the measurement is based. The influence of the measuring instrument on the object whose property is being measured can cause a methodological error. The error due to the threshold discrepancy between the model and the object is also a methodological error. Instrumental systematic errors are errors caused by the imperfections of the measuring instrument (e.g., imprecise calibration of the instrument scale). Setup errors, i.e., errors stemming from the arrangement of measuring instruments, and the effects of one instrument on another, are also instrumental errors. Most often, the additional and dynamic errors are also systematic errors. Note, though, that an unstable input signal can also cause random errors. Personal systematic errors are systematic errors connected with the individual characteristics of the observer.

A systematic error that remains constant and is therefore repeated in each observation or measurement is called a constant systematic error; for example, such an error will be present in measurements performed using balances, resistors, etc. The personal errors made by experienced experimenters can also be classified as constant (personal errors made by inexperienced experimenters, however, are considered random). Errors that increase or decrease throughout the measurement time are called progressing errors. Errors that vary with a definite period are called periodic errors.

It is very difficult to identify systematic errors; for example, variable systematic errors can be identified by using statistical methods, correlation, and regression analysis. Systematic errors can also be identified by measuring the same quantity using two different instruments (methods) or by measuring periodically a known (instead of an unknown) quantity. If a systematic error has been identified, then it can usually be estimated and eliminated. However, making rational estimates of the magnitude of the residual systematic errors and, in particular, assigning consistent levels of confidence to these residual errors is an extremely difficult task. In practice, therefore, *residual systematic errors are assumed to follow a continuous uniform distribution, within ranges that are conservatively estimated based on experience and expert judgment.*

III.D.2. Errors and Uncertainties in Multiple Direct Measurements

Multiple direct measurements are usually needed for measuring the average value of some parameter; for reducing the effects of random errors associated with the measuring instruments; for investigating a new phenomenon (to determine relationships between the quantities characterizing the respective phenomenon and connections to other physical quantities); and for developing new measuring instruments. Under certain restrictions on the measurement data, the methods of mathematical statistics provide means for analyzing observations and estimating measurement errors from multiple measurements. Thus, direct multiple measurements that are free of systematic errors (i.e., in which only random errors occur) can be analyzed by statistical methods directly, without additional considerations. In many practical situations, though, the mathematical restrictions required by mathematical statistics are not entirely fulfilled; hence, it is often necessary to develop practical methods for analyzing such situations individually, case by case.

When the random character of the observational results is caused by measurement errors, the respective observations are assumed to have a normal distribution. This assumption rests on two premises, namely: (i) since measurement errors consist of many components, the central limit theorem implies a normal distribution for such errors; and (ii) measurements are performed under controlled conditions, so that the distribution function of their error is actually bounded. Hence, approximating a bounded distribution by a normal distribution (for which the random quantity can take any real value) is a conservative procedure since such an approximation leads to larger confidence intervals than would be obtained if the true bounded distribution were known. Nevertheless, the hypothesis that the distribution of the observations is normal must be verified, since the measured results do not always correspond to a normal distribution. For example, when the measured quantity is an average value, the distribution of the observations can have any form.

In general, both the systematic and random components of the error must be estimated. Although repeating the measurements yields information about the random components of the error, information about the systematic component cannot be extracted from the measurements themselves. Hence, the systematic errors are estimated from information about the properties of the measuring instruments employed, the method of measurement, and the conditions under which the measurements are performed. Although the random error can only be estimated *a posteriori*, the systematic error can also be estimated *a priori*. Note that the random components of all conditionally constant errors become, *a posteriori*, part of the random error of the respective measurement. Thus, the remaining parts of conditionally constant errors in multiple measurements become purely systematic errors. However, the values of these errors can vary in repeated measurements of the same quantity, even if the measurements are performed by the same method.

For easy reference, the main steps usually required for estimating the uncertainties of direct measurements are summarized below:

1. Analyze the measurement problem to establish the purpose of the measurement and the required measurement accuracy. Construct a model for the parameter to be measured by taking into account the physical quantities characterizing the surrounding environment and affecting the size of the measured parameter. Determine how these quantities are to be measured if the measurement is being planned (or were measured, if the measurement has already been performed); in particular, estimate their nominal values and range of variation.
2. Establish the important metrological properties of the measuring instruments chosen or used for the measurement.
3. Compile a list of all possible elementary errors in the given measurement and estimate their limits.
4. Whenever possible, obtain point estimates, particularly for the dominant elementary errors, for the purpose of correcting the respective errors. For each correction thus introduced, estimate its limits of inaccuracy and add these limits to the list of elementary errors.
5. Assess dependencies among elementary errors. For example, if two errors δ_1 and δ_2 (arising from two distinct causes), depend on a third physical quantity, then δ_1 and δ_2 depend on each other. To eliminate this dependence, it often suffices to introduce a new elementary error that reflects the effect of the third quantity on the result of measurement. Consequently, the original dependent elementary errors, δ_1 and δ_2, will be replaced by two new elementary errors, δ_1^{new} and δ_2^{new}, which can now be regarded as being independent from each other.

6. Divide all elementary errors into conditionally and absolutely constant errors, and highlight those with limits unsymmetrical relative to the result of measurement.
7. For multiple measurements, determine whether the respective error is random or quasi-random, and estimate the respective confidence limits.
8. Estimate the possible change in the intrinsic error of the instruments over the time period elapsed since they were last calibrated. If there are reasons to assume that the intrinsic error could have exceeded permissible values, then the respective instruments must be verified prior to performing the measurement and, if necessary, adjusted or recalibrated.
9. The result of a single measurement is often obtained directly from the indication of the measuring instrument. In other cases, however, the indication must be multiplied by a scale factor or corrected.
10. The result of multiple measurements is obtained from the arithmetic mean of the results from the component measurements, unless the definition of the measured quantity requires that a different algorithm be used.
11. *A priori* estimation of error or uncertainty is usually made for the least favorable case. For example, if a multiple measurement is planned, then the value of the standard deviation is often taken based on expert recommendations.
12. Perform *a posteriori* estimation of error and/or uncertainty, with the respective confidence interval.
13. Present the results of measurement together with their uncertainties or errors in the forms indicated in Sec. II.J.

III.E. INDIRECT MEASUREMENTS

An *indirect measurement* is a measurement in which the value of the unknown quantity is calculated by using matched measurements of other quantities, called *measured arguments* or, briefly, *arguments,* which are related through a known relation to the measured quantity. From the perspective of conducting a measurement, the indirect measurements can also be divided into two classes, namely *single indirect measurements* (in which all arguments are measured once only) and *multiple indirect measurements* (in which all arguments are measured several times).

In an indirect measurement, the *true but unknown value of the measured quantity* or *response*, denoted by R, is related to the *true but unknown values of arguments*, denoted as $(\alpha_1,...,\alpha_k)$, by a *known relationship (i.e., function) f.* This relationship is called the *measurement equation*, and can be generally represented in the form

$$R = f(\alpha_1,\ldots,\alpha_k). \tag{III.E.1}$$

The specific forms of measurement equations can be considered as mathematical models of specific indirect measurements.

In practice, the *nominal parameter values*, $(\alpha_1^0,\ldots,\alpha_k^0)$, are known together with their *uncertainties or errors*, $(\delta\alpha_1,\ldots,\delta\alpha_k)$. The nominal parameter values are given by their respective expectations, while the associated errors and/or uncertainties are given by their respective standard deviations. Processing the experimental data obtained in an indirect measurement is performed with the same objectives as for direct measurements, namely to calculate the expected value, $E(R)$, of the measured response R, and to calculate the error and/or uncertainty, including confidence intervals, associated with $E(R)$. The higher-order moments of the distribution of R are also of interest, if they can be calculated.

It is very important to note here that the measurement equation, namely Eq. (III.E.1), can be interpreted to represent not only results of indirect measurements but also **results of calculations.** *In this interpretation, $(\alpha_1,\ldots,\alpha_k)$ are considered to be the parameters underlying the respective calculation, R is considered to represent the result or response of the calculation, while f represents not only the explicit relationships between parameters and response but also represents implicitly the relationships among the parameters and the independent and dependent variables comprising the respective mathematical model.*

Calculating the expected value, $E(R)$, of the measured and/or calculated response R, and calculating the various moments (variances and covariances, skewness, kurtosis) and confidence intervals associated with $E(R)$ are the objectives of a procedure referred to as the method of *propagation of moments* or *propagation of errors*, which is presented in the next section.

III.F. PROPAGATION OF MOMENTS (ERRORS)

The measured or calculated system response R (i.e., the result of an indirect measurement or the result of a calculation) is considered to be a real-valued function of k system parameters, denoted as $(\alpha_1,\ldots,\alpha_k)$; without loss of generality, these parameters can be considered to be real scalars. As discussed in the previous section, the *true values* of the parameters $(\alpha_1,\ldots,\alpha_k)$ are *not* known; only their *nominal values*, $(\alpha_1^0,\ldots,\alpha_k^0)$, are known together with *their uncertainties or errors*, $(\delta\alpha_1,\ldots,\delta\alpha_k)$. Usually, the nominal parameter values are taken to be the expected parameter values, while the associated errors and/or uncertainties are given by their respective standard deviations. The relative uncertainties $\delta\alpha_i/\alpha_i^0$ are usually symmetrical around α_i^0, and smaller than unity. Thus, the true parameter values $(\alpha_1,\ldots,\alpha_k)$ can be expressed in vector form as

$$\boldsymbol{\alpha} = \boldsymbol{\alpha}^0 + \delta\boldsymbol{\alpha} = \left(\alpha_1^0 + \delta\alpha_1,\ldots,\alpha_k^0 + \delta\alpha_k\right). \tag{III.F.1}$$

As noted in the previous section, the response is related to the parameters via the measurement equation (or computational model), which is traditionally written in the simpler form

$$R = R(\alpha_1,\ldots,\alpha_k) = R\left(\alpha_1^0 + \delta\alpha_1,\ldots,\alpha_k^0 + \delta\alpha_k\right). \tag{III.F.2}$$

In the functional relation above, R is used in the dual role of both a (random) function and the numerical realization of this function, which is consistent with the notation used for random variables and function. Expanding $R\left(\alpha_1^0 + \delta\alpha_1,\ldots,\alpha_k^0 + \delta\alpha_k\right)$ in a Taylor series around the nominal values $\boldsymbol{\alpha}^0 = \left(\alpha_1^0,\ldots,\alpha_k^0\right)$ and retaining only the terms up to the nth order in the variations $\delta\alpha_i \equiv \left(\alpha_i - \alpha_i^0\right)$ around α_i^0 gives:

$$\begin{aligned}
R(\alpha_1,\ldots,\alpha_k) &\equiv R\left(\alpha_1^0 + \delta\alpha_1,\ldots,\alpha_k^0 + \delta\alpha_k\right) \\
&= R\left(\boldsymbol{\alpha}^0\right) + \sum_{i_1=1}^{k}\left(\frac{\partial R}{\partial \alpha_{i_1}}\right)_{\boldsymbol{\alpha}^0} \delta\alpha_{i_1} \\
&+ \frac{1}{2}\sum_{i_1,i_2=1}^{k}\left(\frac{\partial^2 R}{\partial \alpha_{i_1}\partial \alpha_{i_2}}\right)_{\boldsymbol{\alpha}^0} \delta\alpha_{i_1}\delta\alpha_{i_2} \\
&+ \frac{1}{3!}\sum_{i_1,i_2,i_3=1}^{k}\left(\frac{\partial^3 R}{\partial \alpha_{i_1}\partial \alpha_{i_2}\partial \alpha_{i_3}}\right)_{\boldsymbol{\alpha}^0} \delta\alpha_{i_1}\delta\alpha_{i_2}\delta\alpha_{i_3} + \cdots \\
&+ \frac{1}{n!}\sum_{i_1,i_2,\ldots,i_n=1}^{k}\left(\frac{\partial^n R}{\partial \alpha_{i_1}\partial \alpha_{i_2}\ldots\partial \alpha_{i_n}}\right)_{\boldsymbol{\alpha}^0} \delta\alpha_{i_1}\ldots\delta\alpha_{i_n}.
\end{aligned} \tag{III.F.3}$$

Using the above Taylor-series expansion, the various moments of the random variable $R(\alpha_1,\ldots,\alpha_k)$, namely its mean, variance, skewness, and kurtosis, are calculated by considering that the system parameters $(\alpha_1,\ldots,\alpha_k)$ are random variables distributed according to a joint probability density function $p(\alpha_1,\ldots,\alpha_k)$, with:

mean values:

$$E(\alpha_i) = \alpha_i^0, \qquad (III.F.4)$$

variances:

$$\operatorname{var}(\alpha_i, \alpha_i) \equiv \sigma_i^2 \equiv \int_{S_\alpha} (\alpha_i - \alpha_i^0)^2 p(\alpha_1, \ldots, \alpha_k) d\alpha_1 d\alpha_2 \ldots d\alpha_k, \quad (III.F.5)$$

covariances:

$$\operatorname{cov}(\alpha_i, \alpha_j) \equiv \int_{S_\alpha} (\alpha_i - \alpha_i^0)(\alpha_j - \alpha_j^0) p(\alpha_1, \ldots, \alpha_k) d\alpha_1 d\alpha_2 \ldots d\alpha_k. \quad (III.F.6)$$

The procedure outlined above is called the *method of propagation of errors* or *propagation of moments*, and the resulting equations for the various moments of $R(\alpha_1, \ldots, \alpha_k)$ are called the **moment propagation equations**.

For large complex systems, with many parameters, it is impractical to consider the nonlinear terms in Eq. (III.F.3). In such cases, the response $R(\alpha_1, \ldots, \alpha_k)$ becomes *a linear function of the parameters* $(\alpha_1, \ldots, \alpha_k)$ of the form

$$R(\alpha_1, \ldots, \alpha_k) = R(\boldsymbol{\alpha}^0) + \sum_{i=1}^{k} \left(\frac{\partial R}{\partial \alpha_i}\right)_{\boldsymbol{\alpha}^0} \delta\alpha_i = R^0 + \sum_{i=1}^{k} S_i \delta\alpha_i, \quad (III.F.7)$$

where $R^0 \equiv R(\boldsymbol{\alpha}^0)$ and $S_i \equiv (\partial R/\partial \alpha_i)_{\boldsymbol{\alpha}^0}$ is *the sensitivity of the response* $R(\alpha_1, \ldots, \alpha_k)$ *to the parameter* α_i. The *mean value* of $R(\alpha_1, \ldots, \alpha_k)$ is obtained from Eq. (III.F.7) as

$$\begin{aligned}
E(R) &\equiv \int_{S_\alpha} \left(\sum_{i=1}^{k} S_i \delta\alpha_i\right) p(\alpha_1, \ldots, \alpha_k) d\alpha_1 d\alpha_2 \ldots d\alpha_k + R^0 \\
&= \sum_{i=1}^{k} S_i \int_{S_\alpha} (\alpha_i - \alpha_i^0) p(\alpha_1, \ldots, \alpha_k) d\alpha_1 d\alpha_2 \ldots d\alpha_k + R^0 \quad (III.F.8) \\
&= R^0.
\end{aligned}$$

The various moments of $R(\alpha_1, \ldots, \alpha_k)$ can be calculated by using Eqs. (III.F.7) and (III.F.8) in conjunction with the definitions previously introduced in Sec.

II.D.1. Thus, the lth central moment $\mu_l(R)$ of $R(\alpha_1,\ldots,\alpha_k)$ is obtained as the following k-fold integral over the domain S_α of the parameters $\boldsymbol{\alpha}$:

$$\mu_l(R) \equiv E\big((R-E(R))^l\big) = \int_{S_\alpha} \left(\sum_{i=1}^{k} S_i \delta\alpha_i\right)^l p(\alpha_1,\ldots,\alpha_k) d\alpha_1 d\alpha_2 \ldots d\alpha_k. \quad \text{(III.F.9)}$$

The variance of $R(\alpha_1,\ldots,\alpha_k)$ is calculated by setting $l = 2$ in Eq. (III.F.9) and by using the result obtained in Eq. (III.F.8); the detailed calculations are as follows:

$$\mu_2(R) \equiv \text{var}(R) \equiv E\left((R - R^0)^2\right)$$

$$= \int_{S_\alpha} \left(\sum_{i=1}^{k} S_i \delta\alpha_i\right)^2 p(\alpha_1,\ldots,\alpha_k) d\alpha_1 d\alpha_2 \ldots d\alpha_k$$

$$= \sum_{i=1}^{k} S_i^2 \int_{S_\alpha} (\delta\alpha_i)^2 p(\alpha_1,\ldots,\alpha_k) d\alpha_1 d\alpha_2 \ldots d\alpha_k$$

$$+ 2 \sum_{i \neq j=1}^{k} S_i S_j \int_{S_\alpha} (\delta\alpha_i)(\delta\alpha_j) p(\alpha_1,\ldots,\alpha_k) d\alpha_1 d\alpha_2 \ldots d\alpha_k$$

$$= \sum_{i=1}^{k} S_i^2 \, \text{var}(\alpha_i) + 2 \sum_{i \neq j=1}^{k} S_i S_j \, \text{cov}(\alpha_i, \alpha_j).$$

The result obtained in the above equation can be written in matrix form as

$$\text{var}(R) = \boldsymbol{S} \boldsymbol{V}_\alpha \boldsymbol{S}^T, \quad \text{(III.F.10)}$$

where the superscript "T" denotes transposition, $\boldsymbol{V}_\alpha$ denotes the *covariance matrix* for the parameters $(\alpha_1,\ldots,\alpha_k)$, with elements defined as

$$(\boldsymbol{V}_\alpha)_{ij} = \begin{cases} \text{cov}(\alpha_i,\alpha_j) = \rho_{ij}\sigma_i\sigma_j, & i \neq j, \ \rho_{ij} \equiv \textit{correlation coefficient} \\ \text{var}(\alpha_i) = \sigma_i^2, & i = j, \end{cases}$$

and the column vector $\boldsymbol{S} = (S_1,\ldots,S_k)$, with components $S_i = (\partial R/\partial \alpha_i)_{\alpha^0}$, denotes the *sensitivity vector*. Equation (III.F.10) is colloquially known as the *sandwich rule*.

If the system parameters are uncorrelated, Eq. (III.F.10) takes on the simpler form

$$\mathrm{var}(R) = \sum_{i=1}^{k} S_i^2 \, \mathrm{var}(\alpha_i) = \sum_{i=1}^{k} S_i^2 \sigma_i^2 \,. \qquad \text{(III.F.11)}$$

The above concepts can be readily extended from a single response to n responses that are functions of the parameters $(\alpha_1,\ldots,\alpha_k)$. In vector notation, the n responses are represented as the column vector

$$\boldsymbol{R} = (R_1,\ldots,R_n). \qquad \text{(III.F.12)}$$

In this case, the vector-form equivalent of Eq. (III.F.7) is the following linear, first-order Taylor expansion of $\boldsymbol{R}(\boldsymbol{\alpha})$:

$$\boldsymbol{R}(\boldsymbol{\alpha}^0 + \delta\boldsymbol{\alpha}) = \boldsymbol{R}(\boldsymbol{\alpha}^0) + \delta\boldsymbol{R} \cong \boldsymbol{R}(\boldsymbol{\alpha}^0) + \boldsymbol{S}\delta\boldsymbol{\alpha}, \qquad \text{(III.F.13)}$$

where $\boldsymbol{S}$ is a rectangular matrix of order $n \times k$ with elements representing the sensitivity of the jth response to the ith system parameter, namely

$$(\boldsymbol{S})_{ji} = \partial R_j / \partial \alpha_i \,. \qquad \text{(III.F.14)}$$

The expectation $E(\boldsymbol{R})$ of $\boldsymbol{R}$ is obtained by following the same procedure as that leading to Eq. (III.F.8), to obtain

$$E(\boldsymbol{R}) = \boldsymbol{R}^0 \,. \qquad \text{(III.F.15)}$$

The covariance matrix $\boldsymbol{V}_R$ for $\boldsymbol{R}$ is obtained by following the same procedure as that leading to Eq. (III.F.10); this yields

$$\boldsymbol{V}_R = E\big(\boldsymbol{S}\delta\boldsymbol{\alpha}(\boldsymbol{S}\delta\boldsymbol{\alpha})^T\big) = \boldsymbol{S}E\big(\delta\boldsymbol{\alpha}\delta\boldsymbol{\alpha}^T\big)\boldsymbol{S}^T = \boldsymbol{S}\boldsymbol{V}_\alpha \boldsymbol{S}^T, \qquad \text{(III.F.16)}$$

where the superscript "T" denotes transposition. Note that Eq. (III.F.16) has the same "sandwich" form as Eq. (III.F.10) for a single response.

The equations for the *propagation of higher-order moments* become increasingly complex and are seldom used in practice. For example, *for a single response* $R(\alpha_1,\ldots,\alpha_k)$ *and uncorrelated parameters* $(\alpha_1,\ldots,\alpha_k)$, the respective propagation of moments equations can be obtained from Eq. (III.F.3), after considerable amount of algebra, as follows:

$$E(R) = R(\alpha_1^0, \ldots, \alpha_k^0) + \frac{1}{2} \sum_{i=1}^{k} \left\{ \frac{\partial^2 R}{\partial \alpha_i^2} \right\}_{\alpha^0} \mu_2(\alpha_i)$$

$$+ \frac{1}{6} \sum_{i=1}^{k} \left\{ \frac{\partial^3 R}{\partial \alpha_i^3} \right\}_{\alpha^0} \mu_3(\alpha_i) + \frac{1}{24} \sum_{i=1}^{k} \left\{ \frac{\partial^4 R}{\partial \alpha_i^4} \right\}_{\alpha^0} \mu_4(\alpha_i) \quad \text{(III.F.17)}$$

$$+ \frac{1}{24} \sum_{i=1}^{k-1} \sum_{j=i+1}^{k} \left\{ \frac{\partial^4 R}{\partial \alpha_i^2 \partial \alpha_j^2} \right\}_{\alpha^0} \mu_2(\alpha_i) \mu_2(\alpha_j);$$

$$\mu_2(R) = \sum_{i=1}^{k} \left\{ \left(\frac{\partial R}{\partial \alpha_i} \right)^2 \right\}_{\alpha^0} \mu_2(\alpha_i) + \sum_{i=1}^{k} \left\{ \frac{\partial R}{\partial \alpha_i} \frac{\partial^2 R}{\partial \alpha_i^2} \right\}_{\alpha^0} \mu_3(\alpha_i)$$

$$+ \frac{1}{3} \sum_{i=1}^{k} \left\{ \frac{\partial R}{\partial \alpha_i} \frac{\partial^3 R}{\partial \alpha_i^3} \right\}_{\alpha^0} \mu_4(\alpha_i) \quad \text{(III.F.18)}$$

$$+ \frac{1}{4} \sum_{i=1}^{k} \left\{ \left(\frac{\partial^2 R}{\partial \alpha_i^2} \right)^2 \right\}_{\alpha^0} \left[\mu_4(\alpha_i) - (\mu_2(\alpha_i))^2 \right];$$

$$\mu_3(R) = \sum_{i=1}^{k} \left\{ \left(\frac{\partial R}{\partial \alpha_i} \right)^3 \right\}_{\alpha^0} \mu_3(\alpha_i)$$

$$+ \frac{3}{2} \sum_{i=1}^{k} \left\{ \left(\frac{\partial R}{\partial \alpha_i} \right)^2 \frac{\partial^2 R}{\partial \alpha_i^2} \right\}_{\alpha^0} \left[\mu_4(\alpha_i) - (\mu_2(\alpha_i))^2 \right]; \quad \text{(III.F.19)}$$

$$\mu_4(R) = \sum_{i=1}^{k} \left\{ \left(\frac{\partial R}{\partial \alpha_i} \right)^4 \right\}_{\alpha^0} \left[\mu_4(\alpha_i) - 3(\mu_2(\alpha_i))^2 \right] + 3[\mu_2(R)]^2. \quad \text{(III.F.20)}$$

In Eqs. (III.F.17-20), the quantities $\mu_l(R), (l = 1, \ldots, 4)$, denote the respective central moments of the response $R(\alpha_1, \ldots, \alpha_k)$, while the quantities $\mu_k(\alpha_i), (i = 1, \ldots, k; k = 1, \ldots, 4)$ denote the respective central moments of the parameters $(\alpha_1, \ldots, \alpha_k)$. Note that $E(R) \neq R^0$ when the response $R(\alpha_1, \ldots, \alpha_k)$ is a nonlinear function of the parameters $(\alpha_1, \ldots, \alpha_k)$. As has been already mentioned, Eqs. (III.F.17-20) are valid for uncorrelated parameters only.

It is important to emphasize that the "propagation of moments" equations are used not only for processing experimental data obtained from indirect measurements, but are also used for performing statistical analysis of computational models. In the latter case, the "propagation of errors" equations provide a systematic way of obtaining the uncertainties in computed results, arising not only from uncertainties in the parameters that enter the respective computational model but also from the numerical approximations themselves. The efficient computation of sensitivities and, subsequently, uncertainties in results produced by various models (algebraic, differential, integrals, etc.) are the objectives of subsequent chapters in this book.

Some simple illustrative examples of using Eq. (III.F.10) to calculate the standard deviation σ_R of a response R will be presented next. As a first example, consider that R is a multiplicative function of two correlated parameters, i.e., $R_1 = \alpha_1 \alpha_2$. Hence, the sensitivities of R_1 to α_1 and α_2 can be readily calculated as

$$S_1 = \frac{\partial R_1}{\partial \alpha_1}\bigg|_{\alpha^0} = \alpha_2^0, \quad S_2 = \frac{\partial R_1}{\partial \alpha_2}\bigg|_{\alpha^0} = \alpha_1^0.$$

Substituting the above sensitivities in Eq. (III.F.10) yields the following result for the relative standard deviation of R_1:

$$\frac{\sigma_{R_1}}{R_1} = \left(\frac{\sigma_1^2}{(\alpha_1^0)^2} + \frac{\sigma_2^2}{(\alpha_2^0)^2} + 2\frac{V_{12}}{\alpha_1^0 \alpha_2^0} \right)^{1/2}. \qquad \text{(III.F.21)}$$

As another example, consider a response that is the ratio of two uncorrelated parameters, i.e.,

$$R_2 = \alpha_1 / \alpha_2.$$

In this case, the respective sensitivities of R_2 to α_1 and α_2 are obtained as

$$S_1 = 1/\alpha_2^0, \quad S_2 = -\alpha_1^0 / (\alpha_2^0)^2.$$

Substituting the above expressions in Eq. (III.F.10) yields:

$$\frac{\sigma_{R_2}}{R_2} = \left(\frac{\sigma_1^2}{(\alpha_1^0)^2} + \frac{\sigma_2^2}{(\alpha_2^0)^2} \right)^{1/2}. \qquad \text{(III.F.22)}$$

GLOSSARY

Absolute error (of a measuring instrument): The difference between the value of the measured quantity obtained by using a measuring instrument and the true (yet unknown) value of the measured quantity.

Absolutely constant elementary error: An elementary error that retains the same value in repeated measurements performed under the same conditions. The value of an absolutely constant error is unknown but its limits can be estimated.

Accuracy of measurement: A qualitative expression of the closeness of the result of a measurement to the true value of the measured quantity.

Accuracy of a measuring instrument: The ability of a measuring instrument to produce measurements whose results are close to the true value of the measured quantity.

Combined measurement: Measurement of several quantities of the same kind, using results from (and/or combinations of) direct measurements.

Conditionally constant elementary error (of a measurement): An elementary error, having definite limits, which varies in repeated measurements performed under the same conditions or with different measuring instruments of the same type. These limits can be calculated or estimated.

Dead band: Maximum interval through which a stimulus may be changed in either direction without producing a change in the response of a measuring instrument.

Drift: A slow variation in time of the output of a measuring instrument, independently of the respective stimulus.

Elementary error (of a measurement): A component of error or uncertainty of a measurement associated with a single source of inaccuracy of the measurement.

Error (of a measurement): The deviation of the result of a measurement from the true value of the measured quantity; the error is expressed in absolute or relative form.

Inaccuracy (of a measurement): A qualitative characteristic of the deviation of a measurement result from the true value of the measured quantity. Quantitatively, inaccuracy can be characterized either as a measurement error or as a measurement uncertainty.

Indirect measurement: A measurement in which the value of the measured quantity is calculated by using measurements of other quantities that are connected to the measured quantity by a known relation.

Intrinsic error: The error of a measuring instrument, determined under reference conditions.

Measurement: The set of experimental operations that are performed using technical products (measuring instruments) for the purpose of finding the value of a physical quantity.

Measuring instrument: A technical product with standardized metrological characteristics.

Measuring standard: A measuring instrument intended to materialize and/or conserve a unit of a physical quantity in order to transmit its value to other measuring instruments.

Metrology: Science of measurement: an applied science that includes knowledge of measurements of physical quantities.

Normal operating conditions: Conditions within which a measuring instrument is designed to operate so that its metrological characteristics lie within specified limits.

Primary standard: A measuring standard that has the highest accuracy (in a country).

Random error (of a measurement): A component of the inaccuracy of a measurement that varies in an unpredictable way in the course of repeated measurements of the same measured quantity under the same conditions.

Relative error: Absolute error divided by the true value of the measured quantity. In practice, the true (but unknown) value is replaced by the measurement result.

Repeatability of a measurement: The closeness of agreement among several consecutive measurements of the same quantity, performed under the same operating conditions with the same measuring instruments, over a short period of time.

Reproducibility of a measurement: The closeness of agreement among repeated measurements for the same measured quantities performed in different locations, under different operating conditions, or over a long period of time.

Resolution: The smallest, still distinguishable interval between two adjacent values of the output signal of a measuring instrument.

Result of measurement: The value obtained by measurement of a quantity. The measurement result is expressed as a product of a numerical value and a proper unit.

Sensitivity of a measuring instrument: The change in the response of a measuring instrument divided by the corresponding change in the stimulus.

Systematic error (of measurement): A component of the inaccuracy of measurement that remains constant or varies in a predictable way in the course of repeated measurements of the same measured quantity.

True value: The value of a measured quantity that (if it were known) would ideally reflect, qualitatively and quantitatively, the respective property of the quantity of interest.

Uncertainty of measurement: An interval within which the true value of a measured quantity would lie with a given probability. Uncertainty is defined with its limits and corresponding confidence probability, and can be expressed in absolute or relative form.

Chapter IV

LOCAL SENSITIVITY AND UNCERTAINTY ANALYSIS OF LINEAR SYSTEMS

A physical system is modeled mathematically in terms of: (a) linear and/or nonlinear equations that relate the system's independent variables and parameters to the system's state (i.e., dependent) variables, (b) inequality and/or equality constraints that delimit the ranges of the system's parameters, and (c) one or several quantities, customarily referred to as system responses (or objective functions, or indices of performance) that are to be analyzed as the parameters vary over their respective ranges. The objective of *local sensitivity analysis* is to analyze the behavior of the system responses locally around a chosen point or trajectory in the combined phase space of parameters and state variables. On the other hand, the objective of *global sensitivity analysis* is to determine all of the system's critical points (bifurcations, turning points, response extrema) in the combined phase space formed by the parameters, state variables, and adjoint variables, and subsequently analyze these critical points by local sensitivity analysis. The concepts underlying local sensitivity analysis will be presented in Chapters IV and V; specifically, Chapter IV presents the mathematical formalism for sensitivity analysis of *linear systems*, while Chapter V presents the sensitivity analysis formalism for nonlinear systems with operator responses in the presence of feedback.

The scope of sensitivity analysis is to calculate exactly and efficiently the sensitivities of the system's response to variations in the system's parameters, around their nominal values. As will be shown in this chapter, these sensitivities are obtained by calculating the first Gâteaux-differential of the system's response at the nominal value of the system's dependent variables (also called state functions) and parameters. Two procedures will be developed for calculating the sensitivities: the Forward Sensitivity Analysis Procedure (*FSAP*) and the Adjoint Sensitivity Analysis Procedure (*ASAP*). As will be shown in this chapter, the *FSAP* is conceptually easier to develop and implement than the *ASAP*; however, the *FSAP* is advantageous to employ only if, in the problem under consideration, the number of different responses of interest exceeds the number of system parameters and/or parameter variations to be considered. However, most problems of practical interest comprise a large number of parameters and comparatively few responses. In such situations, it is by far more advantageous to employ the *ASAP*. Once the sensitivities to all parameters are available, they can subsequently be used for various purposes, such as for ranking the respective parameters in order of their relative importance to the response, for assessing changes in the response due to parameter variations, or for performing uncertainty analysis by using the propagation of errors (moments) formulas presented in Section III.F. It is important to note that the physical systems considered in this chapter are *linear in the dependent (state) variables*. This important feature makes it possible to solve the adjoint equations underlying the ASAP independently of the original (forward) equations. This fact is in contradistinction to physical systems in which the underlying operators act nonlinearly on the dependent variables; in such cases, as will be seen in Chapter V, the adjoint equations underlying the ASAP cannot be solved independently of the original (forward) equations.

A physical system that is *linear in the dependent (state) variables* can be modeled mathematically by a system of K coupled operator equations of the form

$$L(\alpha)u = Q[\alpha(x)], \quad x \in \Omega, \tag{IV.1}$$

where:

1. $x = (x_1, \ldots, x_J)$ denotes the phase-space position vector; $x \in \Omega \subset \mathbb{R}^J$, where Ω is a subset of the J-dimensional real vector space $\mathbb{R}^J$;

2. $u(x) = [u_1(x), \ldots, u_K(x)]$ denotes the state vector; $u(x) \in \mathscr{E}_u$, where $\mathscr{E}_u$ is a normed linear space over the scalar field $\mathscr{F}$ of real numbers;

3. $\alpha(x) = [\alpha_1(x), \ldots, \alpha_I(x)]$ denotes the vector of system parameters; $\alpha \in \mathscr{E}_\alpha$, where $\mathscr{E}_\alpha$ is also a normed linear space; in usual applications, $\mathscr{E}_\alpha$ may be one of the Hilbert spaces $\mathscr{L}_2$ or ℓ_2; occasionally, the components of α may simply be a set of real scalars, in which case $\mathscr{E}_\alpha$ is $\mathbb{R}^I$;

4. $Q[\alpha(x)] = [Q_1(\alpha), \ldots, Q_K(\alpha)]$ denotes a (column) vector whose elements represent inhomogeneous source terms that depend either linearly or nonlinearly on α; $Q \in \mathscr{E}_Q$, where $\mathscr{E}_Q$ is a normed linear space; the components of Q may be nonlinear operators (rather than just functions) acting on $\alpha(x)$ and x.

The components of the (column) vector $L = [L_1(\alpha), \ldots, L_K(\alpha)]$ are operators (including: differential, difference, integral, distributions, finite or infinite matrices) that *act linearly on the state vector* $u(x)$, *and depend nonlinearly* (in general) *on the vector of system parameters* $\alpha(x)$.

In view of the definitions given above, L represents the mapping $L: \mathscr{D} \subset \mathscr{E} \to \mathscr{E}_Q$, where $\mathscr{D} = \mathscr{D}_u \times \mathscr{D}_\alpha$, $\mathscr{D}_u \subset \mathscr{E}_u$, $\mathscr{D}_\alpha \subset \mathscr{E}_\alpha$, and $\mathscr{E} = \mathscr{E}_u \times \mathscr{E}_\alpha$. Note that an arbitrary element $e \in \mathscr{E}$ is of the form $e = (u, \alpha)$.

Even though in most practical applications, $\mathscr{E}$ and $\mathscr{E}_Q$ will be Hilbert spaces (e.g., the space $\mathscr{L}_2$, the Sobolev spaces $\mathscr{H}^m$), this restriction is not imposed at this stage, for the sake of generality. The domain $\mathscr{D}$ of L is specified by the characteristics of its component operators. Thus, if differential operators appear in Eq. (IV.1), then a corresponding set of boundary and/or initial conditions (which are essential to define $\mathscr{D}$) must also be given. This set can be represented in operator form as

$$[B(\alpha)u - A(\alpha)]_{\partial\Omega} = 0, \quad x \in \partial\Omega, \qquad (IV.2)$$

where A and B are operators and $\partial\Omega$ is the boundary of Ω. *The operator B acts linearly on u, but can depend nonlinearly on α.* The operator $A(\alpha)$ represents all inhomogeneous boundary terms, which can, in general, be nonlinear functions of α. The function $u(x)$ is considered to be the unique nontrivial solution of the physical problem described by Eqs. (IV.1) and (IV.2). This requirement is fulfilled (or is assumed to be fulfilled, when rigorous existence and uniqueness proofs are lacking) in most problems of practical interest.

In general, the *system response* (i.e., performance parameter) R associated with the problem modeled by Eqs. (IV.1) and (IV.2) can be represented mathematically by an operator that *acts nonlinearly on the system's state vector u and parameters α*, namely:

$$R(e): \mathscr{D}_R \subset \mathscr{E} \to \mathscr{E}_R, \qquad (IV.3)$$

where $\mathscr{E}_R$ is a normed vector space.

In practice, the exact values of the parameters α are not known; usually, only their nominal (mean) values, α^0, and their covariances, $\text{cov}(\alpha_i, \alpha_j)$, are available (in exceptional cases, higher moments may also be available). The nominal values, u^0, of the dependent variables u are obtained by solving Eq. (IV.1) and (IV.2) for the nominal values α^0. Thus, $e^0 = (u^0, \alpha^0)$ is the "base case" solution of the so-called "base-case system"

$$L(\alpha^0)u^0 = Q[\alpha^0(x)], \quad x \in \Omega, \qquad (IV.4)$$

$$[B(\alpha^0)u^0 - A(\alpha^0)]_{\partial\Omega} = 0, \quad x \in \partial\Omega. \qquad (IV.5)$$

Once the nominal values $e^0 = (u^0, \alpha^0)$ have been obtained, the nominal value $R(e^0)$ of the response $R(e)$ is obtained by evaluating Eq. (IV.3) at $e^0 = (u^0, \alpha^0)$.

The most general and fundamental concept for the definition of the sensitivity of a response to variations in the system parameters is the Gâteaux-(G)-differential, as defined in Eq. (I.D.1) in Section I.D. Applying that definition to Eq. (IV.3) yields the G-differential $\delta R(e^0; h)$ of the response $R(e)$ at e^0 with increment h, as

$$\delta R(e^0; h) \equiv \left\{ \frac{d}{dt} \left[R(e^0 + th) \right] \right\}_{t=0} = \lim_{t \to 0} \frac{R(e^0 + th) - R(e^0)}{t}, \quad \text{(IV.6)}$$

for $t \in \mathscr{T}$, and all (i.e., arbitrary) vectors $h \in \mathscr{E}$; here $h = (h_u, h_\alpha)$, since $\mathscr{E} = \mathscr{E}_u \times \mathscr{E}_\alpha$. As indicated by Eq. (I.D.2), the G-differential $\delta R(e^0; h)$ is related to the total variation $[R(e^0 + th) - R(e^0)]$ of R at e^0 through the relation

$$R(e^0 + th) - R(e^0) = \delta R(e^0; h) + \Delta(h), \text{ with } \lim_{t \to 0} [\Delta(th)]/t = 0. \quad \text{(IV.7)}$$

It is important to note that R need not be continuous in u and/or α for $\delta R(e^0; h)$ to exist at $e^0 = (u^0, \alpha^0)$, and that $\delta R(e^0; h)$ is not necessarily linear in h. Thus, by defining $\delta R(e^0; h)$ to be *the sensitivity of the response R to variations h*, the concept of sensitivity can also be used for responses that involve discontinuities. The objective of *local sensitivity analysis* is to evaluate $\delta R(e^0; h)$. To achieve this objective, two alternative formalisms are developed and discussed in the following, namely the "Forward Sensitivity Analysis Procedure" (*FSAP*) and the "Adjoint Sensitivity Analysis Procedure" (*ASAP*).

IV.A. THE FORWARD SENSITIVITY ANALYSIS PROCEDURE (FSAP)

In practice, we know at the outset the vector, h_α, of parameter variations around the nominal values α^0. However, since the system's state vector u and parameters α are related to each other through Eqs. (IV.1) and (IV.2), it follows that h_u and h_α are also related to each other. Therefore, the sensitivity $\delta R(e^0; h)$ of $R(e)$ at e^0 can only be evaluated after determining the vector of variations

h_u (in the state functions u around u^0), in terms of the vector of parameter variations h_α. The first-order relationship between h_u and h_α is obtained by taking the G-differentials of Eqs. (IV.1) and (IV.2). Thus, applying the definition given in Eq. (1.D.1) to Eqs. (IV.1) and (IV.2), respectively, gives

$$L(\alpha^0)h_u + [L'_\alpha(\alpha^0)u^0]h_\alpha - \delta Q(\alpha^0;h_\alpha) = 0, \quad x \in \Omega, \qquad \text{(IV.A.1)}$$

and

$$\{B(\alpha^0)h_u + [B'_\alpha(\alpha^0)u^0]h_\alpha - \delta A(\alpha^0;h_\alpha)\}_{\partial\Omega} = 0, \quad x \in \partial\Omega, \quad \text{(IV.A.2)}$$

where $L'_\alpha(\alpha^0)$ and $B'_\alpha(\alpha^0)$ denote, respectively, the partial G-derivatives at α^0 of L and B with respect to α. For a given vector of parameter variations h_α around α^0, the system of equations represented by Eqs. (IV.A.1) and (IV.A.2) is solved to obtain h_u. Once h_u is available, it is in turn used in Eq. (IV.6) to calculate the sensitivity $\delta R(e^0;h)$ of $R(e)$ at e^0, for a given vector of parameter variations h_α.

Equations (IV.A.1) and (IV.A.2) are called the "*Forward Sensitivity Equations*" (*FSE*), and the direct calculation of the response sensitivity $\delta R(e^0;h)$ by using the (h_α-dependent) solution h_u of these equations constitutes the *Forward Sensitivity Analysis Procedure* (*FSAP*). From the standpoint of computational costs and effort, the *FSAP* is advantageous to employ only if, in the problem under consideration, the number of different responses of interest exceeds the number of system parameters and/or parameter variations to be considered. This is very rarely the case in practice, however, since most problems of practical interest are characterized by many parameters (i.e., α has many components) and comparatively few responses. In such situations, it is not economical to employ the *FSAP* to answer all sensitivity questions of interest, since it becomes prohibitively expensive to solve the h_α-dependent *FSE* repeatedly to determine h_u for all possible vectors h_α. Hence, it is desirable to devise, whenever possible, an alternative procedure to evaluate $\delta R(e^0;h)$ so as to circumvent the necessity for solving the *FSE* repeatedly.

IV.B. THE ADJOINT SENSITIVITY ANALYSIS PROCEDURE (ASAP)

The practical motivation underlying the development of an alternative method for sensitivity analysis is to avoid the need for repeatedly solving the *FSE*, i.e., Eqs. (IV.A.1) and (IV.A.2). This goal could be achieved if all unknown values of h_u were eliminated from the expression of $\delta R(e^0;h)$. As will be shown in this Section, this elimination can indeed be accomplished by constructing an adjoint system that is (a) uniquely defined, (b) independent of the vectors h_u and h_α, and (c) such that its solution can be used to eliminate all unknown values of h_u from the expression of $\delta R(e^0;h)$.

As has been discussed in Section I.C, adjoint operators can be introduced uniquely only for densely defined linear operators in Banach spaces. Furthermore, since the lack of an inner product in a general Banach space gives rise to significant conceptual distinctions between the adjoint of a linear operator on a Banach space and the adjoint of a linear operator on a Hilbert space, the choice of space becomes important for subsequent derivations. Since all problems of practical importance are ultimately solved numerically in discrete Hilbert spaces, the subsequent developments in this Section will also be set forth in Hilbert spaces. Specifically, the spaces $\mathcal{E}_u$ and $\mathcal{E}_Q$ will henceforth be considered to be real Hilbert spaces denoted by $\mathcal{H}_u$ and $\mathcal{H}_Q$, respectively. The inner product on $\mathcal{H}_u$ and $\mathcal{H}_Q$ will be denoted by $\langle \bullet,\bullet \rangle_u$ and $\langle \bullet,\bullet \rangle_Q$, respectively, since these inner products need not be identical.

To define the formal adjoint L^+ of L, we recall from the geometry of Hilbert spaces that the following relationship holds for a vector $\psi \in \mathcal{H}_Q$:

$$\langle \psi, L(\alpha^0) h_u \rangle_Q = \langle L^+(\alpha^0)\psi, h_u \rangle_u + \{P[h_u,\psi]\}_{\partial\Omega} \qquad \text{(IV.B.1)}$$

In the above equation, therefore, the operator $L^+(\alpha^0)$ is the $K \times K$ matrix

$$L^+(\alpha^0) = \left[L^+_{ji}(\alpha^0) \right], \quad (i,j = 1,\ldots,K), \qquad \text{(IV.B.2)}$$

obtained by transposing the formal adjoints of the operators $L_{ij}(\alpha^0)$, while $\{P[h_u,\psi]\}_{\partial\Omega}$ is the associated bilinear form [recall Eq. (I.B.26)] evaluated on $\partial\Omega$. The domain of L^+ is determined by selecting appropriate adjoint boundary conditions, represented here in operator form as

$$\{B^+(\psi;\alpha^0) - A^+(\alpha^0)\}_{\partial\Omega} = 0, \quad x \in \partial\Omega. \tag{IV.B.3}$$

These boundary conditions for L^+ are obtained by requiring that:

(a) They must be independent of h_u, h_α, and G-derivatives with respect to α;
(b) The substitution of Eqs. (IV.A.2) and (IV.B.3) into the expression of $\{P[h_u,\psi]\}_{\partial\Omega}$ must cause all terms containing unknown values of h_u to vanish.

This selection of the boundary conditions for L^+ reduces $\{P[h_u,\psi]\}_{\partial\Omega}$ to a quantity that contains boundary terms involving only known values of h_α, ψ, and, possibly, α^0; this quantity will be denoted by $\hat{P}(h_\alpha,\psi;\alpha^0)$. In general, $\hat{P}$ does not automatically vanish as a result of these manipulations, although it may do so in particular instances. In principle, $\hat{P}$ could be forced to vanish by considering extensions of $L(\alpha^0)$, in the operator sense, based on Eqs. (IV.A.1) and (IV.A.2). In practice, however, it is more convenient to avoid such complications, since $\hat{P}$ will ultimately appear only as a readily computable quantity in the final expression of the response sensitivity $\delta R(e^0;h)$.

Using the adjoint boundary conditions for L^+ [i.e., Eq. (IV.B.3)] together with Eq. (IV.A.2) in Eq. (IV.B.1) reduces the latter to

$$\langle \psi, L(\alpha^0)h_u \rangle_Q = \langle L^+(\alpha^0)\psi, h_u \rangle_u + \hat{P}(h_\alpha,\psi;\alpha^0). \tag{IV.B.4}$$

Using now Eq. (IV.A.1) to replace the quantity $L(\alpha^0)h_u$ in Eq. (IV.B.4) transforms the latter equation into the form

$$\langle L^+(\alpha^0)\psi, h_u \rangle_u = \langle \psi, \delta Q(\alpha^0; h_\alpha) - [L'_\alpha(\alpha^0)u^0]h_\alpha \rangle_Q \\ - \hat{P}(h_\alpha,\psi;\alpha^0). \tag{IV.B.5}$$

At this stage in the development of the *ASAP*, we first examine the special case when $\mathscr{E}_R$ is simply the field of real scalars $\mathscr{F}$, so that the system response simply becomes a nonlinear *functional* $R: \mathscr{D}_R \to \mathscr{F}$. Subsequently, we will consider the general case of a nonlinear *operator* $R: \mathscr{D}_R \to \mathscr{E}_R$.

IV.B.1. System Responses: Functionals

When $R: \mathcal{D}_R \to \mathcal{F}$, the sensitivity $\delta R(e^0; h)$ is also a functional that takes values in $\mathcal{F}$. Note, also, that the functional on the right-side of Eq. (IV.B.5) does not contain any values of h_u. Thus, if the h_u-dependence in the functional $\delta R(e^0; h)$ could be separated from the h_α-dependence and, furthermore, if the quantity containing this h_u-dependence could be expressed in terms of the left-side of Eq. (IV.B.5), then the *ASAP* would have attained its objective. At this stage, however, the functional $\left\langle L^+(\alpha^0)\psi, h_u \right\rangle_u$ is linear in h_u, while, in general, $\delta R(e^0; h)$ is not. For $\delta R(e^0; h)$ to be linear in h and consequently in h_u, it becomes apparent that $R(e)$ must satisfy (see Section I.D) a weak Lipschitz condition at e^0, and must also satisfy the relation

$$R(e^0 + th_1 + th_2) - R(e^0 + th_1) - R(e^0 + th_2) + R(e^0) = o(t). \quad \text{(IV.B.6)}$$

In this case, the G-differential $\delta R(e^0; h)$ will be linear in h [and will therefore be denoted by $DR(e^0; h)$], and can be written as

$$DR(e^0; h) = R'_u(e^0) h_u + R'_\alpha(e^0) h_\alpha, \quad \text{(IV.B.7)}$$

where $R'_u(e^0)$ and $R'_\alpha(e^0)$ denote, respectively, the partial G-derivatives at e^0 of $R(e)$ with respect to u and α. As indicated by Eq. (IV.B.7), the h_u-dependence in $DR(e^0; h)$ has thus been separated from the h_α-dependence. Since the quantity $R'_\alpha(e^0) h_\alpha$ can be calculated directly at this stage, it is called the *direct effect term*. On the other hand, the quantity $R'_u(e^0) h_u$ cannot be evaluated directly at this stage (since it depends on the unknown function h_u), and is therefore called the *indirect effect term*. This terminology reflects the fact that the response can be considered to depend on the parameters α both "directly" and "indirectly," through the state vector u.

Since the functional $R'_u(e^0) h_u$ is linear in h_u and since Hilbert spaces are self-dual, the Riesz representation theorem (see Section I.C) ensures that there exists a unique vector $\nabla_u R(e^0) \in \mathcal{H}_u$ such that

$$R'_u(e^0) h_u = \left\langle \nabla_u R(e^0), h_u \right\rangle_u, \quad h_u \in \mathcal{H}_u. \quad \text{(IV.B.8)}$$

The right-side of Eq. (IV.B.8) and the left-side of Eq. (IV.B.5) can now be required to represent the same functional, which is accomplished by imposing the relationship

$$L^+(\alpha^0)\psi = \nabla_u R(e^0). \qquad (IV.B.9)$$

Furthermore, the Riesz representation theorem ensures that the above relationship, where ψ satisfies the adjoint boundary conditions given in Eq. (IV.B.3), holds uniquely. The construction of the requisite *adjoint system*, consisting of Eqs. (IV.B.3) and (IV.B.9), has thus been accomplished. It is very important to note that this adjoint system is independent not only of the functions h_u but also of the nominal values u^0 of u. This means that *the adjoint system*, namely Eqs. (IV.B.3) and (IV.B.9), *can be solved independently of the solution* u^0 *of the original equations*. In turn, this fact simplifies considerably the choice of numerical methods for solving the adjoint system. It is also important to note that *this advantageous situation arises iff the original equations that model the physical problem*, namely Eqs. (IV.1) and (IV.2), *are linear in the state-variables* u.

Replacing now Eqs. (IV.B.8), (IV.B.9), and (IV.B.5) in Eq. (IV.B.7) yields the following expression for response sensitivity $DR(e^0;h)$

$$DR(e^0;h) = R'_\alpha(e^0)h_\alpha + \left\langle \psi, \delta Q(e^0;h_\alpha) - [L'_\alpha(\alpha^0)u^0]h_\alpha \right\rangle_Q - \hat{P}(h_\alpha,\psi;e^0). \qquad (IV.B.10)$$

As indicated by the above expression, the desired elimination of the unknown values of h_u from the expression giving the sensitivity $DR(e^0;h)$ of $R(e)$ at e^0 to variations h_α in the system parameters α has been accomplished. Thus, once the single calculation to determine the adjoint function ψ from Eqs. (IV.B.3) and (IV.B.9) has been carried out, ψ can be used in Eq. (IV.B.10) to obtain efficiently the sensitivity $DR(e^0;h)$ of $R(e)$.

IV.B.2. System Responses: Operators

The analysis of the necessary and sufficient conditions underlying the validity of the *ASAP* presented in the previous section for responses that are functionals can also be employed to establish the guidelines for treating operator responses. When the response $R(e)$ is an operator, the sensitivity $\delta R(e^0;h)$ is also an operator, defined on the same domain, and with the same range as $R(e)$. Based on the considerations presented in the previous Section, the guidelines for developing the *ASAP* for operator responses can be formulated as follows:

(a) Isolate the h_u-dependence of $\delta R(e^0;h)$ from the functional dependence of $\delta R(e^0;h)$ on the remaining quantities;

(b) Express the quantity containing the h_u-dependence, isolated as described in (a) above, in the form of linear combinations of functionals that are themselves linear in h_u; and

(c) Employ the *ASAP for functionals*, as developed in the previous Section, to evaluate the functionals determined in item (b) above.

The implementation of the guidelines formulated above into a rigorous formalism will necessarily involve the use of adjoint operators. Since adjoint operators in Hilbert spaces are in practice more convenient to use than adjoint operators in Banach spaces, the subsequent developments are facilitated by taking advantage of the simplifying geometrical properties of Hilbert spaces while still retaining sufficient generality for practical applications. Henceforth, therefore, the spaces $\mathcal{E}_u$, $\mathcal{E}_Q$, and $\mathcal{E}_R$ are considered to be Hilbert spaces and denoted as $\mathcal{H}_u(\Omega)$, $\mathcal{H}_Q(\Omega)$, and $\mathcal{H}_R(\Omega_R)$, respectively. The elements of $\mathcal{H}_u(\Omega)$ and $\mathcal{H}_Q(\Omega)$ are, as before, vector-valued functions defined on the open set $\Omega \subset \mathbb{R}^J$, with smooth boundary $\partial\Omega$. The elements of $\mathcal{H}_R(\Omega_R)$ are vector or scalar functions defined on the open set $\Omega_R \subset \mathbb{R}^m, 1 \le m \le J$, with a smooth boundary denoted as $\partial\Omega_R$. Of course, if $J=1$, then $\partial\Omega$ merely consists of two endpoints; similarly, if $m=1$, then $\partial\Omega_R$ consists of two endpoints only. The inner products on $\mathcal{H}_u(\Omega)$, $\mathcal{H}_Q(\Omega)$, and $\mathcal{H}_R(\Omega_R)$ are denoted by $\langle\bullet,\bullet\rangle_u$, $\langle\bullet,\bullet\rangle_Q$, and $\langle\bullet,\bullet\rangle_R$, respectively.

In view of the guidelines (a) and (b) above, it becomes apparent that further progress is possible only if $\delta R(e^0;h)$ is linear in h, that is, if $R(e)$ satisfies a weak Lipschitz condition at e^0, and

$$R(e^0 + th_1 + th_2) - R(e^0 + th_1) - R(e^0 + th_2) + R(e^0) = o(t);$$
$$h_1, h_2 \in \mathcal{H}_u \times \mathcal{H}_\alpha; t \in \mathcal{F}.$$
(IV.B.11)

Therefore, the response sensitivity $\delta R(e^0;h)$ will henceforth be assumed to be linear in h, and will consequently be denoted as $DR(e^0;h)$. This means that $R(e)$ admits a total G-derivative at $e^0 = (u^0, \alpha^0)$, such that the relationship

$$DR(e^0;h) = R'_u(e^0)h_u + R'_\alpha(e^0)h_\alpha$$
(IV.B.12)

holds, where $R'_u(e^0)$ and $R'_\alpha(e^0)$ are the partial G-derivatives at e^0 of $R(e)$ with respect to u and α.

With the derivation of Eq. (IV.B.12), the task outlined in guideline (a) above has been completed, and Eq. (IV.B.11) gives the necessary and sufficient conditions that make the accomplishment of this guideline possible. Note also that $R'_u(e^0)$ is a linear operator, on h_u, from $\mathcal{H}_u$ into $\mathcal{H}_R$, i.e., $R'_u(e^0) \in L(\mathcal{H}_u(\Omega), \mathcal{H}_R(\Omega_R))$. By analogy to the particular case when the response is a functional, it is still convenient to refer to the quantities $R'_u(e^0)h_u$ and $R'_\alpha(e^0)h_\alpha$ appearing in Eq. (IV.B.12) as the "indirect effect term" and the "direct effect term," respectively.

The direct effect term can be evaluated efficiently at this stage. To proceed with the evaluation of the indirect effect term, we consider that the orthonormal set $\{p_s\}_{s \in \mathcal{S}}$, where s runs through an index set $\mathcal{S}$, is an orthonormal basis of $\mathcal{H}_R(\Omega_R)$. Then, since $R'_u(e^0)h_u \in \mathcal{H}_R(\Omega_R)$, it follows that $R'_u(e^0)h_u$ can be represented as the Fourier series

$$R'_u(e^0)h_u = \sum_{s \in \mathcal{S}} \left\langle R'_u(e^0)h_u, p_s \right\rangle_R p_s. \quad \text{(IV.B.13)}$$

The notation $\sum_{s \in \mathcal{S}}$ is used to signify that in the above sum only an at most countable number of elements are different from zero, and the series extended upon the nonzero elements converges unconditionally. According to customary terminology, the functionals $\left\langle R'_u(e^0)h_u, p_s \right\rangle_R$ are called the Fourier coefficients [in this case, of $R'_u(e^0)h_u$] with respect to the basis $\{p_s\}$. These functionals are linear in h_u since $R(e)$ was required to satisfy the conditions stated in Eq. (IV.B.11). Thus, the task outlined in guideline (b) above is now completed with the derivation of Eq. (IV.B.13).

To accomplish the task outlined in guideline (c) above, we recall that the *ASAP* for functionals required that the indirect effect term be represented as an inner product of h_u with an appropriately defined vector in $\mathcal{H}_u$. This indicates that further progress can be made only if each of the functionals in Eq. (IV.B.13) is expressed as an inner product of h_u with a uniquely defined vector in $\mathcal{H}_u(\Omega)$, which remains to be determined in the sequel.

The construction of the previously mentioned inner products can be accomplished with the help of the operator adjoint to $R'_u(e^0)$. Since $R'_u(e^0) \in L(\mathcal{H}_u(\Omega), \mathcal{H}_R(\Omega_R))$, and since Hilbert spaces are self-dual, the adjoint of $R'_u(e^0)$ is the operator $M(e^0) \in L(\mathcal{H}_R(\Omega_R), \mathcal{H}_u(\Omega))$ defined by means of relationship

$$\left\langle R'_u(e^0) h_u, p_s \right\rangle_R = \left\langle M(e^0) p_s, h_u \right\rangle_u, \quad s \in \mathcal{S}. \qquad \text{(IV.B.14)}$$

The operator $M(e^0)$ is unique if $R'_u(e^0)$ is densely defined. The *ASAP* developed in the previous section for responses that are functionals can now be used to construct the adjoint system whose solution will subsequently enable the elimination of unknown values of h_u from the expression of each functional $\left\langle h_u, M(e^0) p_s \right\rangle_u$, $s \in \mathcal{S}$. This adjoint system is constructed based on the same necessary and sufficient conditions as those underlying the validity of Eq. (IV.B.1) so that, for every vector $\psi_s \in \mathcal{H}_Q, s \in \mathcal{S}$, the following relationship holds:

$$\left\langle \psi_s, L(\alpha^0) h_u \right\rangle_Q = \left\langle L^+(\alpha^0) \psi_s, h_u \right\rangle_u + \{P(h_u; \psi_s)\}_{\partial\Omega}, \quad s \in \mathcal{S}, \qquad \text{(IV.B.15)}$$

where $L^+(\alpha^0)$ is the operator formally adjoint to $L(\alpha^0)$ and $\{P(h_u; \psi_s)\}_{\partial\Omega}$ is the associated bilinear form evaluated on $\partial\Omega$. The adjoint boundary conditions which determine the domain of $L^+(\alpha^0)$ are obtained as in the previous section, by requiring that they satisfy the same criteria as the criteria satisfied by the adjoint boundary conditions given in Eq. (IV.B.3). From this requirement and from the fact that Eq. (IV.B.15) and (IV.B.1) have formally identical mathematical expressions, it follows that the desired adjoint boundary conditions are formally identical to the boundary conditions given in Eq. (IV.B.3) and can therefore be expressed as

$$\{B^+(\psi_s; \alpha^0) - A^+(\alpha^0)\}_{\partial\Omega} = 0, \quad s \in \mathcal{S}. \qquad \text{(IV.B.16)}$$

As before, selecting the adjoint boundary conditions given in Eq. (IV.B.16) reduces the bilinear form $\{P(h_u; \psi_s)\}_{\partial\Omega}$ appearing in Eq. (IV.B.15) to $\hat{P}(h_\alpha, \psi_s; \alpha^0)$. Using this fact together with Eq. (IV.A.1) reduces Eq. (IV.B.15) to

$$\left\langle L^+(\alpha^0) \psi_s, h_u \right\rangle_u = \left\langle \psi_s, \delta Q(\alpha^0; h_\alpha) - [L'_\alpha(\alpha^0) u^0] h_\alpha \right\rangle_Q \\ - \hat{P}(h_\alpha, \psi_s; \alpha^0), \quad s \in \mathcal{S}. \qquad \text{(IV.B.17)}$$

The left-side of Eq. (IV.B.17) and the right-side of Eq. (IV.B.14) are now required to represent the same functional; this is accomplished by imposing the relation

$$L^+(\alpha^0)\psi_s = M(\alpha^0)p_s, \quad s \in \mathcal{S}, \qquad \text{(IV.B.18)}$$

which holds uniquely in view of the Riesz representation theorem. This last step completes the construction of the desired adjoint system, which consists of Eq. (IV.B.18) and the adjoint boundary conditions given in Eq. (IV.B.16). Furthermore, Eqs. (IV.B.12-18) can now be used to obtain the following expression for the sensitivity $DR(e^0;h)$ of $R(e)$ at e^0:

$$DR(e^0;h) = R'_\alpha(e^0)h_\alpha$$
$$+ \sum_{s \in \mathcal{S}}\left[\left\langle \psi_s, \delta Q(\alpha^0;h_\alpha) - [L'_\alpha(\alpha^0)u^0]h_\alpha \right\rangle_Q - \hat{P}(h_\alpha,\psi_s;\alpha^0)\right]p_s.$$
(IV.B.19)

As Eq. (IV.B.19) indicates, the desired elimination of all unknown values of h_u from the expression of the sensitivity $DR(e^0;h)$ of $R(e)$ at e^0 has thus been accomplished. Note that Eq. (IV.B.19) includes the particular case of functional-type responses, in which case the summation $\sum_{s \in \mathcal{S}}$ would contain a single term $(s=1)$ only, and the derivations presented in this section would reduce to those presented in the previous section.

To evaluate the sensitivity $DR(e^0;h)$ by means of Eq. (IV.B.19), one needs to compute as many adjoint functions ψ_s from Eqs. (IV.B.18) and (IV.B.16) as there are nonzero terms in the representation of $R'_u(e^0)h_u$ given in Eq. (IV.B.13). Although the linear combination of basis elements p_s given in Eq. (IV.B.13) may, in principle, contain infinitely many terms, obviously only a finite number of the corresponding adjoint functions ψ_s can be calculated in practice. Therefore, special attention is required to select the Hilbert space $\mathcal{H}_R(\Omega_R)$, a basis $\{p_s\}_{s \in \mathcal{S}}$ for this space, and a notion of convergence for the representation given in Eq. (IV.B.13) to best suit the problem at hand. This selection is guided by the need to represent the indirect effect term $R'_u(e^0)h_u$ as accurately as possible with the smallest number of basis elements; a related consideration is the viability of deriving bounds and/or asymptotic expressions for the remainder after truncating Eq. (IV.B.13) to the first few terms.

IV.C. ILLUSTRATIVE EXAMPLE: SENSITIVITY ANALYSIS OF LINEAR ALGEBRAIC MODELS

As a simple illustrative example of using the *FSAP* and *ASAP*, as developed in Sections IV.A and B above, we consider in this Section a system response of the form

$$R = cx, \qquad \text{(IV.C.1)}$$

where x is the solution of the system of linear simultaneous equations

$$Ax = b. \qquad \text{(IV.C.2)}$$

In the above equations, the $n \times n$ matrix $A = (a_{ij})$, $(i, j = 1, \ldots, n)$, together with the vectors $b = (b_1, \ldots, b_n)$ and $c = (c_1, \ldots, c_n)$ are considered to stem from experiments and/or calculations, so they are not known exactly. As is often the case in practice, only the respective nominal (i.e., mean) values $A^0 = (a_{ij}^0)$, $(i, j = 1, \ldots, n)$, $b^0 = (b_1^0, \ldots, b_n^0)$, $c^0 = (c_1^0, \ldots, c_n^0)$, and the respective uncertainties are known. In principle, the response R can represent either the result of an indirect measurement or the result of a calculation.

The nominal solution x^0 is obtained by solving Eq. (IV.C.2) for the nominal parameter values A^0 and b^0:

$$A^0 x^0 = b^0. \qquad \text{(IV.C.3)}$$

The above equation is called "the base-case system." The nominal solution x^0 is used together with the nominal parameter values c^0 to calculate the nominal value, R^0, of the response by using Eq. (IV.C.1), to obtain

$$R^0 = c^0 x^0. \qquad \text{(IV.C.4)}$$

As discussed in Chapter III, measurements and/or calculations give rise to errors $\delta A \equiv (\delta a_{ij})$, where $\delta a_{ij} \equiv a_{ij} - a_{ij}^0$, $(i, j = 1, \ldots, n)$; $\delta b \equiv (\delta b_1, \ldots, \delta b_n)$, where $\delta b_i \equiv b_i - b_i^0$, and $\delta c \equiv (\delta c_1, \ldots, \delta c_n)$, where $\delta c_i \equiv c_i - c_i^0$, $(i = 1, \ldots, n)$, in the system parameters A, b, c. These errors are mathematically expressed by means of parameter covariances $\text{cov}(a_{ij}, a_{lm})$, $\text{cov}(b_i, b_j)$, $\text{cov}(c_i, c_j)$, $\text{cov}(a_{ij}, b_l)$, $\text{cov}(a_{ij}, c_l)$, and $\text{cov}(b_i, c_j)$, $(i, j, l, m = 1, \ldots, n)$. The aim of uncertainty analysis is to obtain the mean value (i.e., expectation), $E(R)$, and the variance, $\text{var}(R)$, of the response R, which arise due to parameter uncertainties. Since the response R considered in Eq. (IV.C.1) depends nonlinearly on some of the system parameters, the propagation of moments equations, presented in Chapter III, must be used to calculate the mean value, variance, and (possibly) higher order moments of R. As shown there, the propagation of errors (moments) equations require knowledge of the respective sensitivities of R to all of the system parameters.

In principle, the sensitivities of R to all system parameters could be calculated by solving the perturbed linear system

$$\left(A^0 + \delta A\right)\left(x^0 + \delta x\right) = b^0 + \delta b, \qquad \text{(IV.C.5)}$$

to obtain

$$x_{new} = x^0 + \delta x, \qquad \text{(IV.C.6)}$$

and then use x_{new} to calculate

$$R_{new} = \left(c^0 + \delta c\right)x_{new} \equiv R^0 + \delta R. \qquad \text{(IV.C.7)}$$

The response sensitivities could then be obtained by dividing R_{new} by the parameter variations $\delta a_{ij} \equiv a_{ij} - a_{ij}^0$, $\delta b_i \equiv b_i - b_i^0$, and $\delta c_i \equiv c_i - c_i^0$. When the system under consideration is large, involving many parameters, the base-case solution x^0 is obtained by iterative methods rather than by inverting the matrix A^0 directly; in practice, therefore, the inverse matrix $\left(A^0\right)^{-1}$ is very rarely available. Furthermore, even if the inverse matrix $\left(A^0\right)^{-1}$ were available, the inverse matrix $\left(A^0 + \delta A\right)^{-1}$ is extremely unlikely to be available, since this inverse matrix would need to be computed anew every time an element of (δA) would change, as would be the case, for example, if a new experimental or computational evaluation of any of its components would result in new uncertainties. In such cases, the response sensitivities could be obtained by employing the *FSAP* to calculate the Gâteaux variation δR of the response R at the point $\left(A^0, x^0, b^0\right)$ along the directions $(\delta A, \delta x, \delta b)$. The Gâteaux variation of Eq. (IV.A.1) is calculated from the respective definition as

$$\left\{\frac{d}{d\varepsilon}\left[\left(R^0 + \varepsilon \delta R\right) - \left(c^0 + \varepsilon \delta c\right)\left(x^0 + \varepsilon \delta x\right)\right]\right\}_{\varepsilon=0} = 0,$$

which yields the relation

$$\delta R = c^0(\delta x) + (\delta c)x^0. \qquad \text{(IV.C.8)}$$

The quantity $(\delta c)x^0$ is the "direct effect" term, which can be calculated already at this stage. On the other hand, the quantity $c^0(\delta x)$ is the "indirect effect" term, which can be calculated only after determining the variation δx. In turn, δx is obtained by solving the Gâteaux-differentiated Eq. (IV.C.2), namely

$$\left\{ \frac{d}{d\varepsilon} \left[(A^0 + \varepsilon \delta A)(x^0 + \varepsilon \delta x) - (b^0 + \varepsilon \delta b) \right] \right\}_{\varepsilon=0} = 0,$$

or, equivalently,

$$A^0(\delta x) = \delta b - (\delta A)x^0. \qquad \text{(IV.C.9)}$$

The quantity δx can be obtained by noting that Eq. (IV.C.9) can be solved by using the same amount of effort as required to solve the base-case system, Eq. (IV.C.3), since it involves inverting (directly or iteratively) the matrix A^0 once only. Note, however, that the right-side (i.e., the source) of Eq. (IV.C.9) involves the variations (δA) and δb. Therefore, if any of the elements of (δA) and δb would change, for example, due to new evaluations, then Eq. (IV.C.9) would need to be solved anew. *In practice, solving Eq. (IV.C.9) repeatedly becomes impractical for large systems with many parameters, because of the large computational resources required by such repeated calculations. Such computationally intensive requirements limit the practical usefulness of the FSAP.*

The alternative procedure, which avoids the need for solving Eq. (IV.C.9) repeatedly, is the *Adjoint Sensitivity Analysis Procedure (ASAP)*. The *ASAP* relies on constructing and using the hermitian adjoint matrix to A^0, and the steps involved in constructing the requisite adjoint are as generally outlined in Section IV.B, namely:

1. Introduce an *n*-component vector $\psi = (\psi_1,\ldots,\psi_n)$, and form the inner (scalar) product of ψ with Eq. (IV.C.9) to obtain

$$\langle \psi, A^0(\delta x) \rangle = \langle \psi, \delta b - (\delta A)x^0 \rangle. \qquad \text{(IV.C.10)}$$

2. Note that the hermitian adjoint of A^0 is simply its transpose, $(A^0)^+$, and transpose the left side of Eq. (IV.C.10) to obtain

$$\langle \psi, A^0(\delta x) \rangle = \langle \delta x, (A^0)^+ \psi \rangle. \tag{IV.C.11}$$

3. Since the vector ψ is still arbitrary at this stage, it is possible to specify it by identifying the right side of Eq. (IV.C.11) with the first term on the right side of Eq. (IV.C.8), to obtain

$$(A^0)^+ \psi = c^0. \tag{IV.C.12}$$

4. Collecting now the results of Eqs. (IV.C.8-12) yields the following succession of equalities:

$$\langle \psi, \delta b - (\delta A)x^0 \rangle = \langle \delta x, (A^0)^+ \psi \rangle = c^0(\delta x) = \delta R - (\delta c)x^0. \tag{IV.C.13}$$

5. Retaining the first and the last terms in Eq. (IV.C.13) yields the expression

$$\delta R = (\delta c)x^0 + \langle \psi, \delta b - (\delta A)x^0 \rangle, \tag{IV.C.14}$$

where the vector ψ is the solution of Eq. (IV.C.12). Since the matrix $(A^0)^+$ is the hermitian adjoint of A^0, Eq. (IV.C.12) is called the *adjoint sensitivity equation,* and the vector ψ is called the *adjoint function.*

Note that the adjoint equation, i.e., Eq. (IV.C.12), is independent of any variations (δA) and δb; it only depends on the response R through the vector c^o, which appears as the source-term for this equation. Furthermore, Eq. (IV.C.12) is linear in ψ and independent of x. Hence, the adjoint sensitivity equation needs to be solved only once in order to obtain the adjoint function ψ.

Note also that the adjoint sensitivity equation can be solved independently of the original equation (IV.C.3), a fact that is characteristic of linear problems as exemplified by Eq. (IV.1) and (IV.C.2). In turn, once the adjoint function has been calculated, it is used in Eq. (IV.C.14) to obtain the sensitivities δR to all system parameters. Thus, the need to solve repeatedly the forward sensitivity equations has been circumvented. In conclusion, the *ASAP* is the most efficient method to use for sensitivity analysis of systems in which the number of parameters exceeds the number of responses under consideration. It is noteworthy that this is, indeed, the case for most practical problems.

IV.D. ILLUSTRATIVE EXAMPLE: LOCAL SENSITIVITY ANALYSIS OF A LINEAR NEUTRON DIFFUSION PROBLEM

This Section presents the sensitivity analysis of a linear neutron diffusion problem. This problem is sufficiently simple to admit an exact solution, yet contains (ordinary) differential operators with boundary conditions, which serve as a nontrivial test-bed for illustrating the application of the *FSAP* and *ASAP* for local sensitivity analysis of representative responses. The reason for choosing a problem that admits analytical solution is to demonstrate that the *FSAP* and *ASAP* yield the same results, but the *ASAP* is considerably more efficient to apply than the *FSAP*, even for simple problems with few parameters.

Consider the diffusion of monoenergetic neutrons due to distributed sources of strength S neutrons/cm$^3 \cdot$s within a slab of material of extrapolated thickness $2a$. The linear neutron diffusion equation that models mathematically this problem is

$$D\frac{d^2\varphi}{dx^2} - \Sigma_a \varphi + S = 0, \quad x \in (-a, a), \qquad \text{(IV.D.1)}$$

where $\varphi(x)$ is the neutron flux, D is the diffusion coefficient, Σ_a is the macroscopic absorption cross section, and S is the distributed source term. Note that, in view of the problem's symmetry, the origin $x = 0$ has been conveniently chosen at the middle (center) of the slab. The boundary conditions for Eq. (IV.D.1) are that the neutron flux must vanish at the extrapolated distance, i.e.,

$$\varphi(\pm a) = 0. \qquad \text{(IV.D.2)}$$

A typical response R for the neutron diffusion problem modeled by Eqs. (IV.D.1) and (IV.D.2) would be the reading of a detector placed within the slab, for example, at a distance b from the slab's midline at $x = 0$. Such a response is given by the reaction rate

$$R(e) \equiv \Sigma_d \varphi(b), \qquad \text{(IV.D.3)}$$

where Σ_d represents the detector's equivalent reaction cross section. The system parameters for this problem are thus the positive constants Σ_a, D, S, and Σ_d, which will be considered to be the components of the vector $\boldsymbol{\alpha}$ of system parameters, defined as

$$\boldsymbol{\alpha} \equiv (\Sigma_a, D, S, \Sigma_d). \tag{IV.D.4}$$

The vector $e(x)$ appearing in the functional dependence of R in Eq. (IV.D.3) denotes the concatenation of $\varphi(x)$ with $\boldsymbol{\alpha}$, and is defined as

$$e \equiv (\varphi, \boldsymbol{\alpha}). \tag{IV.D.5}$$

The nominal value $\varphi^0(x)$ of the flux is determined by solving Eqs. (IV.D.1) and (IV.D.2) for the nominal parameter values $\boldsymbol{\alpha}^0 = (\Sigma_a^0, D^0, S^0, \Sigma_d^0)$, to obtain

$$\varphi^0(x) = \frac{S^0}{\Sigma_a^0}\left(1 - \frac{\cosh xk}{\cosh ak}\right), \quad k \equiv \sqrt{\Sigma_a^0/D^0}, \tag{IV.D.6}$$

where $k \equiv \sqrt{\Sigma_a^0/D^0}$ is the nominal value of the reciprocal diffusion length for our illustrative example. Using Eq. (IV.D.6) together with the nominal value Σ_d^0 in Eq. (IV.D.3) gives the nominal response

$$R(e^0) = \frac{S^0 \Sigma_d^0}{D^0}\left(1 - \frac{\cosh bk}{\cosh ak}\right), \quad e^0 \equiv (\varphi^0, \boldsymbol{\alpha}^0). \tag{IV.D.7}$$

Note that even though Eq. (IV.D.1) is *linear* in φ, the solution $\varphi(x)$ depends *nonlinearly* on $\boldsymbol{\alpha}$, as evidenced by Eq. (IV.D.6). The same is true of the response $R(e)$. Even though $R(e)$ is *linear separately* in φ and in $\boldsymbol{\alpha}$, as shown in Eq. (IV.D.3), R is not *simultaneously linear* in φ and $\boldsymbol{\alpha}$, which leads to a *nonlinear dependence of $R(e)$ on $\boldsymbol{\alpha}$*. This fact is confirmed by the explicit expression of $R(e)$ given in Eq. (IV.D.7).

Consider now that the system parameters $\boldsymbol{\alpha}$ are allowed to vary (because of uncertainties, external influences, etc.) from their nominal values $\boldsymbol{\alpha}^0$ by amounts

$$\boldsymbol{h}_\alpha \equiv (\delta\Sigma_a, \delta D, \delta S, \delta\Sigma_d). \tag{IV.D.8}$$

Then, the *sensitivities* of the response $R(e)$ to the variations h_α are given by the *G-differential* $\delta R(e^0;h)$ of $R(e)$ at e^0 to variations

$$h \equiv (h_\varphi, h_\alpha). \qquad \text{(IV.D.9)}$$

Applying the definition

$$\delta R(e^0;h) \equiv \frac{d}{d\varepsilon}\{R(e^0 + \varepsilon h)\}_{\varepsilon=0} \qquad \text{(IV.D.10)}$$

of the G-differential to the response defined by Eq. (IV.D.3) gives

$$\begin{aligned}\delta R(e^0;h) &\equiv \frac{d}{d\varepsilon}\{(\Sigma_d^0 + \varepsilon \delta \Sigma_d)[\varphi^0(b) + \varepsilon h_\varphi(b)]\}_{\varepsilon=0} \\ &= R'_\alpha(e^0)h_\alpha + R'_\varphi(e^0)h_\varphi,\end{aligned} \qquad \text{(IV.D.11)}$$

where the "direct-effect" term $R'_\alpha h_\alpha$ is defined as

$$R'_\alpha(e^0)h_\alpha \equiv \delta \Sigma_d \, \varphi^0(b), \qquad \text{(IV.D.12)}$$

while the "indirect-effect" term $R'_\varphi h_\varphi$ is defined as

$$R'_\varphi(e^0)h_\varphi \equiv \Sigma_d^0 h_\varphi(b). \qquad \text{(IV.D.13)}$$

As indicated by Eq. (IV.D.11), the operator $\delta R(e^0;h)$ is linear in h; in particular, $R'_\varphi h_\varphi$ is a *linear operator* on h_φ. This linear dependence of $\delta R(e^0;h)$ on h is underscored by writing henceforth $DR(e^0;h)$ to denote the sensitivity of $R(e)$ at e^0 to variations h. The "direct-effect" term $R'_\alpha h_\alpha$ can be evaluated at this stage by replacing Eq. (IV.D.6) into Eq. (IV.D.12), to obtain

$$R'_\alpha(e^0)h_\alpha = \delta \Sigma_d \frac{S^0}{\Sigma_a^0}\left(1 - \frac{\cosh bk}{\cosh ak}\right). \qquad \text{(IV.D.14)}$$

The "indirect-effect" term $R'_\varphi h_\varphi$, though, cannot be evaluated at this stage, since $h_\varphi(x)$ is not yet available. The first-order (in $\|h_\alpha\|$) approximation to the exact value of $h_\varphi(x)$ is obtained by calculating the *G-differentials* of Eqs. (IV.D.1) and (IV.D.2) and solving the resulting equations. Thus, calculating the

G-differentials of Eqs. (IV.D.1) and (IV.D.2) yields the "forward sensitivity equations" *(FSE)*

$$L(\alpha^0)h_\varphi + [L'_\alpha(\alpha^0)\varphi^0]h_\alpha = O(\|h_\alpha\|^2), \tag{IV.D.15}$$

together with the boundary conditions

$$h_\varphi(\pm a) = 0. \tag{IV.D.16}$$

In Eq. (IV.D.15), the operator $L(\alpha^0)$ is defined as

$$L(\alpha^0) \equiv D^0 \frac{d^2}{dx^2} - \Sigma_a^0, \tag{IV.D.17}$$

while the quantity

$$[L'_\alpha(\alpha^0)\varphi^0]h_\alpha \equiv \delta D \frac{d^2\varphi^0}{dx^2} - \delta\Sigma_a \varphi^0 + \delta S, \tag{IV.D.18}$$

which is the partial G-differential of $L\varphi$ at α^0 with respect to α, contains all of the first-order parameter variations h_α. Solving the *FSE* yields the solution

$$\begin{aligned}h_\varphi(x) = &\, C_1(\cosh xk - \cosh ak) \\ &+ C_2(x \sinh xk \cosh ak - a \sinh ak \cosh xk),\end{aligned} \tag{IV.D.19}$$

where the constants C_1 and C_2 are defined as

$$C_1 \equiv \frac{(\delta\Sigma_a S^0/\Sigma_a^0 - \delta S)}{\Sigma_a^0 (\cosh ak)}, \tag{IV.D.20}$$

and, respectively,

$$C_2 \equiv \frac{(\delta D/D^0 - \delta\Sigma_a/\Sigma_a^0)S^0}{2\sqrt{D^0 \Sigma_a^0}(\cosh ak)^2}. \tag{IV.D.21}$$

Evaluating Eq. (IV.D.19) at $x = b$ and replacing the resulting expression in Eq. (IV.D.13) gives the "indirect-effect" term as

$$R'_\varphi(e^0)h_\varphi = \Sigma_d^0 [C_1(\cosh bk - \cosh ak) + C_2(b\sinh bk \cosh ak - a\sinh ak \cosh bk)]. \tag{IV.D.22}$$

In practice, though, it would not be possible to solve the *FSE* [i.e., Eqs. (IV.D.15) and (IV.D.16)] analytically, as has been done above; instead, the solution $h_\varphi(x)$ would be obtained by solving the *FSE* numerically. Therefore, the *FSE* would need to be solved anew for each parameter variation $\mathbf{h}_\alpha$.

As generally shown in Section IV.B, the need to solve repeatedly the *FSE* can be circumvented by using the *Adjoint Sensitivity Analysis Procedure (ASAP)*. The first prerequisite for applying the *ASAP* is that the "indirect-effect" term $R'_\varphi(e^0)h_\varphi$ be expressible as a *linear functional of* h_φ. An examination of Eq. (IV.D.13) readily reveals that $R'_\varphi(e^0)h_\varphi$ is indeed a linear functional of h_φ. Therefore, $R'_\varphi(e^0)h_\varphi$ can be represented as an inner product in an appropriately defined Hilbert space $\mathcal{H}_u$, as discussed generally in Section IV.B. For our illustrative example, we chose $\mathcal{H}_u$ to be the real Hilbert space $\mathcal{H}_u \equiv \mathcal{L}_2(\Omega)$, with $\Omega \equiv (-a, a)$, equipped with the inner product

$$\langle f(x), g(x) \rangle \equiv \int_{-a}^{a} f(x)g(x)dx, \tag{IV.D.23}$$
$$\text{for } f, g \in \mathcal{H}_u \equiv \mathcal{L}_2(\Omega), \quad \Omega \equiv (-a, a).$$

In $\mathcal{H}_u \equiv \mathcal{L}_2(\Omega)$, the linear functional $R'_\varphi(e^0)h_\varphi$ defined in Eq. (IV.D.13) can be represented as the inner product

$$R'_\varphi(e^0)h_\varphi \equiv \Sigma_d^0 h_\varphi(b) = \int_{-a}^{a} \Sigma_d^0 h_\varphi(x)\delta(x-b)dx = \langle \Sigma_d^0 \delta(x-b), h_\varphi \rangle. \tag{IV.D.24}$$

The next step underlying the *ASAP* is the construction of the operator $L^+(\alpha^0)$ that is *formally adjoint* to $L(\alpha^0)$. Using Eq. (I.B.20) shows that the formal adjoint of $L(\alpha^0)$ is the operator

$$L^+(\alpha^0) \equiv D^0 \frac{d^2}{dx^2} - \Sigma_a^0. \tag{IV.D.25}$$

Note that $L^+(\alpha^0)$ and $L(\alpha^0)$ are *formally self-adjoint*. The qualifier "formally" must still be kept at this stage, since the boundary conditions for $L^+(\alpha^0)$ have not been determined yet. The boundary conditions for $L^+(\alpha^0)$ are derived by applying Eq. (IV.B.1) to the operators $L(\alpha^0)$ and $L^+(\alpha^0)$, to obtain

$$\int_{-a}^{a} \psi(x)\left[D^0 \frac{d^2 h_\varphi}{dx^2} - \Sigma_a^0 h_\varphi(x)\right] dx = \int_{-a}^{a}\left[D^0 \frac{d^2 \psi(x)}{dx^2} - \Sigma_a^0 \psi(x)\right] h_\varphi(x) dx + \left\{P[h_\varphi, \psi]\right\}_{x=-a}^{x=a}.$$

(IV.D.26)

Note that the function $\psi(x)$ is still arbitrary at this stage, except for the requirement that $\psi \in \mathcal{H}_Q = \mathcal{L}_2(\Omega)$; note also that the Hilbert spaces $\mathcal{H}_u$ and $\mathcal{H}_Q$ have now both become the same space, i.e., $\mathcal{H}_u = \mathcal{H}_Q = \mathcal{L}_2(\Omega)$.

Integrating the left-side of Eq. (IV.D.26) by parts twice and canceling terms yields the following expression for the bilinear boundary form:

$$\left\{P[h_\varphi, \psi]\right\}_{-a}^{a} = D^0 \left[\psi \frac{dh_\varphi}{dx} - h_\varphi \frac{d\psi}{dx}\right]_{-a}^{a}.$$

(IV.D.27)

The boundary conditions for $L^+(\alpha^0)$ can now be selected by applying to Eq. (IV.D.27) the general principles (a) and (b) outlined immediately following Eq. (IV.B.3). Since h_φ is known at $x = \pm a$ from Eq. (IV.D.16), the application of these principles to Eq. (IV.D.27) leads to the choice

$$\psi(\pm a) = 0,$$

(IV.D.28)

for the adjoint function $\psi(x)$. This choice of boundary conditions for ψ ensures that unknown values of h_φ, such as the derivatives $\left\{dh_\varphi/dx\right\}_{-a}^{a}$, would be eliminated from the bilinear form $\left\{P[h_\varphi,\psi]\right\}_{-a}^{a}$ in Eq. (IV.D.27). Note that the implementation of both Eqs. (IV.D.28) and (IV.D.16) into Eq. (IV.D.27) actually causes $\left\{P[h_\varphi,\psi]\right\}_{-a}^{a}$ to vanish; hence, the quantity $\hat{P}(h_\alpha, \psi; \alpha^0)$ appearing in Eq. (IV.B.4) is identically zero for this illustrative example.

Since the boundary conditions selected in Eq. (IV.D.28) for the adjoint function $\psi(x)$ are *the same as* the boundary conditions for $h_\varphi(x)$ in Eq. (IV.D.16), and since the operators $L^+(\alpha^0)$ and $L(\alpha^0)$ are formally self-adjoint, we can at this stage drop the qualifier "formally," and *can now conclude that the operators $L^+(\alpha^0)$ and $L(\alpha^0)$ are indeed self-adjoint.*

The last step in the construction of the adjoint system is the identification of the source term, which is done by applying Eq. (IV.B.9) to Eqs. (IV.D.24) and (IV.D.25). This identification readily shows that

$$\nabla_\varphi R(e^0) = \Sigma_d^0 \delta(x-b), \tag{IV.D.29}$$

so that the complete adjoint system becomes

$$L^+(\alpha^0)\psi \equiv D^0 \frac{d^2\psi}{dx^2} - \Sigma_a^0 \psi(x) = \Sigma_d^0 \delta(x-b), \tag{IV.D.30}$$

where the adjoint function $\psi(x)$ is subject to the boundary conditions $\psi(\pm a) = 0$, as shown in Eq. (IV.D.28).

Finally, the sensitivity $DR(e^0;h)$ of $R(e)$ at e^0 to parameter variations h_α can be expressed in terms of the adjoint function $\psi(x)$ by applying Eq. (IV.B.10) to our illustrative example. Thus, recalling that $\delta Q = 0$ and $\hat{P} = 0$ for our example, and using Eq. (IV.D.18) gives the following expression for the "indirect-effect" term $R'_\varphi(e^0)h_\varphi$:

$$R'_\varphi(e^0)h_\varphi = -\int_{-a}^{a} \psi(x)\left[\delta D \frac{d^2\varphi^0}{dx^2} - \delta\Sigma_a \varphi^0(x) + \delta S\right]dx, \tag{IV.D.31}$$

where $\psi(x)$ is the solution of the adjoint sensitivity system defined by Eqs. (IV.D.30) and (IV.D.28).

As expected, the adjoint sensitivity system is independent of parameter variations h_α, so it needs to be solved only once to obtain the adjoint function $\psi(x)$. Very important, too, is the fact (characteristic of linear systems) that the adjoint system is independent of the original solution $\varphi^0(x)$, and can therefore be solved directly, without any knowledge of the neutron flux $\varphi(x)$. Of course, the adjoint system depends on the response, which provides the source term, as shown in Eq. (IV.D.30). Solving the adjoint system for our illustrative example yields the following expression for the adjoint function $\psi(x)$:

$$\psi(x) = \frac{\Sigma_d^0}{\sqrt{\Sigma_a^0 D^0}}\left[\frac{\sinh(b-a)k}{\sinh 2ak}\sinh(x+a)k + H(x-b)\sinh(x-b)k\right], \tag{IV.D.32}$$

where $H(x-b)$ is the Heaviside-step functional defined as

$$H(x) = \begin{cases} 0, & \text{for } x < 0 \\ 1, & \text{for } x \geq 0 \end{cases}. \qquad \text{(IV.D.33)}$$

Using Eq. (IV.D.32) in Eq. (IV.D.31) and carrying out the respective integrations over x yields, as expected, the same expression for the "indirect-effect" term $R'_\varphi(e^0)h_\varphi$ as obtained in Eq. (IV.D.22). Finally, the local sensitivity $DR(e^0;h)$ of $R(e)$ at e^0 to variations h_α in the system parameters is obtained from Eqs. (IV.D.11), (IV.D.14), and either Eq. (IV.D.22) provided by the *FSAP* or, alternatively, using the adjoint function in Eq. (IV.D.31), as provided by the *ASAP*; either way, the final expression of the sensitivity $DR(e^0;h)$ is

$$DR(e^0;h) = \delta\Sigma_d \frac{S^0}{\Sigma_a^0}\left(1 - \frac{\cosh bk}{\cosh ak}\right)$$
$$+ \Sigma_d^0[C_1(\cosh bk - \cosh ak) \qquad \text{(IV.D.34)}$$
$$+ C_2(b\sinh bk \cosh ak - a\sinh ak \cosh bk)].$$

It is instructive to compare the expression of the local sensitivity $DR(e^0;h)$ with the expression of the exact variation

$$(\Delta R)_{exact} \equiv R(e^0 + h) - R(e^0), \qquad \text{(IV.D.35)}$$

which would be induced in the response $R(e)$ by parameter variations h_α. The exact variation $(\Delta R)_{exact}$ is readily obtained from Eq. (IV.D.7) as

$$(\Delta R)_{exact} = \frac{S^0 + \delta S}{D^0 + \delta D}(\Sigma_d^0 + \delta\Sigma_d)\left(1 - \frac{\cosh bk_p}{\cosh ak_p}\right) - R(e^0), \qquad \text{(IV.D.36)}$$

where

$$k_p \equiv \sqrt{(\Sigma_a^0 + \delta\Sigma_a)/(D^0 + \delta D)}. \qquad \text{(IV.D.37)}$$

On the other hand, we can solve *exactly* the perturbed equation

$$L(\alpha^0 + h_\alpha)[\varphi^0 + h_\varphi^{exact}(x)] + (S^0 + \delta S) = 0, \qquad \text{(IV.D.38)}$$

subject to the boundary conditions given by Eq. (IV.D.16), to obtain

$$h_\varphi^{exact}(x) = \frac{S^0 \delta\Sigma_a/\Sigma_a^0 - \delta S}{(\Sigma_a^0 + \delta\Sigma_a)\cosh ak_p}(\cosh xk_p - \cosh ak_p) \\ + S^0 \frac{\cosh(ak_p)\cosh(xk) - \cosh(ak)\cosh(xk_p)}{\Sigma_a^0 \cosh ak_p \cosh ak}.$$ (IV.D.39)

Comparing Eq. (IV.D.39) to Eq. (IV.D.19) readily confirms that the solution $h_\varphi(x)$ of the *FSE* is the first-order, in $\|h_\alpha\|$, approximation of $h_\varphi^{exact}(x)$, i.e.,

$$h_\varphi^{exact}(x) = h_\varphi(x) + O(\|h_\alpha\|^2).$$ (IV.D.40)

Similarly, comparing Eq. (IV.D.36) to Eq. (IV.D.34) confirms, as expected, that the local sensitivity $DR(e^0;h)$ is the first-order, in $\|h_\alpha\|$, approximation of the exact response variation, namely:

$$R(e^0 + h) = R(e^0) + DR(e^0;h) + O(\|h_\alpha\|^2).$$ (IV.D.41)

Actually, it can be shown that the functional Taylor-series of $R(e^0 + h)$ contains three terms only, namely

$$R(e^0 + h) = R(e^0) + DR(e^0;h) + \frac{1}{2}D^2R(e^0;h),$$ (IV.D.42)

where $D^2R(e^0;h) = 2(\delta\Sigma_d)h_\varphi(b)$.

IV.E. ILLUSTRATIVE EXAMPLE: SENSITIVITY ANALYSIS OF EULER'S ONE-STEP METHOD APPLIED TO A FIRST-ORDER LINEAR ODE

This Section presents the application of the *FSAP* and *ASAP* to a first-order linear ordinary equation, that also admits a readily obtainable closed-form, exact solution. For this problem, though, the *FSAP* and *ASAP* are applied both to the differential equation and to the difference equation that would result by applying a numerical method to solve the respective differential equation. For illustrative purposes, this difference equation is obtained by applying Euler's one-step method to the first-order differential equation. This way, we illustrate not only the parallels between the applications of the *FSAP* and *ASAP* to the differential equation and its discretized counterpart, but also the respective distinctions, showing, in particular, that the operations of Gâteaux-differentiation and (Euler) discretization do *not* commute. Thus, this illustrative example highlights the importance of ensuring that the sensitivity equations, both forward and adjoint, are derived in a consistent manner in the course of preparing them for numerical solution.

Consider the first-order linear ordinary differential equation

$$\begin{cases} \dfrac{du}{dt} + au = b, \\ u(0) = u_0, \end{cases} \qquad\qquad \text{(IV.E.1)}$$

where a, b, and u_0 are real scalars.

The solution of Eq. (IV.E.1) is:

$$u(t) = u_0 e^{-at} + \frac{b}{a}\left(1 - e^{-at}\right). \qquad\qquad \text{(IV.E.2)}$$

Along with Eq. (IV.E.1), consider a response of the form:

$$R \equiv \int_0^{t_f} \gamma(t)\, d(t)\, u(t)\, dt, \qquad\qquad \text{(IV.E.3)}$$

where $\gamma(t)$ is a weight function to be used for numerical quadrature and/or time-step sensitivity analysis, while $d(t)$ allows flexibility in modeling several types of responses: $d(t)$ can be the Dirac functional $\delta(t - t_0)$, the Heaviside functional $H(t - t_0)$, etc. For illustrative purposes, the vector $\boldsymbol{\alpha}$ of system parameters for this problem is defined as $\boldsymbol{\alpha} \equiv (a, b, u_0, \gamma)$, but $d(t)$ is fixed, unaffected by uncertainties.

Hence, the sensitivity DR of R to parameter variations $\boldsymbol{h}_\alpha$ around the nominal value $\boldsymbol{\alpha}^0$ is obtained by applying the definition given in Eq. (IV.6) to Eq. (IV.E.3); this yields:

$$DR = \int_0^{t_f} \gamma^0(t)\, d(t)\, h(t)\, dt + \int_0^{t_f} \delta\gamma(t)\, d(t)\, u^0(t)\, dt. \qquad \text{(IV.E.4)}$$

In the above expression, the variation $h(t)$ around the nominal solution value u^0 denotes the solution of the forward sensitivity equation (*FSE*) obtained by taking the G-differential of Eq. (IV.E.1); thus, $h(t)$ is the solution of

$$\begin{cases} \dfrac{dh}{dt} + ah = \delta b - u^0 \delta a, \\ h(0) = \delta u^0 \equiv h^0, \end{cases} \quad \text{(IV.E.5)}$$

where u^0 represents the base-case solution.

The solution $h(t)$ of Eq. (IV.E.5) can be readily obtained as

$$h(t) = \frac{b}{a^2}\left(tae^{-at} + e^{-at} - 1\right)\delta a + \frac{1-e^{-at}}{a}\delta b + \delta u^0 e^{-at}, \quad \text{(IV.E.6)}$$

where a and b are evaluated at their base-case values but where the superscript "zero" has been omitted for notational simplicity.

Applying the Adjoint Sensitivity Analysis Procedure (*ASAP*) to Eq. (IV.E.5) yields the following adjoint sensitivity equation for the adjoint function $\psi(t)$:

$$\begin{cases} -\dfrac{d\psi}{dt} + a\psi = \gamma(t)d(t) \\ \psi(t_f) = 0. \end{cases} \quad \text{(IV.E.7)}$$

Equation (IV.E.7) can also be solved to obtain the adjoint function explicitly:

$$\psi(t) = e^{at}\int_t^{t_f} \gamma(t)d(t)e^{-at}dt. \quad \text{(IV.E.8)}$$

In terms of the adjoint function $\psi(t)$, the sensitivity DR, i.e., Eq. (IV.E.4), becomes

$$DR = \int_0^{t_f}\psi(t)[\delta b - u(t)\delta a]dt + \psi(0)\delta u^0 + \int_0^{t_f}\delta\gamma(t)d(t)u(t)dt. \quad \text{(IV.E.9)}$$

To calculate the response R numerically, the integral in Eq. (IV.E.3) must be discretized. Any numerical quadrature used for this purpose can be represented in the general form

$$R = \int_0^{t_f = N\Delta t_s^q} \gamma(t) d(t) u(t) dt = \sum_{s=0}^{N} \Delta t_s^q \gamma^s d^s u^s, \qquad \text{(IV.E.10)}$$

where $N = t_f / \Delta t_s^q$ denotes the number of steps in the quadrature, and the superscript "q" refers to "quadrature step." Taking the *G-differential* of Eq. (IV.E.10) yields the following *discretized* form for the response sensitivity DR:

$$DR = \sum_{s=0}^{N} \gamma^s \Delta t_s^q d^s h^s + \sum_{s=0}^{N} \delta \gamma^s \Delta t_s^q d^s u^s. \qquad \text{(IV.E.11)}$$

We now use *Euler's one-step method* to discretize both the original and the adjoint systems. Recall that Euler's one-step method applied to a first-order differential equation of the form

$$\frac{dy}{dt} = f(t, y) \qquad \text{(IV.E.12)}$$

leads to a discrete algebraic system of the form

$$\begin{aligned} y^{n+1} &= y^n + \Delta t_n f(t^n, y^n), \\ y^n &= y(t_n); \quad t_n = t_0 + n\Delta t_n, \quad (n = 0, 1, \ldots, N). \end{aligned} \qquad \text{(IV.E.13)}$$

Recall also that Euler's one-step method produces a local truncation error $= O((\Delta t)^2)$ and a global truncation error $= O(\Delta t)$. Applying Euler's one-step method with $f(u) = (b - au)$ to Eq. (IV.E.1) gives

$$\begin{cases} u^0 = u(0), \\ u^{n+1} = u^n + \Delta t_n^e (b^0 - a^0 u^n), \quad (n = 0, 1, \ldots, N-1), \end{cases} \qquad \text{(IV.E.14)}$$

where the superscript "e" in Δt_n^e denotes "equation," in contradistinction to the superscript "q" in Eq. (IV.E.10), which denotes "quadrature." We also apply Euler's method to Eq. (IV.E.5) to obtain the "**Discretized Forward Sensitivity Equations**" (*DFSE*)

$$\begin{cases} h^0 = \delta u^0, \\ h^{n+1} = h^n + \Delta t_n^e \left(-a^0 h^n + \delta b - \left(u^0\right)^n \delta a \right) \\ \phantom{h^{n+1}} = h^n \left(1 - a^0 \Delta t_n^e\right) + \Delta t_n^e \left(\delta b - u^n \delta a\right), \quad (n = 0, 1, \ldots, N-1). \end{cases}$$ (IV.E.15)

In the same manner, Euler's method is applied to Eq. (IV.E.7) to discretize the adjoint sensitivity system. There are two ways to discretize the adjoint system, namely by using either

(i) Euler's *forward* method: $y(t_n + \Delta t_n) = y(t_n) + \Delta t_n \left(\dfrac{dy}{dt}\right)_{t_n} + \cdots$

to obtain

$$\begin{cases} \psi^n = \dfrac{\psi^{n+1} + \Delta t_n^e \gamma^n d^n}{1 + a^0 \Delta t_n^e} \\ = \psi^{n+1}\left(1 - a^0 \Delta t_n^e\right) + \Delta t_n^e \gamma^n d^n + O\left[\left(\Delta t_n^e\right)^2\right], \quad (n = N-1, \ldots, 1, 0), \\ \psi^N = \begin{cases} 0, \text{ or} \\ \Delta t_n^q \gamma^N d^N, \end{cases} \end{cases}$$

(IV.E.16)

or by using

(ii) Euler's *backward* method: $y(t_{n+1} - \Delta t_n) = y(t_{n+1}) - \Delta t_n \left(\dfrac{dy}{dt}\right)_{t_{n+1}} + \cdots$

to obtain

$$\begin{cases} \psi^n = \psi^{n+1} - \Delta t_n^e \left(a^0 \psi^{n+1} - \gamma^{n+1} d^{n+1}\right) \\ = \psi^{n+1}\left(1 - a^0 \Delta t_n^e\right) + \Delta t_n^e \gamma^{n+1} d^{n+1}, \quad (n = N-1, \ldots, 0) \\ \psi^N = \begin{cases} 0, \text{ or} \\ \Delta t_n^q \gamma^N d^N. \end{cases} \end{cases}$$ (IV.E.17)

The *FSAP* is now applied to the *Euler-discretized original system* represented by Eq. (IV.E.14): applying the definition of the G-differential to Eq. (IV.E.14) yields

$$\begin{cases} \dfrac{d}{d\varepsilon}\left[u^0 + \varepsilon g^0\right]_{\varepsilon=0} = \dfrac{d}{d\varepsilon}\left[u(0) + \varepsilon \delta u^0\right]_{\varepsilon=0}, \\ \dfrac{d}{d\varepsilon}\left[u^{n+1} + \varepsilon g^{n+1}\right]_{\varepsilon=0} = \dfrac{d}{d\varepsilon}\left\{u^n + \varepsilon g^n + \left[\Delta t_n^e + \varepsilon \delta\left(\Delta t_n^e\right)\right] \times \right. \\ \left. \left[b^0 + \varepsilon\,\delta b - \left(a^0 + \varepsilon\,\delta a\right)\left(u^n + \varepsilon g^n\right)\right]\right\}_{\varepsilon=0}, \end{cases}$$

which leads, after obvious simplifications, to the following algebraic system

$$\begin{cases} g^0 = \delta u^0 \\ g^{n+1} = g^n\left(1 - a^0 \Delta t_n^e\right) + \Delta t_n^e\left(\delta b - u^n \delta a\right) + \delta\left(\Delta t_n^e\right)\left(b^0 - a^o u^n\right), \\ (n = 0, 1, \ldots, N-1). \end{cases} \quad \text{(IV.E.18)}$$

It is useful to refer to Eq. (IV.E.18) as the **Discrete Sensitivity System**, as opposed to the "*discretized sensitivity system*" represented by Eq. (IV.E.15), to underscore the conceptual as well as mathematical differences between the two systems. Note also that *the symbol g^n has been used in Eq. (IV.E.18), as opposed to the symbol h^n that was used in Eq. (IV.E.15), also to underscore the fact that the two systems are not identical.* In other words, the system obtained by G-differentiating the discretized original system is not identical to the result obtained by discretizing the G-differentiated original system, even if the same discretization procedure is used in both cases. *Intuitively speaking, the operations of G-differentiation and discretization do not commute.* The *Discrete Sensitivity System* makes it possible to analyze the sensitivity of the response to the step-size and truncation error, Δt_n^e; the respective contribution is represented by the last term on the right-side of Eq. (IV.E.18). Note that setting $\delta\left(\Delta t_n^e\right) = 0$ in Eq. (IV.E.18) reduces it to Eq. (IV.E.15); furthermore, taking the limit as $\Delta t_n^e \to 0$ in both Eq. (IV.E.18) and Eq. (IV.E.15) leads to Eq. (IV.E.5). This fact indicates that, for the one-step Euler Method, the *Discrete Sensitivity System* and the *Discretized Forward Sensitivity Equations* lead to difference equations for g^n and h^n, respectively, which are both *consistent* with the *FSE* for h given in Eq. (IV.E.5).

In preparation for applying the *Discrete Adjoint Sensitivity Analysis Procedure (DASAP)*, we note that the *inner product* of two functions $u(t)$ and $v(t)$ is discretized by the use of a quadrature formula of the general form:

$$\int_0^{t_f} \gamma(t)\,u(t)\,v(t)\,dt = \sum_{s=0}^{N} \gamma^s \Delta t_s^q u^s v^s. \qquad \text{(IV.E.19)}$$

Writing the one-step Euler method, namely Eq. (IV.E.14), in the form

$$\begin{cases} u^0 = u(0) \\ u^{n+1} = u^n + \Delta t_n^e f(u^n), \quad (n = 0,\ldots,N-1), \end{cases} \qquad \text{(IV.E.20)}$$

and calculating the G-differential of Eq. (IV.E.20) gives the following form for the *Discretized Forward Sensitivity Equations (DFSE)*:

$$\begin{cases} g^0 = g(0) = \delta u^0 \\ g^{n+1} = g^n + \Delta t_n^e \left(\dfrac{\partial f}{\partial u}\right)^n g^n + q^{n+1}, \quad (n = 0,\ldots,N-1), \end{cases} \qquad \text{(IV.E.21)}$$

where $q^{n+1} \equiv \Delta t_n^e \sum_{i=1}^{I} \left(\dfrac{\partial f}{\partial \alpha_i}\right)^n \delta\alpha_i + \delta(\Delta t_n^e) f(u^n)$. The above *DFSE* can be written in matrix form as

$$A\,g = q, \qquad \text{(IV.E.22)}$$

with:

$$A \equiv \begin{bmatrix} 1 & 0 & \cdots & 0 & 0 \\ -\left[1+\Delta t_0^e \left(\dfrac{\partial f}{\partial u}\right)^0\right] & 1 & \cdots & 0 & 0 \\ 0 & -\left[1+\Delta t_1^e \left(\dfrac{\partial f}{\partial u}\right)^1\right] & \cdots & 0 & 0 \\ \cdots & \cdots & \cdots & \cdots & \cdots \\ 0 & 0 & \cdots & 1 & 0 \\ 0 & 0 & \cdots & -\left[1+\Delta t_{N-1}^e \left(\dfrac{\partial f}{\partial u}\right)^{N-1}\right] & 1 \end{bmatrix};$$

$$g \equiv \begin{bmatrix} g^0 & g^1 & g^2 & \cdots & g^{N-1} & g^N \end{bmatrix}^T;$$

$$q \equiv \left[\delta u^0, \Delta t_0^e \left(\delta b - u^0 \delta a\right) + \delta\left(\Delta t_0^e\right)\left(b^0 - u^0 a^0\right), \ldots \right.$$
$$\left. \ldots, \Delta t_{N-1}^e \left(\delta b - u^{N-1} \delta a\right) + \delta\left(\Delta t_{N-1}^e\right)\left(b^0 - u^{N-1} a^0\right)\right]^T ;$$

and where the superscript "T" denotes "transposition"; note that the components of the vector q have been written for the particular form taken on by $f(u)$ for this illustrative example.

Taking the G-differential of the discretized response R given in Eq. (IV.E.10) yields the "*discrete response sensitivity*" (DDR) as

$$(DDR) = \sum_{s=0}^{N} \Delta t_s^q \gamma^s d^s g^s + \sum_{s=0}^{N} \Delta t_s^q \left(\delta \gamma^s\right) d^s u^s$$
$$+ \sum_{s=0}^{N} \delta\left(\Delta t_s^q\right) \gamma^s d^s u^s .$$
(IV.E.23)

The last sum in Eq. (IV.E.23) contains the explicit contribution from the sensitivity to the "quadrature time-step"; comparing the expression of (DDR) given in Eqs. (IV.E.23) to the expression of (DR) given in Eq. (IV.E.11) shows that the explicit contribution from the "time-step" sensitivity appears as an additional term.

The *Discrete Adjoint Sensitivity Equations* (*Discrete ASE*) are obtained by multiplying Eq. (IV.E.22) with a vector $\psi^T \equiv \left(\psi^0, \ldots, \psi^N\right)$, at this stage arbitrary but later to become the "*discrete adjoint function*," and the resulting equation is subsequently transposed to obtain

$$g^T A^+ \psi = q^T \psi .$$
(IV.E.24)

Defining the vector r as

$$r = \left(r^0, \ldots, r^N\right)$$
$$\equiv \left(\gamma^0 \Delta t_0^q d^0, \ldots, \gamma^N \Delta t_N^q d^N\right),$$
(IV.E.25)

and requiring the discrete adjoint function ψ to be the solution of the *Discrete ASE*

$$A^+ \psi = r ,$$
(IV.E.26)

reduces Eq. (IV.E.24) to

$$g^T r = q^T \psi. \tag{IV.E.27}$$

Replacing Eq. (IV.E.27) in Eq. (IV.E.23) gives

$$\begin{aligned}(DDR) &= \sum_{n=0}^{N} q^n \psi^n + \sum_{s=0}^{N}\left[\Delta t_s^q (\delta\gamma^s) + \delta(\Delta t_s^q)\gamma^s\right]d^s u^s \\ &= \psi^0 \delta u^0 + \sum_{n=0}^{N-1} \psi^{n+1}\left[\Delta t_n^e(\delta b - u^n \delta a) + \delta(\Delta t_n^e)(b^0 - u^n a^0)\right] \\ &+ \sum_{n=0}^{N}\left[\Delta t_n^q(\delta\gamma^n) + \delta(\Delta t_n^q)\gamma^n\right]d^n u^n. \end{aligned} \tag{IV.E.28}$$

Recalling that

$$A^+ \equiv \begin{bmatrix} 1 & -\left[1+\Delta t_0^e\left(\frac{\partial f}{\partial u}\right)\right]^0 & \cdots & 0 & 0 \\ 0 & 1 & \cdots & 0 & 0 \\ \cdots & \cdots & \cdots & \cdots & \cdots \\ 0 & 0 & \cdots & 1 & -\left[1+\Delta t_{N-1}^e\left(\frac{\partial f}{\partial u}\right)\right]^{N-1} \\ 0 & 0 & \cdots & 0 & 1 \end{bmatrix}$$

reduces Eq. (IV.E.26) to

$$\begin{cases} \psi^n - \left[1+\Delta t_n^e\left(\frac{\partial f}{\partial u}\right)\right]^n \psi^{n+1} = r^n \left(= \gamma^n \Delta t_n^q d^n\right), & (n = N-1,\ldots,1,0), \\ \psi^N = r^N \left(= \gamma^N \Delta t_N^q d^N\right). \end{cases} \tag{IV.E.29}$$

Recalling also that $(\partial f/\partial u)^n = -a^0$ (for the example under consideration) further simplifies Eq. (IV.E.29) to

$$\begin{cases} \psi^N = \gamma^N \Delta t_N^q d^N \\ \psi^n = \psi^{n+1}\left(1 - a^0 \Delta t_n^e\right) + \gamma^n \Delta t_n^q d^n, & (n = N-1,\ldots,0). \end{cases} \tag{IV.E.30}$$

Comparing Eq. (IV.E.30) to Eqs. (IV.E.16) and (IV.E.17) reveals the following features, within $O((\Delta t)^2)$:

(i) if $\Delta t_n^q \neq \Delta t_n^e$, then Eq. (IV.E.30) is not identical to either Eq. (IV.E.16) or Eq. (IV.E.17), which means that the "Discrete ASE" produced by the DASAP are **not identical** (and may even be inconsistent) with the Euler-discretized "Differential ASE" produced by the ASAP formalism.

(ii) if $\Delta t_n^q = \Delta t_n^e$, then Eq. (IV.E.30) is consistent with Eq. (IV.E.16), which means that the Discrete ASE are consistent with the Differential ASE discretized using the **forward** one-step Euler Method in a **backward** stepping procedure.

Note that Eq. (IV.E.30) can be obtained directly in component form from Eq. (IV.E.21) by multiplying the latter by ψ^{n+1}, summing over n from 0 to $N-1$, and rearranging the summations, as follows:

$$\sum_{n=0}^{N-1}\psi^{n+1}g^{n+1} = \sum_{n=0}^{N-1}\psi^{n+1}g^n\left[1+\Delta t_n^e\left(\frac{\partial f}{\partial u}\right)^n\right] + \sum_{n=0}^{N-1}\psi^{n+1}q^{n+1}.$$

Rearranging once again the sums over g^n gives

$$\sum_{n=0}^{N-1}g^n\left\{\psi^n - \psi^{n+1}\left[1+\Delta t_n^e\left(\frac{\partial f}{\partial u}\right)^n\right]\right\} = \psi^0 g^0 - \psi^N g^N + \sum_{n=0}^{N-1}\psi^{n+1}q^{n+1}.$$

(IV.E.31)

Using Eq. (IV.E.29) to replace the terms within the braces on the left-side of Eq. (IV.E.31) by $\gamma^n \Delta t_n^q d^n$ $(n=0,\ldots,N)$, and using the relation $\delta u^0 = q^0$, leads to

$$\sum_{n=0}^{N}g^n\gamma^n\Delta t_n^q d^n = \psi^0\delta u^0 + \sum_{n=0}^{N-1}\psi^{n+1}q^{n+1}\left(=\sum_{n=0}^{N}\psi^n q^n\right). \quad \text{(IV.E.32)}$$

Comparing the right-side of Eq. (IV.E.32) to Eq.(IV.E.23) shows that

$$(DDR) = \psi^0 \delta u^0 + \sum_{n=0}^{N-1} \psi^{n+1} q^{n+1} + \sum_{n=0}^{N} \left[\Delta t_n^q (\delta \gamma^n) + \delta(\Delta t_n^q) \gamma^n \right] d^n u^n$$

$$= \psi^0 \delta u^0 + \sum_{n=0}^{N-1} \psi^{n+1} \left[\Delta t_n^e \sum_{i=1}^{I} \left(\frac{\partial f}{\partial \alpha_i} \right)^n \delta \alpha_i + \delta(\Delta t_n^e) f(u^n) \right] \quad \text{(IV.E.33)}$$

$$+ \sum_{n=0}^{N} \left[\Delta t_n^q (\delta \gamma^n) + \delta(\Delta t_n^q) \gamma^n \right] d^n u^n.$$

On the other hand, applying a quadrature rule to discretize the sensitivity DR expressed in integral form by Eq. (IV.E.9) would yield

$$DR = \sum_{n=0}^{N} \psi^n \left[\delta b - u^n \delta a \right] \Delta t_n^q + \psi^0 \delta u^0$$

$$+ \sum_{n=0}^{N} \delta \gamma^n \Delta t_n^q d^n u^n, \quad \text{(IV.E.34)}$$

where $\delta b - u^n \delta a = \left(\frac{\partial f}{\partial u} \right)^n$.

Comparing Eq. (IV.E.34) with Eq. (IV.E.33) reveals that they are *not* identical. Consistency between the expressions of DR and DDR would require the condition:

$$\Delta t_n^q = \Delta t_n^e, \quad \text{(IV.E.35)}$$

just as was necessary to require in order to ensure that Eq. (IV.E.30) is consistent with Eq. (IV.E.16). Note also that Eq. (IV.E.34) contains the terms $\psi^n \left(\frac{\partial f}{\partial u} \right)^n \Delta t_n^q$, whereas the corresponding sum in Eq. (IV.E.33) contains the terms $\psi^{n+1} \left(\frac{\partial f}{\partial u} \right)^n \Delta t_n^q$, which are not identical to one another. Since

$$\psi^n = \psi^{n+1} + O(\Delta t_n^q), \quad \text{(IV.E.36)}$$

it follows that

$$DR = (DDR) + O(\Delta t_n^q). \quad \text{(IV.E.37)}$$

Of course, the summation $\sum_{n=0}^{N-1} \psi^{n+1} q^{n+1}$ in Eq. (IV.E.33) also contains the terms $\delta(\Delta t_n^e) f(u^n)$, but these terms are of $O\left[(\Delta t_n^q)^2\right] = O\left[(\Delta t_n^e)^2\right]$, which is consistent with the second-order truncation error in Eq. (IV.E.34).

The example presented in this Section illustrates the importance of consistency between the differential form of the sensitivity equations, the respective adjoint equations, and their discretized counterparts. As long as consistency is ensured, it is possible to use either the *DASAP* or the discretized equations produced by the *ASAP*. If this consistency is not ensured, then it is recommended to use the differential and/or integral formulation of the *ASAP*, and subsequently discretize the resulting adjoint equations, in order to solve them numerically.

IV.F. ILLUSTRATIVE EXAMPLE: SENSITIVITY AND UNCERTAINTY ANALYSIS OF A SIMPLE MARKOV CHAIN SYSTEM

This Section presents a paradigm that illustrates the main steps underlying *sensitivity and uncertainty* analysis of any system. Thus, consider a pump that is to start operating on demand at time $t = 0$. The probability of success (i.e., the pump starts up) at $t = 0$ is $s, 0 < s < 1$, while the probability of failure (i.e., the pump remains "down") at $t = 0$ is $(1-s)$. If started successfully, the pump operates until it fails; the time to failure, t_f, can be considered a random variable with probability distribution function $p(t_f)$. When the pump fails, it will be repaired and then restored to service; the repair time, t_r, can also be considered a random variable, with probability distribution $p(t_r)$. The pump continues to function on the "up/down" (i.e., in operation/under repair) cycle until it fails irreparably at a final time $t = T$, at which time the pump is decommissioned. A typical response of interest for such a pump is the limiting interval availability, namely the fraction of time the pump operates over a very long mission time T.

For illustrative purposes, consider that the time dependent behavior of the pump (during the period $0 < t < T$) is described by the following Markov model:

$$\begin{cases} \dfrac{dU(t)}{dt} = -\lambda U(t) + \mu D(t) \\ \dfrac{dD(t)}{dt} = \lambda U(t) - \mu D(t) \\ U(0) = U_s \\ D(0) = D_s \end{cases} \qquad \text{(IV.F.1)}$$

where $U(t)$ denotes the probability that the pump is operational ("up"), $D(t)$ denotes the probability that the pump is in the failed state ("down"), λ denotes the instantaneous failure rate, μ denotes the instantaneous repair rate, and U_s is the probability that the pump starts up, on demand, at $t=0$. Note that $U(t)+D(t)=1$, for all $t \in [0,T]$. Consider that only one trial is performed at $t=0$, so that this trial is described by the Bernoulli probability distribution $p(U_s) = s^{U_s}(1-s)^{1-U_s}$, $(U_s = 0,1)$, and $p(U_s) = 0$ otherwise. For illustrative purposes, consider that $\lambda = 1/t_f$, where the probability distribution $p(t_f)$ of the time to failure, t_f, is the *gamma distribution* $p(t_f) = (t_f)^{\alpha-1} e^{-(t_f/\beta)} / [\beta^\alpha \Gamma(\alpha)]$, with $\Gamma(\alpha) = \int_0^\infty t^{\alpha-1} e^{-t} dt$, and consider that $\mu = 1/t_r$, where the probability distribution $p(t_r)$ of the time to repair, t_r, is the *log-normal distribution* $p(t_r) = \frac{1}{\sqrt{2\pi\sigma^2}} \frac{1}{t_r} \exp\left(-\frac{(\log t_r - \log v)^2}{2\sigma^2}\right)$. Consider also that the random vector of parameters $\boldsymbol{a} \equiv (\mu, \lambda, U_s, D_s)$ is distributed according to a multivariate probability distribution function $p(\boldsymbol{a})$, defined on a set $\mathcal{S}_a$.

The limiting interval availability response is represented mathematically as

$$U_{av} \equiv \frac{1}{T} \int_0^T U(\tau) d\tau. \tag{IV.F.2}$$

The typical question to be answered by sensitivity and uncertainty analysis is to determine the moments of the distribution function of the response, U_{av}, arising from the uncertainties in the vector of system parameters $\boldsymbol{a} \equiv (\mu, \lambda, U_s, D_s)$.

The remainder of this Section will illustrate in detail the steps involved in calculating the mean value and variance in the response U_{av}, commencing with the calculation of the mean values and variances of each parameter, followed by the calculation of the covariance matrix for the random vector $\boldsymbol{a} \equiv (\mu, \lambda, U_s, D_s)$, and the calculation of the requisite sensitivities of U_{av} to variations in the components of $\boldsymbol{a} \equiv (\mu, \lambda, U_s, D_s)$.

Recall that the mean value, $E(x)$, and variance, $\text{var}(x)$, of a random variable x defined on a set of values S_x and with probability distribution function $p(x)$

are defined as $E(x) \equiv \int_{S_x} x\, p(x)\, dx$, and $\text{var}(x) = \int_{S_x} [x - E(x)]^2\, p(x)\, dx$, respectively.

Thus, for the Bernoulli distribution, the mean and variance of U are, respectively:

$$U_s^0 \equiv E(U_s) \equiv \sum_{U_s=0}^{1} U_s\, s^{U_s} (1-s)^{1-U_s} = s. \qquad \text{(IV.F.3)}$$

$$\text{var}(U_s) = \sum_{U_s=0}^{1} (U_s - s)^2\, s^{U_s} (1-s)^{1-U_s}$$
$$= s^2(1-s) + (1-s)^2 s = s(1-s). \qquad \text{(IV.F.4)}$$

The mean value, t_f^0, and variance, $\text{var}(t_f)$, of the random variable t_f (distributed according to the gamma distribution) are, respectively:

$$t_f^0 \equiv E(t_f) \equiv (MTTF) = \int_0^\infty t_f (t_f)^{\alpha-1} e^{-t_f/\beta}\, \frac{dt_f}{\beta^\alpha \Gamma(\alpha)}$$

$$= \frac{\beta}{\Gamma(\alpha)} \underbrace{\int_0^\infty \left(\frac{t_f}{\beta}\right)^\alpha e^{-t_f/\beta} \left(\frac{dt_f}{\beta}\right)}_{\Gamma(\alpha+1)} \qquad \text{(IV.F.5)}$$

$$= \frac{\beta\, \Gamma(\alpha+1)}{\Gamma(\alpha)} = \alpha\beta.$$

$$\text{var}(t_f) \equiv \int_0^\infty (t_f - \alpha\beta)^2\, t_f^{\alpha-1} e^{-t_f/\beta}\, \frac{dt_f}{\beta^\alpha \Gamma(\alpha)}$$

$$= \frac{\beta^2}{\Gamma(\alpha)} \underbrace{\int_0^\infty \left(\frac{t_f}{\beta}\right)^{\alpha+1} e^{-t_f/\beta} \left(\frac{dt_f}{\beta}\right)}_{\Gamma(\alpha+2)} \qquad \text{(IV.F.6)}$$

$$- \frac{2\alpha\beta^2}{\Gamma(\alpha)} \underbrace{\int_0^\infty \left(\frac{t_f}{\beta}\right)^\alpha e^{-t_f/\beta} \left(\frac{dt_f}{\beta}\right)}_{\Gamma(\alpha+1)} + \frac{\alpha^2\beta^2}{\Gamma(\alpha)} \Gamma(\alpha) = \alpha\beta^2.$$

The mean value, t_r^0, and variance, $\text{var}(t_r)$, of t_r are obtained as

$$t_r^0 \equiv E(t_r) \equiv (MTTR) = \int_0^\infty t_r \frac{1}{\sqrt{2\pi\sigma^2}} \frac{1}{t_r} \exp\left[-\frac{(\log t_r/v)^2}{2\sigma^2}\right] dt_r \quad \text{(IV.F.7)}$$
$$\text{var}(t_r) = E(t_r^2) - [E(t_r)]^2.$$

Although $E(t_r)$ and $E(t_r^2)$ could be calculated directly, the respective calculations are very cumbersome, and can be avoided by using the relationships between the log-normal and the normal distribution. For this purpose, recall that these distributions are related to each other by means of the transformation

$$p(x)dx = p(t_r)dt_r, \quad \text{(IV.F.8)}$$

where

$$p(x) = \frac{1}{\sqrt{2\pi\sigma^2}} \exp\left[-\frac{(x-m_0)^2}{2\sigma^2}\right]; \quad t_r = \exp(x); \quad m_0 = \log(v). \quad \text{(IV.F.9)}$$

Recall also that the moment generating function (*MGF*) for the normal distribution $p(x)$ is

$$M_x(t) \equiv \int_{-\infty}^\infty e^{tx} p(x)dx = \exp\left(m_0 t + \frac{\sigma^2 t^2}{2}\right). \quad \text{(IV.F.10)}$$

On the other hand, the central moments of order k of t_r are defined as

$$E(t_r^k) \equiv \int_{-\infty}^\infty t_r^k \, p(t_r) dt_r. \quad \text{(IV.F.11)}$$

Using Eqs. (IV.F.8) through (IV.F.10) in Eq. (IV.F.11) gives:

$$E(t_r^k) = \int_{-\infty}^\infty t_r^k \, p(x)dx = \int_{-\infty}^\infty e^{kx} p(x)dx = M_x(t)\big|_{t=k}. \quad \text{(IV.F.12)}$$

Thus, setting $k = 1$ in Eq. (IV.F.12) yields

$$E(t_r) = M_x(t)\big|_{t=1} = \exp(m_0 + \sigma^2/2) = v \exp\left(\frac{\sigma^2}{2}\right). \quad \text{(IV.F.13)}$$

Setting $k = 2$ in Eq. (IV.F.12) yields

$$E(t_r^2) = M_x(t)\big|_{t=2} = \exp\left(2m_0 + \frac{4\sigma^2}{2}\right) = v^2 \exp(2\sigma^2). \quad \text{(IV.F.14)}$$

The variance $\text{var}(t_r)$ can now be calculated by replacing Eqs. (IV.F.14) and (IV.F.13) into Eq. (IV.F.7) to obtain

$$\text{var}(t_r) = v^2 \exp(2\sigma^2) - v^2 \exp(\sigma^2). \quad \text{(IV.F.15)}$$

To calculate the mean value and variance of $\lambda \equiv (t_f)^{-1}$ it is convenient to use the "propagation of moments" method, applied to λ as a function of the random variable t_f.

$$\lambda(t_f) = \lambda(t_f^0) + (\partial \lambda / \partial t_f)_0 \delta t_f + \frac{1}{2}(\partial^2 \lambda / \partial t_f^2)_0 (\delta t_f)^2 + \text{higher order terms}.$$

Since only the second moment [in this case, $\text{var}(t_f)$] is considered to be available for this paradigm problem (as would be the case in large-scale practical problems), the above Taylor-series is truncated at the second order terms. Taking the expectation value of the above Taylor-series expansion gives:

$$\lambda^0 \equiv E(\lambda) = E[\lambda(t_f^0)] + \frac{1}{2}\left(\frac{\partial^2 \lambda}{\partial t_f^2}\right)_{t_f^0} \text{var}(t_f) + \text{higher order terms}$$

$$= (t_f^0)^{-1} + \frac{1}{2}\left[2(t_f^0)^{-3}\right]\text{var}(t_f)$$

$$= (t_f^0)^{-1}\left[1 + (t_f^0)^{-2} \text{var}(t_f)\right] = (\alpha\beta)^{-1}\left[1 + \underbrace{(\alpha\beta)^{-2}}_{\text{from Eq.}} \underbrace{(\alpha\beta^2)}_{\text{from Eq.}}\right] \quad \text{(IV.F.16)}$$
$$\phantom{= (t_f^0)^{-1}\left[1 + (t_f^0)^{-2} \text{var}(t_f)\right] = (\alpha\beta)^{-1}}\text{\scriptsize (IV.F.5) (IV.F.6)}$$

$$= (\alpha\beta)^{-1}\left[1 + \alpha^{-1}\right]$$

Similarly:

$$\text{var}(\lambda) = E(\lambda^2) - [E(\lambda)]^2$$

$$= E\left\{\left[\lambda(t_f^0) + \left(\frac{\partial \lambda}{\partial t_f}\right)_{t_f^0} \delta t_f + \frac{1}{2}\left(\frac{\partial^2 \lambda}{\partial t_f^2}\right)_{t_f^0}(\delta t_f)^2\right]^2\right\}$$

$$-\left\{E[\lambda(t_f^0)] + \frac{1}{2}\left(\frac{\partial^2 \lambda}{\partial t_f^2}\right)_{t_f^{0o}}\text{var}(t_f)\right\}^2 \qquad \text{(IV.F.17)}$$

$$= \left(\frac{\partial \lambda}{\partial t_f}\right)_{t_f^0}^2 \text{var}(t_f) - \frac{1}{4}\left(\frac{\partial^2 \lambda}{\partial t_f^2}\right)_{t_f^0}^2 [\text{var}(t_f)]^2.$$

Using Eqs. (IV.F.5) and (IV.F.6) in Eq. (IV.F.17) gives

$$\text{var}(\lambda) = \left[-(t_f^0)^{-2}\right]^2 \text{var}(t_f) - \frac{1}{4}\left[2(t_f^0)^{-3}\right]^2 [\text{var}(t_f)]^2 \qquad \text{(IV.F.18)}$$
$$= (\alpha\beta)^{-4}\alpha\beta^2 - (\alpha\beta)^{-6}(\alpha\beta^2)^2 = \alpha^{-3}\beta^{-2}(1-\alpha^{-1}).$$

The mean and variance of $\mu \equiv (t_r)^{-1}$ are calculated using the same procedures as were used above for calculating the mean and variance of λ, to obtain

$$\mu^0 \equiv E(\mu) = (t_r^0)^{-1}\left[1 + (t_r^0)^{-2}\text{var}(t_r)\right] \qquad \text{(IV.F.19)}$$

and

$$\text{var}(\mu) = \left[-(t_r^0)^{-2}\right]^2 \text{var}(t_r) - \frac{1}{4}\left[2(t_r^0)^{-3}\right]^2 [\text{var}(t_r)]^2. \qquad \text{(IV.F.20)}$$

Using now Eqs. (IV.F.14) and (IV.F.15) in Eqs. (IV.F.19) and (IV.F.20) yields:

$$\mu^0 \equiv E(\mu) = \left[v\exp\left(-\frac{\sigma^2}{2}\right)\right]^{-1}\left\{1 + [v\exp(\sigma^2/2)]^{-2}[v^2\exp(2\sigma^2) - v^2\exp(\sigma^2)]\right\}$$

$$= v^{-1}\exp\left(\frac{\sigma^2}{2}\right)$$

and, respectively,

$$\text{var}(\mu) = (t_r^0)^{-4} \text{var}(t_r) \left[1 - (t_r^0)^{-2} \text{var}(t_r)\right]$$

$$= \frac{v^2 e^{2\sigma^2} - v^2 e^{\sigma^2}}{v^4 e^{2\sigma^2}} \left[1 - \frac{v^2 e^{2\sigma^2} - v^2 e^{\sigma^2}}{v^2 e^{\sigma^2}}\right]$$

$$= v^{-2}\left(1 - e^{-\sigma^2}\right)\left(-e^{\sigma^2}\right) = v^{-2}\left[3 - \exp(\sigma^2) - \exp(-\sigma^2)\right].$$

To obtain the covariance matrix, V_a, for the random vector (μ, λ, U_s, D_s) in terms of $[\text{var}(\mu), \text{var}(\lambda), \text{var}(U_s)]$, it is convenient to consider that (μ, λ, U_s, D_s) plays the role of a response, while the vector $(\mu, \lambda, U_s)^T$ is considered to be the vector of parameters. Then, the covariance V_a is obtained from the "sandwich" formula given in Eq. (III.F.16), i.e.,

$$V_a = S \begin{pmatrix} \text{var}(\mu) & 0 & 0 \\ 0 & \text{var}(\lambda) & 0 \\ 0 & 0 & \text{var}(U_s) \end{pmatrix} S^T, \qquad \text{(IV.F.21)}$$

where S is the matrix of sensitivities defined as

$$S \equiv \begin{pmatrix} \frac{\partial \mu}{\partial \mu} & \frac{\partial \mu}{\partial \lambda} & \frac{\partial \mu}{\partial U_s} \\ \frac{\partial \lambda}{\partial \mu} & \frac{\partial \lambda}{\partial \lambda} & \frac{\partial \lambda}{\partial U_s} \\ \frac{\partial U_s}{\partial \mu} & \frac{\partial U_s}{\partial \lambda} & \frac{\partial U_s}{\partial U_s} \\ \frac{\partial D_s}{\partial \mu} & \frac{\partial D_s}{\partial \lambda} & \frac{\partial D_s}{\partial U_s} \end{pmatrix} = \begin{pmatrix} 1 & 0 & 0 \\ 0 & 1 & 0 \\ 0 & 0 & 1 \\ 0 & 0 & 0 \end{pmatrix}. \qquad \text{(IV.F.22)}$$

Replacing Eq. (IV.F.22) in Eq. (IV.F.21) leads to

$$V_a = \begin{pmatrix} \text{var}(\mu) & 0 & 0 & 0 \\ 0 & \text{var}(\lambda) & 0 & 0 \\ 0 & 0 & \text{var}(U_s) & -\text{var}(U_s) \\ 0 & 0 & -\text{var}(U_s) & \text{var}(U_s) \end{pmatrix}. \qquad \text{(IV.F.23)}$$

The propagation of errors (moments) method is now applied to U_{av}. For this purpose, U_{av} is considered to be a function of the problem's parameters $\boldsymbol{a} \equiv (\mu, \lambda, U_s, D_s)$, and is expanded in a functional Taylor series around the mean values $\boldsymbol{a}^0 \equiv (\mu^0, \lambda^0, U_s^0, D_s^0)$ of the problem's parameters. The formal expression of this Taylor-series, to first order in the respective sensitivities, is

$$U_{av}(\mu, \lambda, U_s, D_s) = U_{av}(\mu^0, \lambda^0, U_s^0, D_s^0)$$
$$+ \left(\frac{\partial U_{av}}{\partial \mu}\right)_{a^0} \delta\mu + \left(\frac{\partial U_{av}}{\partial \lambda}\right)_{a^0} \delta\lambda$$
$$+ \left(\frac{\partial U_{av}}{\partial U_s}\right)_{a^0} \delta U_s + \left(\frac{\partial U_{av}}{\partial D_s}\right)_{a^0} \delta D_s$$
$$+ \text{ higher order terms}.$$

(IV.F.24)

Calculating the mean value U_{av}^0 of Eq. (IV.F.24) gives

$$U_{av}^0 \equiv \int_{S_a} U_{av}(\mu^0, \lambda^0, U_s^0, D_s^0) p(\boldsymbol{a}) d\boldsymbol{a}$$
$$+ \int_{S_a} \left(\frac{\partial U_{av}}{\partial \mu}\right)_{a^0} \delta\mu\, p(\boldsymbol{a}) d\boldsymbol{a} + \int_{S_a} \left(\frac{\partial U_{av}}{\partial \lambda}\right)_{a^0} \delta\lambda\, p(\boldsymbol{a}) d\boldsymbol{a}$$
$$+ \int_{S_a} \left(\frac{\partial U_{av}}{\partial U_s}\right)_{a^0} \delta U_s\, p(\boldsymbol{a}) d\boldsymbol{a} + \int_{S_a} \left(\frac{\partial U_{av}}{\partial D_s}\right)_{a^0} \delta D_s\, p(\boldsymbol{a}) d\boldsymbol{a}$$
$$= U_{av}(\mu^0, \lambda^0, U_s^0, D_s^0),$$

(IV.F.25)

since the other terms vanish.

The variance, $\text{var}(U_{av})$, is obtained by using the formula

$$\text{var}(U_{av}) = E(U_{av}^2) - [E(U_{av})]^2.$$

(IV.F.26)

Calculating the first term on the right-side of Eq. (IV.F.26) gives

$$E(U_{av}^2) = \int_{S_a} \left[U_{av}^0 + \left(\frac{\partial U_{av}}{\partial \mu}\right)_{a^0} \delta\mu + \left(\frac{\partial U_{av}}{\partial \lambda}\right)_{a^0} \delta\lambda \right.$$

$$\left. + \left(\frac{\partial U_{av}}{\partial U_s}\right)_{a^0} \delta U_s + \left(\frac{\partial U_{av}}{\partial D_s}\right)_{a^0} \delta D_s \right]^2 p(\mathbf{a}) d\mathbf{a}$$

$$= \int_{S_a} p(\mathbf{a}) d\mathbf{a} \left\{ (U_{av}^0)^2 + 2U_{av}^0 \left[\left(\frac{\partial U_{av}}{\partial \mu}\right)_{a^0} \delta\mu + \left(\frac{\partial U_{av}}{\partial \lambda}\right)_{a^0} \delta\lambda \right. \right.$$

$$\left. + \left(\frac{\partial U_{av}}{\partial U_s}\right)_{a^0} \delta U_s + \left(\frac{\partial U_{av}}{\partial D_s}\right)_{a^0} \delta D_s \right] + \left(\frac{\partial U_{av}}{\partial \mu}\right)_{a^0}^2 (\delta\mu)^2 + \left(\frac{\partial U_{av}}{\partial \lambda}\right)_{a^0}^2 (\delta\lambda)^2$$

$$+ \left(\frac{\partial U_{av}}{\partial U_s}\right)_{a^0}^2 (\delta U_s)^2 + \left(\frac{\partial U_{av}}{\partial D_s}\right)_{a^0}^2 (\delta D_s)^2$$

$$+ 2\left(\frac{\partial U_{av}}{\partial \mu}\right)_{a^0} \left(\frac{\partial U_{av}}{\partial \lambda}\right)_{a^0} \delta\mu\,\delta\lambda + 2\left(\frac{\partial U_{av}}{\partial \mu}\right)_{a^0} \left(\frac{\partial U_{av}}{\partial U_s}\right)_{a^0} \delta\mu\,\delta U_s$$

$$+ 2\left(\frac{\partial U_{av}}{\partial \mu}\right)_{a^0} \left(\frac{\partial U_{av}}{\partial D_s}\right)_{a^0} \delta\mu\,\delta D_s + 2\left(\frac{\partial U_{av}}{\partial \lambda}\right)_{a^0} \left(\frac{\partial U_{av}}{\partial U_s}\right)_{a^0} \delta\lambda\,\delta U_s$$

$$\left. + 2\left(\frac{\partial U_{av}}{\partial \lambda}\right)_{a^0} \left(\frac{\partial U_{av}}{\partial D_s}\right)_{a^0} \delta\lambda\,\delta D_s + 2\left(\frac{\partial U_{av}}{\partial U_s}\right)_{a^0} \left(\frac{\partial U_{av}}{\partial D_s}\right)_{a^0} \delta U_s\,\delta D_s \right\}$$

$$= (U_{av}^0)^2 + 2U_{av}^0 \times 0 + \left(\frac{\partial U_{av}}{\partial \mu}\right)_{a^0}^2 \text{var}(\mu) + \left(\frac{\partial U_{av}}{\partial \lambda}\right)_{a^0}^2 \text{var}(\lambda)$$

$$+ \left(\frac{\partial U_{av}}{\partial U_s}\right)_{a^0}^2 \text{var}(U_s) + \left(\frac{\partial U_{av}}{\partial D_s}\right)_{a^0}^2 \text{var}(D_s)$$

$$+ 2\left(\frac{\partial U_{av}}{\partial \mu}\right)_{a^0} \left(\frac{\partial U_{av}}{\partial \lambda}\right)_{a^0} \underbrace{\text{cov}(\mu,\lambda)}_{0} + 2\left(\frac{\partial U_{av}}{\partial \mu}\right)_{a^0} \left(\frac{\partial U_{av}}{\partial U_s}\right)_{a^0} \underbrace{\text{cov}(\mu,U_s)}_{0}$$

$$+ 2\left(\frac{\partial U_{av}}{\partial \mu}\right)_{a^0} \left(\frac{\partial U_{av}}{\partial D_s}\right)_{a^0} \underbrace{\text{cov}(\mu,D_s)}_{0} + 2\left(\frac{\partial U_{av}}{\partial \lambda}\right)_{a^0} \left(\frac{\partial U_{av}}{\partial U_s}\right)_{a^0} \underbrace{\text{cov}(\lambda,U_s)}_{0}$$

$$+ 2\left(\frac{\partial U_{av}}{\partial \lambda}\right)_{a^0} \left(\frac{\partial U_{av}}{\partial D_s}\right)_{a^0} \underbrace{\text{cov}(\lambda,D_s)}_{0} + 2\left(\frac{\partial U_{av}}{\partial U_s}\right)_{a^0} \left(\frac{\partial U_{av}}{\partial D_s}\right)_{a^0} \underbrace{\text{cov}(U_s,D_s)}_{-\text{var}(U_s)}$$

$$= (U_{av}^0)^2 + \mathbf{S}_a \mathbf{V}_a \mathbf{S}_a^T$$

(IV.F.27)

where the column vector S_a containing the sensitivities of the response U_{av} to the parameters $a \equiv (\mu, \lambda, U_s, D_s)$ is defined as

$$S_a \equiv \left[\left(\frac{\partial U_{av}}{\partial \mu}\right)_{a^0}, \left(\frac{\partial U_{av}}{\partial \lambda}\right)_{a^0}, \left(\frac{\partial U_{av}}{\partial U_s}\right)_{a^0}, \left(\frac{\partial U_{av}}{\partial D_s}\right)_{a^0} \right]^T.$$

Replacing now Eqs. (IV.F.27) and (IV.F.25) into Eq. (IV.F.26) gives

$$\text{var}(U_{av}) = \left(U_{av}^0\right)^2 + S_a V_a S_a^T - \left(U_{av}^0\right)^2 = S_a V_a S_a^T. \qquad \text{(IV.F.28)}$$

The above equation is the customary "sandwich" formula for calculating the response variance in terms of parameter covariances. The next step, therefore, is the calculation of the sensitivity vector S_a.

The explicit expression of U_{av}^0 is obtained by first solving Eq. (IV.F.1) to determine $U^0(t)$. Among the various methods that can be used for this purpose, the Laplace-transform method is quite convenient and will be used in the following. Define, therefore, the Laplace-transforms $U^0(z)$ and $D^0(z)$ as

$$U^0(z) \equiv \int_0^\infty e^{-zt} U^0(t) dt,$$

$$D^0(z) \equiv \int_0^\infty e^{-zt} D^0(t) dt. \qquad \text{(IV.F.29)}$$

Applying the above Laplace transforms to Eq. (IV.F.1), evaluated at a^0, yields

$$\begin{cases} (z + \lambda^0) U^0(z) - \mu^0 D^0(z) = U_s^0 \\ -\lambda^0 U^0(z) + (z + \mu^0) D^0(z) = D_s^0 \end{cases}.$$

Solving the above system of two algebraic equations gives $U^0(z) = (zU_s^0 + \mu^0)/[z(z + \lambda^0 + \mu^0)]$, which can be inverted by using the "residue theorem" to obtain

$$U^0(t) \equiv \mathcal{L}^{-1}[U^0(z)] = \lim_{z \to 0} \frac{ze^{zt}(\mu^0 + zU_s^0)}{z(z + \lambda^0 + \mu^0)}$$
$$+ \lim_{z \to -(\lambda^0 + \mu^0)} \frac{(z + \lambda^0 + \mu^0)e^{zt}(\mu^0 + zU_s^0)}{z(z + \lambda^0 + \mu^0)} \qquad \text{(IV.F.30)}$$
$$= \frac{\mu^0}{\lambda^0 + \mu^0} + \left(U_s^0 - \frac{\mu^0}{\lambda^0 + \mu^0}\right)e^{-(\lambda^0 + \mu^0)t}.$$

Using the above expression in Eq. (IV.F.2) gives the mean value U_{av}^0 of the response U_{av} as

$$U_{av}^0 = \frac{1}{T}\int_0^T U^0(\tau)d\tau = \frac{1}{T}\int_0^T \left[\frac{\mu^0}{\mu^0 + \lambda^0} + \left(U_s^0 - \frac{\mu^0}{\mu^0 + \lambda^0}\right)e^{-(\mu^0 + \lambda^0)\tau}\right]d\tau$$
$$= \frac{\mu^0}{\mu^0 + \lambda^0} + \left(U_s^0 - \frac{\mu^0}{\mu^0 + \lambda^0}\right)\frac{1 - e^{-(\mu^0 + \lambda^0)T}}{(\mu^0 + \lambda^0)T}. \qquad \text{(IV.F.31)}$$

The expressions of $D^0(z)$ and $D^0(t)$ can be similarly obtained as

$$D^0(z) = \frac{zD_s^0}{z(z + \mu^0 + \lambda^0)}, \text{ and}$$

$$D^0(t) \equiv \mathcal{L}^{-1}[D^0(z)] = \frac{\lambda^0}{\lambda^0 + \mu^0} + \left(D_s^0 - \frac{\lambda^0}{\lambda^0 + \mu^0}\right)e^{-(\lambda^0 + \mu^0)t}. \qquad \text{(IV.F.32)}$$

The sensitivities of the response U_{av} are given by $\delta U_{av}(e^0; h)$, which is obtained by using Eq. (IV.F.2), as follows:

$$\delta U_{av}(e^0; h) \equiv \left\{\frac{d}{d\varepsilon}[U_{av}(e^0 + \varepsilon h)]\right\}_{\varepsilon=0}$$
$$= \frac{1}{T}\int_0^T \left\{\frac{d}{d\varepsilon}[U^0(\tau) + \varepsilon \delta U(\tau)]\right\}_{\varepsilon=0} d\tau = \frac{1}{T}\int_0^T \delta U(\tau) d\tau. \qquad \text{(IV.F.33)}$$

The variation $\delta U(t)$ is obtained as the solution of the G-differentiated Markov-model; in turn the G-differentiated Markov-model is obtained as

$$\left\{\frac{d}{d\varepsilon}\left[\frac{d}{dt}\left(U^0+\varepsilon\delta U\right)-\left(\lambda^0+\varepsilon\delta\lambda\right)\left(U^0+\varepsilon\delta U\right)-\left(\mu^0+\varepsilon\delta\mu\right)\left(D^0+\varepsilon\delta D\right)\right]\right\}_{\varepsilon=0}=0$$

$$\left\{\frac{d}{d\varepsilon}\left[\frac{d}{dt}\left(D^0+\varepsilon\delta D\right)-\left(\lambda^0+\varepsilon\delta\lambda\right)\left(U^0+\varepsilon\delta U\right)+\left(\mu^0+\varepsilon\delta\mu\right)\left(D^0+\varepsilon\delta D\right)\right]\right\}_{\varepsilon=0}=0$$

$$\left\{\frac{d}{d\varepsilon}\left[U^0+\varepsilon\delta U\right]\right\}_{\varepsilon=0}=\left\{\frac{d}{d\varepsilon}\left[U_s^0+\varepsilon\delta U_s\right]\right\}_{\varepsilon=0}$$

$$\left\{\frac{d}{d\varepsilon}\left[D^0+\varepsilon\delta D\right]\right\}_{\varepsilon=0}=\left\{\frac{d}{d\varepsilon}\left[D_s^0+\varepsilon\delta D_s\right]\right\}_{\varepsilon=0}.$$

Carrying out the operations involving ε reduces the above system to the *Forward Sensitivity Equations*, which can be written in matrix form as

$$\begin{pmatrix}\dfrac{d}{dt}+\lambda^0 & -\mu^0 \\ -\lambda^0 & \dfrac{d}{dt}+\mu^0\end{pmatrix}\begin{pmatrix}\delta U \\ \delta D\end{pmatrix}=\begin{pmatrix}D^0(\delta\mu)-U^0(\delta\lambda) \\ U^0(\delta\lambda)-D^0(\delta\mu)\end{pmatrix} \qquad\text{(IV.F.34)}$$

subject to the initial conditions

$$\delta U(0)=\delta U_s \text{ and } \delta D(0)=\delta D_s=-\delta U_s. \qquad\text{(IV.F.35)}$$

Note that Eq. (IV.F.34) can be written in the form $L(a^0)h_u = L'_a(e^0)h_a$, where

$$L(a^0)\equiv\begin{pmatrix}\dfrac{d}{dt}+\lambda^0 & -\mu^0 \\ -\lambda^0 & \dfrac{d}{dt}+\mu^0\end{pmatrix}; \; h_u\equiv\begin{pmatrix}\delta U \\ \delta D\end{pmatrix},$$

which makes it obvious that the operator $L(a^0)$ acts linearly on h_u.

From the definition of the response, it is apparent that the *inner product* for two vector-valued functions $f\equiv(f_1,f_2)^T$ and $g\equiv(g_1,g_2)^T$ for this problem is given by

$$\langle f,g\rangle\equiv\sum_{i=1}^{2}\int_0^T f_i(t)g_i(t)\,dt, \; f_i,g_i\in\mathscr{L}_2(0,T),\; i=1,2.$$

To apply the *ASAP*, we form the inner product of $\Phi \equiv (\Phi^U, \Phi^D)$ with Eq. (IV.F.34), to obtain

$$\int_0^T \left\{ (\Phi^U \quad \Phi^D) \begin{pmatrix} \dfrac{d}{dt} + \lambda^0 & -\mu^0 \\ -\lambda^0 & \dfrac{d}{dt} + \mu^0 \end{pmatrix} \begin{pmatrix} \delta U \\ \delta D \end{pmatrix} \right\} dt \qquad \text{(IV.F.36)}$$

$$= \int_0^T \left\{ (\Phi^U \quad \Phi^D) \begin{pmatrix} D^0(\delta\mu) - U^0(\delta\lambda) \\ U^0(\delta\lambda) - D^0(\delta\mu) \end{pmatrix} \right\} dt .$$

Performing the matrix-vector multiplications on the left-side of the above equation gives

$$\int_0^T \Phi^U \frac{d}{dt}(\delta U) dt + \int_0^T \Phi^U \lambda^0 (\delta U) dt$$

$$- \int_0^T \mu^0 \Phi^U (\delta D) dt + \int_0^T \Phi^D \left[-\lambda^0 (\delta U) \right] dt$$

$$+ \int_0^T \Phi^D \frac{d}{dt}(\delta D) dt + \int_0^T \Phi^D \mu^0 (\delta U) dt .$$

Transferring all operations on (δU) and (δD) in the above expression to operations on Φ^U and Φ^D, integrating once by parts where necessary, leads to

$$\left\{ \Phi^U(t)[\delta U(t)] + \Phi^D(t)[\delta D(t)] \right\}_{t=0}^{t=T} + \int_0^T \left\{ (\delta U \quad \delta D) \underbrace{\begin{pmatrix} -\dfrac{d\Phi^U}{dt} + \lambda^0 \Phi^U - \lambda^0 \Phi^D \\ -\mu^0 \Phi^D - \dfrac{d\Phi^D}{dt} + \mu^0 \Phi^U \end{pmatrix}}_{L^+(a^0)\Phi} \right\} dt .$$

(IV.F.37)

Thus, the formal adjoint operator $L^+(a^o)$ has the expression indicated above. Using now Eq. (IV.F.37) to replace the left-side of Eq. (IV.F.36) yields

$$\left\{\Phi^U(t)[\delta U(t)] + \Phi^D(t)[\delta D(t)]\right\}_{t=0}^{t=T} + \int_0^T (\delta U \quad \delta D)\left[L^+(a^0)\Phi\right]dt$$
$$= \int_0^T \left\{\Phi^U(t)\left[D^0(\delta\mu) - U^0(\delta\lambda)\right] + \Phi^D(t)\left[U^0(\delta\lambda) - D^0(\delta\mu)\right]\right\}dt. \tag{IV.F.38}$$

Identifying the second term on the left-side of Eq. (IV.F.38) with δU_{av} of Eq. (IV.F.33) yields:

$$\begin{cases} -\dfrac{d\Phi^U}{dt} + \lambda^0\Phi^U - \lambda^0\Phi^D = 1 \\ -\dfrac{d\Phi^D}{dt} + \mu^0\Phi^D - \mu^0\Phi^U = 0 \end{cases} \tag{IV.F.39}$$

and

$$T(\delta U_{av}) = -\left\{\Phi^U(t)[\delta U(t)] + \Phi^D(t)[\delta D(t)]\right\}_{t=0}^{t=T}$$
$$+ \int_0^T \left\{\Phi^U(t)\left[D^0(\delta\mu) - U^0(\delta\lambda)\right] + \Phi^D(t)\left[U^0(\delta\lambda) - D^0(\delta\mu)\right]\right\}dt. \tag{IV.F.40}$$

In Eq. (IV.F.40), the values $\delta U(t)$ and $\delta D(t)$ are known for $t=0$ from the *Forward Sensitivity Equations* [namely Eq. (IV.F.32)], but are not known for $t=T$. Therefore the unknown values $\delta U(T)$ and $\delta D(T)$ are eliminated from Eq. (IV.F.40) by imposing the conditions

$$\Phi^U(T) = 0; \; \Phi^D(T) = 0, \text{ at } t = T. \tag{IV.F.41}$$

This selection completes the construction of the Adjoint Sensitivity System, which consists of Eqs. (IV.F.39) and (IV.F.41), and also reduces the G-differential δU_{av} of the response, namely Eq. (IV.F.40), to

$$T[\delta U_{av}] = \left[\Phi^U(0) - \Phi^D(0)\right]\delta U_s$$
$$+ \int_0^T \left\{\Phi^U(t)\left[D^0(\delta\mu) - U^0(\delta\lambda)\right] + \Phi^D(t)\left[U^0(\delta\lambda) - D^0(\delta\mu)\right]\right\}dt.$$

It follows from the above equation that the sensitivities of U_{av} to the parameters (μ, λ, U_s, D_s) are

$$\frac{\partial U_{av}}{\partial \mu} = \frac{1}{T}\int_0^T \left[\Phi^U(t) - \Phi^D(t)\right]D^0(t)dt, \quad \text{(IV.F.42)}$$

$$\frac{\partial U_{av}}{\partial \lambda} = \frac{1}{T}\int_0^T \left[\Phi^D(t) - \Phi^U(t)\right]U^0(t)dt, \quad \text{(IV.F.43)}$$

$$\frac{\partial U_{av}}{\partial U_s} = -\frac{\partial U_{av}}{\partial D_s} = \frac{1}{T}\left[\Phi^U(0) - \Phi^D(0)\right]. \quad \text{(IV.F.44)}$$

The *Adjoint Sensitivity Equations*, namely Eqs. (IV.F.39) and (IV.F.41), can be solved by several methods, too; again, the Laplace-transform technique will be used here, since it is very simple to apply. For this purpose, Eqs. (IV.F.39) and (IV.F.41) will first be transformed from a final-time-value problem, to an initial-value problem, by transforming the independent variable t to

$$\tau\Big|_T^0 \equiv T - t\Big|_0^T, \text{ so that } \frac{d}{dt} \to -\frac{d}{d\tau}. \quad \text{(IV.F.45)}$$

Introducing the transformation (IV.F.45) into Eqs. (IV.F.39) and (IV.F.41) yields:

$$\begin{cases} \dfrac{d\Phi^U}{d\tau} + \lambda^0 \Phi^U(\tau) - \lambda^0 \Phi^D(\tau) = 1 \\[4pt] \dfrac{d\Phi^D}{d\tau} + \mu^0 \Phi^D(\tau) - \mu^0 \Phi^U(\tau) = 0 \\[4pt] \Phi^D(\tau)\Big|_{\tau=0} = 0; \; \Phi^U(\tau)\Big|_{\tau=0} = 0 \end{cases} \quad \text{(IV.F.46)}$$

Laplace-transforming Eq. (IV.F.46) leads to

$$\begin{cases} (z + \lambda^0)\Phi^U(z) - \lambda^0 \Phi^D(z) = \dfrac{1}{z} \\[4pt] -\mu^0 \Phi^U(z) + (z + \mu^0)\Phi^D(z) = 0 \end{cases} \quad \text{(IV.F.47)}$$

where

$$\Phi^U(z) \equiv \int_0^\infty e^{-z\tau}\Phi^U(\tau)d\tau \; ; \; \Phi^D(z) \equiv \int_0^\infty e^{-z\tau}\Phi^D(\tau)d\tau.$$

The solution of Eq. (IV.F.47) reads

$$\Phi^U(z) = \frac{z + \mu^0}{z^2(z + \lambda^0 + \mu^0)};$$

$$\Phi^D(z) = \frac{\mu^0}{z^2(z + \lambda^0 + \mu^0)}.$$
(IV.F.48)

Taking the inverse Laplace transforms of Eq. (IV.F.48) gives

$$\Phi^U(\tau) \equiv \mathcal{L}^{-1}\left[\Phi^U(z)\right] = \frac{\mu^0 \tau}{\mu^0 + \lambda^0} + \frac{\lambda^0}{(\mu^0 + \lambda^0)^2}\left[1 - e^{-(\mu^0 + \lambda^0)\tau}\right], \quad \text{(IV.F.49.a)}$$

$$\Phi^D(\tau) \equiv \mathcal{L}^{-1}\left[\Phi^D(z)\right] = \frac{\mu^0 \tau}{\mu^0 + \lambda^0} - \frac{\mu^0}{(\mu^0 + \lambda^0)^2}\left[1 - e^{-(\mu^0 + \lambda^0)\tau}\right]. \quad \text{(IV.F.49.b)}$$

Reverting back, from τ to t, by using Eq. (IV.F.45) in Eqs. (IV.F.49.a-b) yields the expression of the adjoint function as

$$\boldsymbol{\Phi}(t) \equiv \begin{pmatrix} \Phi^U(t) \\ \Phi^D(t) \end{pmatrix} = \begin{pmatrix} \dfrac{\mu^0(T-t)}{\mu^0 + \lambda^0} + \dfrac{\lambda^0}{(\mu^0 + \lambda^0)^2}\left[1 - e^{-(\mu^0 + \lambda^0)(T-t)}\right] \\ \dfrac{\mu^0(T-t)}{\mu^0 + \lambda^0} - \dfrac{\mu^0}{(\mu^0 + \lambda^0)^2}\left[1 - e^{-(\mu^0 + \lambda^0)(T-t)}\right] \end{pmatrix}. \quad \text{(IV.F.50)}$$

The sensitivities of U_{av} can now be obtained by replacing Eq. (IV.F.50) in Eqs. (IV.F.42-44). Note that

$$\Phi^U(t) - \Phi^D(t) = \frac{1}{\mu^0 + \lambda^0}\left[1 - e^{-(\mu^0 + \lambda^0)(T-t)}\right]. \quad \text{(IV.F.51)}$$

Evaluating Eq. (IV.F.51) at $t = 0$ and inserting the resulting expression in Eq. (IV.F.44) yields

$$\frac{\partial U_{av}}{\partial U_s} = -\frac{\partial U_{av}}{\partial D_s} = \frac{1 - e^{-(\mu^0 + \lambda^0)T}}{T(\mu^0 + \lambda^0)}. \quad \text{(IV.F.52)}$$

The sensitivity $\partial U_{av}/\partial \mu$ can be obtained by inserting Eqs. (IV.F.32) and (IV.F.51) into Eq. (IV.F.42) and performing the integration over t; this sequence of operations gives

$$\frac{\partial U_{av}}{\partial \mu} = \frac{1}{T}\int_0^T \frac{dt}{\mu^0+\lambda^0}\left[1-e^{-(\mu^0+\lambda^0)(T-t)}\right]\left[\frac{\lambda^0}{\mu^0+\lambda^0}+\left(D_s^0-\frac{\lambda^0}{\mu^0+\lambda^0}\right)e^{-(\mu^0+\lambda^0)t}\right]$$

$$= \frac{1}{T(\mu^0+\lambda^0)}\left\{\int_0^T \frac{\lambda^0}{\mu^0+\lambda^0}dt + \int_0^T\left(D_s^0-\frac{\lambda^0}{\mu^0+\lambda^0}\right)e^{-(\mu^0+\lambda^0)t}dt\right.$$

$$\left.-\frac{\lambda^0}{\mu^0+\lambda^0}e^{-(\mu^0+\lambda^0)T}\int_0^T e^{(\mu^0+\lambda^0)t}dt - \left(D_s^0-\frac{\lambda^0}{\mu^0+\lambda^0}\right)e^{-(\mu^0+\lambda^0)T}\int_0^T dt\right\}$$

$$= \frac{\lambda^0}{(\mu^0+\lambda^0)^2}+\frac{1-e^{-(\mu^0+\lambda^0)T}}{T(\mu^0+\lambda^0)^2}\left(D_s^0-\frac{\lambda^0}{\mu^0+\lambda^0}\right)-\frac{\lambda^0}{T(\mu^0+\lambda^0)^3}\left[1-e^{-(\mu^0+\lambda^0)T}\right]$$

$$-\frac{1}{\mu^0+\lambda^0}\left(D_s^0-\frac{\lambda^0}{\mu^0+\lambda^0}\right)e^{-(\mu^0+\lambda^0)T}.$$

(IV.F.53)

Similarly, the sensitivity $\partial U_{av}/\partial \lambda$ can be obtained by inserting Eqs. (IV.F.30) and (IV.F.51) into Eq. (IV.F.43) and performing the respective integration over t. Carrying out these operations yields:

$$\frac{\partial U_{av}}{\partial \lambda} = \frac{1}{T}\int_0^T \frac{dt}{\mu^0+\lambda^0}\left[e^{-(\mu^0+\lambda^0)(T-t)}-1\right]\left[\frac{\mu^0}{\mu^0+\lambda^0}+\left(U_s^0-\frac{\mu^0}{\mu^0+\lambda^0}\right)e^{-(\mu^0+\lambda^0)t}\right]$$

$$= \frac{1}{T(\mu^0+\lambda^0)}\left\{\frac{\mu^0}{\mu^0+\lambda^0}e^{-(\mu^0+\lambda^0)T}\int_0^T e^{(\mu^0+\lambda^0)t}dt - \frac{\mu^0}{\mu^0+\lambda^0}\int_0^T dt\right.$$

$$\left.+\left(U_s^0-\frac{\mu^0}{\mu^0+\lambda^0}\right)e^{-(\mu^0+\lambda^0)T}\int_0^T dt - \left(U_s^0-\frac{\mu^0}{\mu^0+\lambda^0}\right)\int_0^T e^{-(\mu^0+\lambda^0)t}dt\right\}$$

$$= \frac{\mu^0}{T(\mu^0+\lambda^0)^3}\left[1-e^{-(\mu^0+\lambda^0)T}\right]-\frac{\mu^0}{(\mu^0+\lambda^0)^2}+\left(U_s^0-\frac{\mu^0}{\mu^0+\lambda^0}\right)e^{-(\mu^0+\lambda^0)T}$$

$$+\frac{1}{T(\mu^0+\lambda^0)^2}\left(U_s^0-\frac{\mu^0}{\mu^0+\lambda^0}\right)\left[e^{-(\mu^0+\lambda^0)T}-1\right].$$

(IV.F.54)

Having thus determined the sensitivity vector S_a, the "sandwich" formula given by Eq. (IV.F.28) is now used to calculate the time-dependent variance $\text{var}(U_{av})$. It is also of interest to note that, in the limit as $T \to \infty$, Eq. (IV.F.52) yields

$$\lim_{T \to \infty} \frac{\partial U_{av}}{\partial U_s} = 0 = \lim_{T \to \infty} \frac{\partial U_{av}}{\partial D_s} \qquad \text{(IV.F.55)}$$

which indicates that the sensitivities to the initial conditions vanish as $T \to \infty$, as they should for a well-posed problem. Furthermore, taking the limit as $T \to \infty$ in Eqs. (IV.F.53) and (IV.F.54) gives

$$\lim_{T \to \infty} \frac{\partial U_{av}}{\partial \mu} = \frac{\lambda^o}{\left(\mu^o + \lambda^o\right)^2} \quad \text{and} \quad \lim_{T \to \infty} \frac{\partial U_{av}}{\partial \lambda} = -\frac{\mu^o}{\left(\mu^o + \lambda^o\right)^2}. \qquad \text{(IV.F.56)}$$

IV.G. ILLUSTRATIVE EXAMPLE: SENSITIVITY AND UNCERTAINTY ANALYSIS OF NEUTRON AND GAMMA RADIATION TRANSPORT

This Section presents an illustrative example of sensitivity and uncertainty analysis of a large-scale problem, described by the Boltzmann transport equation for neutrons and gamma rays, and involving many parameters (in particular, nuclear cross section data). This illustrative example is based on work by Lillie et al., *Nucl. Sci. Eng.*, 100, 105, 1988 (see Ref. 41, with permission), and underscores the typical uses and physical interpretations of sensitivities and uncertainties.

Note that the *Boltzmann transport equation is linear in the neutron and gamma fluxes*. This fact allows the adjoint transport equation, which underlies the application of the *ASAP*, to be solved independently of the forward transport equation. The response in this neutron and gamma transport problem is the free-in-air (FIA) tissue kermas due to prompt neutrons, secondary gammas, and prompt gammas following the nuclear explosion at Nagasaki. The computation of the (FIA) tissue kerma uncertainties due to cross section uncertainties requires the use of sensitivity analysis to obtain relative sensitivity coefficients. These sensitivity coefficients yield the percent change in kerma due to a 1% change in the various cross sections, and are obtained from two dimensional air-over-ground (A/G) radiation transport calculations that yield the forward and adjoint neutron and gamma fluences. Folding the sensitivity coefficients with the covariance data for a particular cross section yields the kerma uncertainty due to the uncertainty in that cross section.

IV.G.1. Sensitivity Analysis

The FIA tissue kerma response, R, can be represented in the form

$$R = \langle \Sigma(\xi)\varphi(\xi) \rangle, \qquad (IV.G.1)$$

where $\Sigma(\xi)$ denotes the FIA tissue kerma response function, $\varphi(\xi)$ denotes the forward fluence, $\xi \equiv (r, E, \Omega)$ denotes the phase space (energy E, solid angle Ω, and position vector r) underlying this problem, and $\langle\ \rangle$ denotes integration over ξ. The forward neutron flux $\varphi(\xi)$ is the solution of transport equation

$$L(\xi)\varphi(\xi) \equiv \Omega \cdot \nabla\varphi + \Sigma_t \varphi - S\varphi = Q(r, E, \Omega) \qquad (IV.G.2)$$

where $\Omega \cdot \nabla\varphi$ is the rate change of the neutron flux, $\Sigma_t \varphi$ is the total interaction rate, $Q(r, E, \Omega)$ is the external source, and

$$S\varphi \equiv \int_{\Omega'} \int_{E'} \sum_{x \neq f} \Sigma_x(r, E' \to E, \Omega' \to \Omega, t)\ \varphi(r, E', \Omega', t) dE'\ d\Omega'$$

is the scattering operator. The A/G fluence is obtained by solving the transport Eq. (IV.G.2) subject to Dirichlet boundary conditions (i.e., the fluence vanishes at large distances from the source). The formal adjoint of the Boltzmann operator $L(\xi)$ is the operator

$$L^+\varphi(\xi) \equiv -\Omega \cdot \nabla\varphi^+ + \Sigma_t \varphi^+ - S^+\varphi^+ \qquad (IV.G.3)$$

where

$$S^+\varphi^+ \equiv \int_{\Omega'} \int_{E'} \sum_{x \neq f} \Sigma_x(r, E \to E', \Omega \to \Omega', t)\ \varphi^+(r, E', \Omega', t) dE'\ d\Omega'.$$

The relative sensitivity represents the fractional change in a response R per fractional change in an input parameter q, and can be written in the form

$$S_{Rq}(\rho) = \frac{dR/dq}{R/q}. \qquad (IV.G.4)$$

This expression above is re-expressed, by applying the *ASAP*, in the equivalent form

$$S_{Rq(\rho)} = \frac{-q\left\langle \varphi^+(\xi)\frac{\partial L(\xi)}{\partial q}\varphi(\xi)\right\rangle}{R}, \qquad (IV.G.5)$$

where the adjoint fluence, $\varphi^+(\xi)$, is the solution of the adjoint Boltzmann equation

$$L^+\varphi(\xi) = \partial R/\partial \varphi. \qquad (IV.G.6)$$

In Eqs. (IV.G.5-6), $\partial R/\partial \varphi$ denotes the partial Gâteaux-derivative of R with respect to φ, while $\partial L/\partial q$ denotes the partial Gâteaux-derivative of L with respect to q, where q represents generically the complete set of partial cross sections used in the forward and adjoint fluence calculations. Finally, note that the direct effect term, $d\Sigma(\xi)/dq$, is not included in Eq. (IV.G.5) because the uncertainty in R due to the uncertainty in the FIA tissue kerma response function $\Sigma(\xi)$ is not considered in the present analysis.

The relative neutron source sensitivities by neutron energy group are defined as

$$S_{Rg} \equiv \frac{dR/dS_g}{R/S_g} = \frac{S_g \varphi_g^+}{R}, \qquad (IV.G.7)$$

where S_{Rg} denotes the relative sensitivity of R with respect to source group S_g, φ_g^+ denotes the adjoint fluence in group g at the source position, and R denotes the response. The expression above reveals that the relative sensitivity of R with respect to source group S_g is simply the fraction of the response due to particles born in source group S_g; the normalized importance adjoint fluence is defined to be the fraction of the adjoint fluence in group g at the source position.

The forward and adjoint fluences needed to calculate the sensitivities using Eq. (IV.G.5) and/or (IV.G.7) were obtained by performing air-over-ground (A/G) radiation transport calculations using the DOT-IV discrete ordinates radiation transport code. The forward fluence is obtained by solving Eq. (IV.G.2), while the adjoint fluence is obtained by solving Eq. (IV.G.6). The A/G geometric model used in these calculations is a right circular cylinder containing air in the upper portion of the cylinder and soil in the lower portion. Six radial zones extending out to a maximum ground range of 2812.5 m and seven axial zones of decreasing air density extending to a height of 1500 m were used to describe the air. The ground was represented by one zone extending downward 50 cm. The entire spatial mesh consisted of 120 radial intervals and 99 axial intervals of which 21 were in the ground and 78 were in the air.

A coupled 46 neutron- and 22 gamma-multigroup cross-section set using a P_3-Legendre angular expansion of the group-to-group transfer coefficients was employed in the fluence calculations. In the adjoint fluence calculations, the FIA response functions were employed as the source terms. The adjoint sources were placed in the radial intervals corresponding to the 700, 1000, and 1500 m ground ranges. For each ground range, two calculations were performed employing the neutron and gamma response functions separately, in order to separate the effects associated with neutrons from those associated with gammas. After obtaining the forward and adjoint fluence, the energy-dependent sensitivity coefficients (called sensitivity profiles) are obtained by folding the sensitivities with the fluences as indicated in Eq. (IV.G.5).

Table IV.1, below, presents the relative sensitivities of the **p**rompt **n**eutron FIA tissue **k**erma (abbreviated in the sequel as PNK) response, and the normalized adjoint fluence of PNK as a function of the energy-group of the source-neutrons. The largest sensitivities are to prompt source neutrons having energies between 1.4 and 7.4 MeV, in neutron source groups 12 through 20. Close to 90% of the PNK response is due to neutrons born in these groups, at all three ground ranges. This effect is due to the large source importance at high-energies, at all three ground ranges. At 700 and 1000 m, source group 17 shows the largest sensitivity whereas at 1500 m the largest sensitivity has shifted from group 17 to group 13. This shift of the largest sensitivity from lower to higher energy at the larger ground range is due to the rapid decrease in adjoint fluence at the lower energies at 1500 m.

The total relative sensitivities (having absolute values >0.1) of tissue kermas to various cross-sections are presented in Table IV.2 for the 700, 1000, and 1500 m ground ranges, for all three responses (prompt neutron, secondary gamma, and prompt gamma). All of the sensitivities listed in Table IV.2 are negative, indicating that an increase in any of the cross sections induces a decrease in the respective response. For the prompt neutron response, the sensitivities with the largest magnitudes arise from the nitrogen elastic scattering cross section. For the secondary and prompt gamma responses, the sensitivities having the largest magnitude arise from the nitrogen gamma incoherent scattering cross sections. For the prompt neutron responses, all sensitivity coefficients increase in magnitude with increasing ground range. This is because the effects of variations in a material become more pronounced as the number of mean free paths a particle has to travel in the respective material increases (i.e., more interaction lengths).

Table IV.1
Nagasaki Prompt Neutron FIA Tissue Kerma (PNK) Relative Sensitivities (>0.001) to Source Neutrons, and the Normalized Adjoint Fluence (>0.1%) as a Function of a Source Energy Group

Neutron Energy Group	Ground Range (m)					
	700		1000		1500	
1	-	-	-	-	-	-
2	-	(6.5)[a]	-	(6.8)	-	(7.2)
3	-	(6.2)	-	(6.6)	-	(6.9)
4	-	(6.1)	-	(6.4)	-	(6.7)
5	0.001	(6.1)	0.002	(6.3)	0.003	(6.6)
6	-	(6.1)	-	(6.4)	-	(6.8)
7	0.003	(6.2)	0.004	(6.5)	0.005	(7.0)
8	0.004	(6.6)	0.005	(7.2)	0.008	(7.9)
9	0.011	(7.3)	0.015	(8.0)	0.024	(9.1)
10	0.017	(7.1)	0.022	(7.7)	0.034	(8.6)
11	0.024	(6.3)	0.031	(6.7)	0.046	(7.2)
12	0.064	(6.7)	0.084	(7.2)	0.128	(7.9)
13	0.135	(5.3)	0.167	(5.4)	0.226	(5.3)
14	0.025	(5.2)	0.032	(5.5)	0.046	(5.6)
15	0.067	(3.1)	0.069	(2.7)	0.064	(1.8)
16	0.137	(3.1)	0.135	(2.5)	0.116	(1.6)
17	0.234	(3.4)	0.231	(2.7)	0.197	(1.7)
18	0.032	(3.1)	0.030	(2.5)	0.024	(1.4)
19	0.124	(2.1)	0.099	(1.4)	0.055	(0.5)
20	0.061	(1.2)	0.040	(0.6)	0.014	(0.1)
21	0.032	(0.7)	0.018	(0.3)	0.005	(0.1)
22	0.009	(0.4)	0.004	(0.2)	-	-
23	0.008	(0.4)	0.004	(0.2)	-	-
24	0.003	(0.2)	0.001	(0.1)	-	-
25	0.003	(0.2)	0.001	-	-	-
26	0.003	(0.2)	-	-	-	-
27	0.001	(0.1)	-	-	-	-
28 to 46	-	-	-	-	-	-

[a]The normalized importance (in %) is given in parentheses.

In all but one of the cases, the sensitivities of the gamma responses to the cross sections listed below in Table IV.2 increase in magnitude with increasing ground range. The only exception is the secondary gamma response sensitivity to the nitrogen total neutron removal cross section. This sensitivity, although negative, decreases in magnitude with increasing ground range. Unlike the other sensitivities, this sensitivity comprises both a negative and positive component.

Table IV.2
Total Relative Sensitivities (>0,1) of Nagasaki FIA Tissue Kermas to Cross-Section

Response/Cross-Section	Ground Range (m)		
	700	1000	1500
Prompt Neutron RESPONSE			
Nitrogen elastic scattering	-1.75	-2.48	-3.57
Nitrogen inelastic scattering	---	---	-0.16
Nitrogen neutron removal	-0.51	-0.73	-1.13
Oxygen elastic scattering	-0.57	-0.80	-1.10
Hydrogen elastic scattering	-0.41	-0.49	-0.57
Secondary Gamma RESPONSE			
Nitrogen elastic scattering	---	-0.24	-0.30
Nitrogen neutron removal	-1.08	-1.02	-0.29
Nitrogen gamma incoherent scattering	-2.18	-2.67	-3.40
Nitrogen pair production	-0.19	-0.26	-0.40
Oxygen gamma incoherent scattering	-0.80	-0.97	-1.23
Oxygen pair production	---	---	-0.15
Prompt Gamma RESPONSE			
Nitrogen gamma incoherent scattering	-2.94	-3.06	-3.13
Oxygen gamma incoherent scattering	-0.94	-1.17	-1.18

The negative component arises because of the loss of a neutron, whereas the positive component is due to the occasional production of a secondary gamma. For the 700, 1000, and 1500 m ground ranges, the positive components of the sensitivity coefficients associated with the gamma production are 1.02, 0.91, and 0.81, respectively. Subtracting these values from the respective values listed in Table IV.2 yields negative components of -2.10, -1.93, and -1.73, respectively. Both components, although opposite in sign, decrease in magnitude with increasing ground range. This behavior is due to the relatively low energy of the neutrons emerging from the source, which causes a substantial portion of the secondary gamma production to occur close to the point of burst. For increasing ground ranges, the adjoint flux (or importance of particles) at the point of burst decreases, which leads to a decrease in the magnitude of the sensitivity of the secondary gamma response to the total neutron removal cross section.

Albeit informative, the sensitivities listed in Table IV.2 are not used in the uncertainty calculations because their energy- and spatial-dependence has been integrated over. Since the covariances available for uncertainty analysis are sufficiently detailed to contain spatial- and energy-dependent information, the respective sensitivities must contain an equal level of detail. Thus, the sensitivities used for calculating the response uncertainties are calculated separately for each energy group, spatial- and angular-interval, and are often referred to as "differential" sensitivities; this is in contradistinction to the sensitivities presented in Table IV.2, which are referred to as "integral" sensitivities (actually, the qualifiers "differential" and "integral" are both misnomers, and their use is not recommended).

Examples of space-dependent, but energy-integrated, sensitivities are presented in Table IV.3. Specifically, this table presents relative sensitivities (with absolute value >0.001) for the prompt neutron FIA tissue kerma response at 700 m ground range as a function of the air zones in the A/G transport model. All of the sensitivities are negative; the largest sensitivity is associated with the nitrogen total cross section in each air zone. The largest negative sensitivities appear in the air zones immediately below the point of burst, i.e., axial zone 3 and radial zone 1, and immediately above the 700 m ground range, i.e., axial zone 1 and radial zone 2; this indicates that the largest sensitivities appear along the ray connecting the source and detector.

Table IV.3
Relative Total Cross-Section FIA Sensitivities (>0.001) for Nagasaki Prompt Neutron FIA Tissue Kerma (PNK) Response at 700 m as Function of the Air Zones in the A/G Transport Model Air Zone

Axial Zone	Height Above Ground (m)	Element	Radial Zone[a]	
			1	2
1	0 to 125	Oxygen	-0.031	-0.167
		Nitrogen	-0.105	-0.513
		Hydrogen	-0.016	-0.070
2	125 to 275	Oxygen	-0.134	-0.083
		Nitrogen	-0.434	-0.257
		Hydrogen	-0.044	-0.039
3	275 to 449	Oxygen	-0.183	-0.024
		Nitrogen	-0.637	-0.081
		Hydrogen	-0.058	-0.015
4	449 to 635	Oxygen	-0.057	-0.003
		Nitrogen	-0.243	-0.015
		Hydrogen	-0.030	-0.004
5	635 to 835	Oxygen	---	---
		Nitrogen	-0.009	-0.002
		Hydrogen	-0.003	-0.001
6	635 to 1095	Oxygen	---	---
		Nitrogen	-0.001	---
		Hydrogen	---	---

[a]Radial zone 1 extends from 0 to 512.5 m and radial zone 2 extends from 512.5 to 1012.5 m.

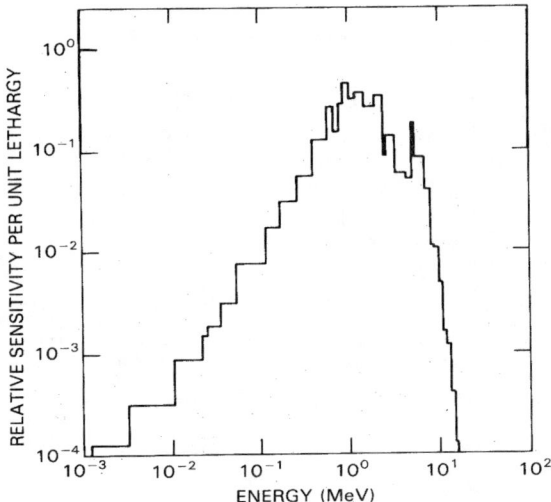

Figure IV.1. Relative sensitivity of Nagasaki prompt neutron FIA tissue kerma at 700-m ground range to the oxygen elastic scattering cross section, integrated over all air zones

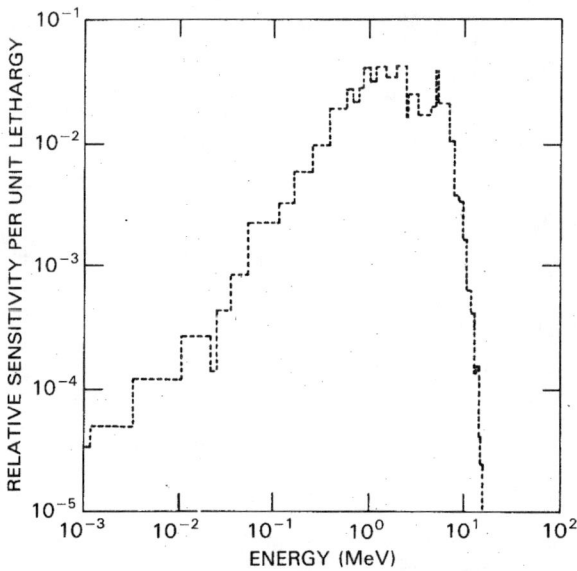

Figure IV.2. Relative sensitivity of Nagasaki prompt neutron FIA tissue kerma at a 700-m ground range to the oxygen elastic scattering cross section, integrated over the ground zone

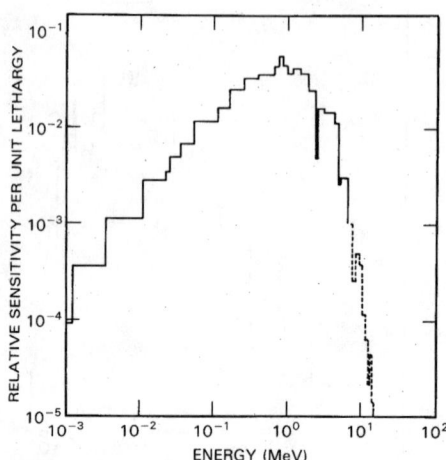

Figure IV.3. Relative sensitivity of Nagasaki prompt neutron FIA tissue kerma at 700-m ground range to the hydrogen elastic scattering cross section, integrated over the ground zone

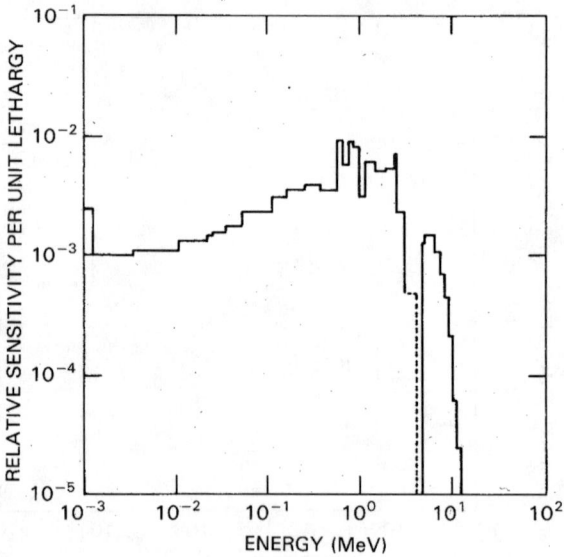

Figure IV.4. Relative sensitivity of Nagasaki secondary gamma FIA tissue kerma at 700-m ground range to the nitrogen elastic scattering cross section, integrated over all air zones

Spatially integrated, but energy-dependent sensitivities per unit lethargy (so-called sensitivity profiles) are shown in Figures IV.1 through IV.4 for the prompt neutron FIA tissue kerma at 700 m ground range. In these figures, solid lines indicate negative sensitivities whereas dashed lines indicate positive sensitivities. The relative sensitivity of the oxygen elastic scattering cross section integrated only over the air zones, on the one hand, and integrated only over the ground zone, on the other hand, is shown in Figures IV.1 and IV.2, respectively. Adding the profiles in these two figures and integrating the resulting profile over the entire energy range yields the value of -0.57 for the integrated sensitivity presented previously in Table IV.2. The sensitivity profile in the air is negative at all energies whereas the sensitivity profile in the ground is positive. In air, the negative profile results because an increase in the oxygen elastic scattering cross section at any energy produces more neutron down-scattering, which increases neutron absorption, so that fewer neutrons reach the detector. In the ground, this negative effect also exists, but it is outweighed by the positive effect associated with the increased probability that a neutron will reenter the air and contribute to the detector response.

The relative sensitivity of the hydrogen elastic scattering cross section integrated only over the ground zone is shown in Figure IV.3. Unlike the all-positive profile for oxygen in the ground, the hydrogen profile is both negative and positive. This behavior is caused by the greater decrease in average neutron energy due to collisions, in hydrogen. For spherically symmetric scattering (i.e., isotropic scattering in the center-of-mass), the average energy loss per collision in hydrogen is approximately 5 times greater than the average energy loss per collision in oxygen. Thus, for changes in the hydrogen elastic scattering cross section in the ground, the positive contribution to the sensitivity associated with the increased probability that a neutron will reenter the air begins to outweigh the corresponding negative contribution only at higher energies.

Figure IV.4 depicts the relative sensitivities of the secondary gamma FIA tissue kerma to the nitrogen elastic scattering cross section integrated over all air zones. The profile is negative at all energies except those between 2.5 and 4 MeV. Neutrons scattering in nitrogen in this energy range have a good chance of landing in the energy range that contains the minimum of the oxygen elastic scattering cross section, which is centered at 2.3 MeV. These neutrons can travel large distances before undergoing further scatterings and thus have a greater chance of producing gammas close to the detector.

IV.G.2. Uncertainty Analysis

The covariance matrices for the Nagasaki FIA tissue kermas due to the uncertainties in the cross sections are obtained by using the "sandwich" formula given in Eq. (III.F.16) to construct the matrix product of the sensitivity profiles, generated as described above for each partial cross section, with the energy-dependent covariance matrix for that cross section. Thus, for N responses and M input parameters, this response covariance is written in matrix notation as

$$V(R) = S_{Rq} V(q) S_{Rq}^T .\qquad \text{(IV.G.8)}$$

In the above expression, the diagonal elements of $V(R)$ represent the variance of the N responses, and its off-diagonal elements represent the covariance between the response uncertainties. The $(ij)^{\text{th}}$-element of the $M \times N$ matrix S_{Rq} of sensitivity coefficients denotes the relative sensitivity of response j to input parameter i. The matrix $V(q)$, of dimension $M \times M$, represents the covariance matrix of the input parameters. For the neutron cross sections, the elements of $V(q)$ were processed from the basic ENDF/B-V data covariance files. The uncertainties used for the nitrogen gamma yields varied by gamma energy group between 5 and 15%. Finally, gamma cross sections are fairly well known, and since the ENDF file does not contain uncertainty data for them, an uncertainty of 1% was assumed.

To obtain the uncertainties due to geometrical and numerical modeling, the respective modeling parameters are treated as fluence bias factors, b_i, defined as

$$b_i \equiv \varphi_i(\xi)/\varphi_{i-1}(\xi) .\qquad \text{(IV.G.9)}$$

For N bias factors, $\varphi_0 \equiv \varphi(\xi)$ is the unbiased flux, and φ_N is the biased flux $\varphi_b(\xi)$. In terms of the unbiased flux, the biased flux is given by

$$\varphi_b(\xi) = \prod_i b_i \varphi(\xi) ,\qquad \text{(IV.G.10)}$$

while the biased response R_b is given by

$$R_b = \left\langle \Sigma(\xi) \prod_i b_i \, \varphi(\xi) \right\rangle ,\qquad \text{(IV.G.11)}$$

which may also be written as

$$R_b = \prod_i b_i \langle \Sigma(\xi) \varphi(\xi) \rangle = \prod_i b_i R .\qquad \text{(IV.G.12)}$$

The constants b_i, as defined by Eq. (IV.G.9), are simply multipliers of the fluence. This procedure yields relative sensitivity coefficients equal to unity for the fluence bias factors. In this manner, a value of unity can be assigned to each

b_i, and the uncertainty in the modeling parameters can then be used to determine the uncertainty in b_i. Table IV.4 presents the nominal values of the modeling parameters considered in this study, together with the uncertainties, representing one standard deviation, associated with each of these modeling parameters. The modeling parameter designated as "A/G transport" in Table IV.4 represents the computational method employed to solve the forward and adjoint transport equations for obtaining the forward and adjoint fluences, respectively. Its inclusion as a modeling parameter facilitates the consideration of uncertainties that arise because of the finite number of energy groups, angular segments, and spatial mesh points employed to represent the continuous energy, angle, and volume in the phase-space associated with the transport operator. This parameter also accounts for uncertainties in computer coding methods, cross-section treatments, and energy-angle regrouping of the source.

Table IV.4
Uncertainty Data (Other Than Cross Sections) Used in the Evaluation of the Nagasaki FIA Tissue Kerma Uncertainty

Modeling Parameter	Nagasaki	
	Nominal Value	Estimated Uncertainty
Weapon yield (kt)	21	±4.3%
Radiation source (particle/kt)	Neutron: 1.6465 + 23 [a] Gamma: 3.7916 + 23	±10% ±25%
Hypocenter (m)	0.0	±6.7%
HOB (m)	503	±3.3%
Air density (g/cm^3)	At HOB: 1.094 − 3 At ground: 1.137 − 3	±0.5% ±0.0%
Water density (g/cm^3)	At HOB: 1.600 − 5 At ground: 2.017 − 5	±5% ±0.0%
Soil density (g/cm^3)	Soil: 1.3 Water: 0.4	±5% ±15%
A/G transport	0 to 200 m [b] 200 to 2000 m	±20% ±10%

[a] Read as $1.6465 \ast 10^{23}$.
[b] Value represents ground range limits for uncertainty.

The uncertainties in the yield, radiation source, and A/G transport modeling parameters propagate directly (i.e., with relative sensitivity of unity) to the fluence and hence to the kerma response. The other modeling parameters, namely the air, water, and soil density parameters, are treated in the same manner as cross section parameters, by using the "sandwich" formula to determine their respective contributions to the kerma uncertainty. The hypocenter and height-of-burst (HOB) modeling parameters require a determination of the variation of the kerma values with range. Variations in both the hypocenter and HOB imply variations in the effective ground range and hence variations in the kerma values. The kerma versus range curves were used to estimate the uncertainties in the kerma due to uncertainties in the location of the hypocenter and HOB.

The kermas due to prompt neutrons, secondary gammas, and prompt gammas were obtained for 700, 1000, and 1500 m ground ranges. Therefore, correlations between the uncertainties in the three responses at each of the three ground ranges had to be determined for the parameters listed in Table IV.4. For the yield, HOB, and air and water densities, full correlations were assumed over all responses at all ground ranges. This assumption is appropriate due to the global nature of changes in these parameters; specifically, a reasonably small variation in each of these parameters produces a constant relative change in the kermas at all locations; hence there is full correlation in the kerma uncertainties due to these parameter uncertainties. A similar situation occurs for the radiation sources in that, for a given response, the correlations by ground range are equal to unity. The prompt neutron and secondary gamma responses due to the radiation source uncertainties are also fully correlated, since the prompt neutrons produce secondary gammas. On the other hand, the correlation between these two responses and the prompt gamma response is zero. The uncertainties due to the soil density uncertainty were considered to be uncorrelated, since the soil density uncertainty simply describes random variations in the soil density at various locations.

The correlations between response type and location cannot be determined accurately for the A/G transport and hypocenter location parameters. Since particle attenuation can be considered to vary nearly exponentially with slant range and, thus, ground range, the correlations for the A/G transport parameter were cast in an exponential form. Finally, the uncertainty correlations as a function of range for the hypocenter location were considered, based on expert opinion, to be small but nonzero; consequently, a correlation value of 0.1 was chosen for the hypocenter location.

The calculated absolute [i.e., normalized per kiloton (kt)] and fractional FIA tissue kermas due to prompt neutrons, secondary gammas, and prompt gammas, as obtained from the forward A/G transport calculations, are presented in Table IV.5. In addition, Table IV.5 also presents the uncertainties associated with each prompt response and with the total prompt response at each ground range. These results indicate that as the ground range increases, the fractional neutron and prompt gamma kermas decrease whereas the secondary gamma kerma increases.

Table IV.5
Calculated Absolute and Fractional FIA Tissue Kermas (Forward A/G Transport Calculations) and Overall Uncertainties at Nagasaki

Ground Range (m)	Response			Total
	Prompt Neutron	Secondary Gamma	Prompt Gamma	
Tissue Kerma (Gy/kt)				
700	3.82-2^a	4.67-1	1.52-1	6.57-1
1000	6.05-3	1.41-1	2.93-2	1.77-1
1500	2.88-4	1.97-2	3.01-3	2.30-2
Fraction of Total Prompt Tissue Kerma				
700	0.0579	0.7101	0.2320	1.00
1000	0.0342	0.7999	0.1659	1.00
1500	0.0125	0.8569	0.1306	1.00
Fractional Standard Deviation (%)				
700	17.03	16.69	27.68	14.71
1000	18.50	16.65	27.75	14.94
1500	22.28	16.69	27.74	15.27

^{a}Read as $3.82 * 10^{-2}$.

The fractional standard deviations are ~20% for the prompt neutron and secondary gamma responses and ~30% for the prompt gamma responses at both cities. The larger uncertainty associated with the prompt gamma responses is due almost entirely to the larger uncertainty on the prompt gamma source. In general, the fractional standard deviations on the secondary and prompt gamma responses remain constant with increasing ground range and increase on the prompt neutron responses with increasing ground range. To understand this behavior, it is necessary to examine the individual contributors to the overall uncertainty.

In Table IV.6, the sources of uncertainty in each response at each ground range have been grouped into three categories. These categories were chosen to allow differentiation between the uncertainties arising from the uncertainties in the basic nuclear data, i.e., cross sections, nuclear densities, and modeling parameters. Since the uncertainties in the responses due to the uncertainties in the air, water vapor, and soil densities and the uncertainties in the nitrogen gamma yields were obtained using the cross section sensitivity coefficients, these contributions were placed in the same category.

The reason for the behavior of the uncertainties in Table IV.5 becomes readily apparent upon examination of the data in Table IV.6. The total uncertainties for all three responses, at all three ground ranges, are dominated by the uncertainties due to the modeling parameters; these uncertainties decrease very slightly with increasing ground range.

Table IV.6
Contributions to Overall Uncertainty in Nagasaki Prompt FIA Tissue Kerma Due to Cross-Section, Density, and Modeling Uncertainties

Ground Range (m)	Fractional of Total Prompt Kerma (%)	Fractional Standard Deviations (%)				Ground Range Correlation Matrices
		Cross Section	Density[a]	Modeling[b]	Total	
Prompt Neutrons						
700	5.79	7.23	2.34	15.24	17.03	1.00 0.78 0.65
1000	3.42	10.21	2.56	15.22	18.50	0.78 1.00 0.76
1500	1.25	15.97	3.17	15.20	22.28	0.65 0.76 1.00
Secondary Gammas						
700	71.01	5.86	3.48	15.24	16.69	1.00 0.78 0.65
1000	79.99	5.84	3.43	15.22	16.65	0.78 1.00 0.70
1500	85.69	5.90	3.59	15.20	16.69	0.65 0.70 1.00
Prompt Gammas						
700	23.20	2.61	1.36	27.52	27.68	1.00 0.92 0.87
1000	16.59	3.28	1.61	27.50	27.75	0.92 1.00 0.89
1500	13.06	3.35	1.49	27.50	27.74	0.87 0.89 1.00
Total						
700	100.0	4.47	2.59	13.77	14.71	1.00 0.78 0.66
1000	100.0	4.98	2.86	13.83	14.94	0.78 1.00 0.71
1500	100.0	5.34	3.17	13.93	15.27	0.66 0.71 1.00

[a]Density represents contributions due to air, water vapor, and soil density uncertainties and the contribution due to nitrogen gamma yield uncertainty.

[b]Modeling represents contributions due to weapon yield, neutron and gamma source, HOB, hypocenter, and A/G transport uncertainties.

This decrease occurs because the sensitivities associated with the weapon yield, neutron and gamma source, hypocenter, and A/G transport are constant over the three ground ranges considered here, but the sensitivity associated with the HOB decreases very slightly with increasing ground range.

Table IV.7
Comparison of Response Uncertainties and FIA Ground Range Correlation Matrices for Nagasaki Prompt FIA Tissue Kerma With and Without Contribution of Neutron and Gamma Source Uncertainties

Ground Range (m)	Fractional Standard Deviations (%)				Ground Range Correlation Matrices	
	Modeling[a]		Total			
	With[b]	Without[c]	With	Without	With	Without
Prompt Neutrons						
700	15.24	11.50	17.03	13.78	1.00 0.78 0.65	1.00 0.68 0.53
1000	15.22	11.47	18.50	15.57	0.78 1.00 0.76	0.68 1.00 0.69
1500	15.20	11.45	22.28	19.91	0.65 0.76 1.00	0.53 0.69 1.00
Secondary Gammas						
700	15.24	11.50	16.69	13.36	1.00 0.78 0.65	1.00 0.65 0.45
1000	15.22	11.47	16.65	13.32	0.78 1.00 0.70	0.65 1.00 0.54
1500	15.20	11.45	16.69	13.37	0.65 0.70 1.00	0.45 0.54 1.00
Prompt Gammas						
700	27.52	11.50	27.68	11.87	1.00 0.92 0.87	1.00 0.56 0.31
1000	27.50	11.47	27.75	12.04	0.92 1.00 0.89	0.56 1.00 0.43
1500	27.50	11.45	27.74	12.02	0.87 0.89 1.00	0.31 0.32 1.00
Total						
700	13.77	9.82	14.71	11.09	1.00 0.78 0.66	1.00 0.64 0.46
1000	13.83	10.19	14.94	11.70	0.78 1.00 0.71	0.64 1.00 0.55
1500	13.93	10.42	15.27	12.12	0.66 0.71 1.00	0.46 0.55 1.00

[a]Modeling represents contributions due to weapon yield, HOB, hypocenter, and A/G transport uncertainties.

[b]**WITH**: represents modeling contributions *with* inclusion of neutron and gamma source uncertainties.

[c]**WITHOUT**: represents modeling contributions *without* inclusion of neutron and gamma source uncertainties.

For the prompt neutron responses, the contributions to the uncertainties from cross section uncertainties are substantial and increase with increasing ground range. For the gamma responses, the contributions to the uncertainties from cross section uncertainties are small and thus only the prompt neutron response uncertainties increase with increasing ground range. Although the cross section uncertainties only contribute substantially to the prompt neutron kerma uncertainties, and although the prompt neutrons contribute only a small portion of the total prompt kerma, their contribution to biological dose is significant.

Because the uncertainties associated with the prompt neutron and prompt gamma sources are very large, especially the 25% uncertainty on the prompt gamma source, the uncertainty analysis was redone without including the source uncertainties in the respective calculations. The results of this analysis are shown in Table IV.7. By not considering the source uncertainties, there was an ~4% decrease in the prompt neutron and secondary gamma response uncertainties, and a 16% decrease in the prompt gamma uncertainties.

Chapter V

LOCAL SENSITIVITY AND UNCERTAINTY ANALYSIS OF NONLINEAR SYSTEMS

This chapter continues the presentation of local sensitivity analysis of system responses around a chosen point or trajectory in the combined phase space of parameters and state variables. In contradistinction to the problems considered in Chapter IV, though, the operators considered in this chapter act nonlinearly not only on the system's parameters but also on the dependent variables, and include linear and/or nonlinear feedback. Consequently, it will no longer be possible to solve the adjoint equations underlying the *ASAP* independently of the original (forward) nonlinear equations. This is the fundamental difference to the problems considered previously in Chapter IV.

V.A. SENSITIVITY ANALYSIS OF NONLINEAR SYSTEMS WITH FEEDBACK AND OPERATOR-TYPE RESPONSES

The physical system considered in this chapter is modeled mathematically by means of K coupled ***nonlinear*** equations represented in operator form as

$$N[u(x),\alpha(x)] = Q[\alpha(x)], \quad x \in \Omega, \qquad (V.A.1)$$

where:

1. $x = (x_1,\ldots,x_J)$ denotes the phase-space position vector; $x \in \Omega \subset \mathbb{R}^J$, where Ω is a subset of the J-dimensional real vector space $\mathbb{R}^J$;

2. $u(x) = [u_1(x),\ldots,u_K(x)]$ denotes the vector of dependent (i.e., state) variables; $u(x) \in \mathscr{E}_u$, where $\mathscr{E}_u$ is a normed linear space over the scalar field $\mathscr{F}$ of real numbers;

3. $\alpha(x) = [\alpha_1(x),\ldots,\alpha_I(x)]$ denotes the vector of system parameters; $\alpha \in \mathscr{E}_\alpha$, where $\mathscr{E}_\alpha$ is also a normed linear space; in usual applications, $\mathscr{E}_\alpha$ may be one of the Hilbert spaces $\mathscr{L}_2$ or l_2; occasionally, the components of α may simply be a set of real scalars, in which case $\mathscr{E}_\alpha$ is $\mathbb{R}^I$;

4. $Q[\alpha(x)] = [Q_1(\alpha),\ldots,Q_K(\alpha)]$ denotes a (column) vector whose elements represent inhomogeneous source terms that depend either

linearly or nonlinearly on α; $Q \in \mathcal{E}_Q$, where $\mathcal{E}_Q$ is again a normed linear space; the components of Q may be operators (rather than just functions) acting on $\alpha(x)$ and x;

5. $N \equiv [N_1(u,\alpha),...,N_K(u,\alpha)]$ is a K-component column vector whose components are, in general, *nonlinear* operators (including differential, difference, integral, distributions, and/or infinite matrices) of u and α.

In view of the definitions given above, N represents the mapping $N : \mathcal{D} \subset \mathcal{E} \to \mathcal{E}_Q$, where $\mathcal{D} = \mathcal{D}_u \times \mathcal{D}_\alpha$, $\mathcal{D}_u \subset \mathcal{E}_u$, $\mathcal{D}_\alpha \subset \mathcal{E}_\alpha$, and $\mathcal{E} = \mathcal{E}_u \times \mathcal{E}_\alpha$. Note that an arbitrary element $e \in \mathcal{E}$ is of the form $e = (u, \alpha)$. Even though in most practical applications $\mathcal{E}$ and $\mathcal{E}_Q$ will be Hilbert spaces (e.g., the space $\mathcal{L}_2$, the Sobolev spaces $\mathcal{H}^m$), this restriction is not imposed at this stage for the sake of generality. If differential operators appear in Eq. (V.A.1), then a corresponding set of boundary and/or initial conditions (which are essential to define $\mathcal{D}$) must also be given. The respective boundary conditions are represented as

$$[B(u,\alpha) - A(\alpha)]_{\partial\Omega} = 0, \quad x \in \partial\Omega, \qquad (V.A.2)$$

where A and B are nonlinear operators while $\partial\Omega$ is the boundary of Ω.

The vector-valued function $u(x)$ is considered to be the unique nontrivial solution of the physical problem described by Eqs. (V.A.1) and (V.A.2). The system response (i.e., performance parameter) $R(u,\alpha)$ associated with the problem modeled by Eqs. (V.A.1) and (V.A.2) is a phase-space dependent mapping that acts nonlinearly on the system's state vector u and parameters α, and is represented in operator form as

$$R(e): \mathcal{D}_R \subset \mathcal{E} \to \mathcal{E}_R, \qquad (V.A.3)$$

where $\mathcal{E}_R$ is a normed vector space.

The nominal parameter values $\alpha^0(x)$ are used in Eqs. (V.A.1) and (V.A.2) to obtain the nominal solution $u^0(x)$ by solving

$$N(u^0, \alpha^0) = Q(\alpha^0), \quad x \in \Omega, \qquad (V.A.4)$$
$$B(u^0, \alpha^0) = A(\alpha^0), \quad x \in \partial\Omega. \qquad (V.A.5)$$

Once the nominal values $u^0(x)$ have been obtained by solving the above equations, they are used together with the nominal parameter values $\alpha^0(x)$ in Eq. (V.A.3) to obtain the nominal response value $R(u^0, \alpha^0)$.

Feedback is introduced into the model by allowing some (or all) of the parameters α to depend on some (or all) of the components of u. Without loss of generality, a feedback mechanism can be specified by adding an operator, $F(u)$, to the nominal parameter values α^0. Thus, in the presence of feedback, the values of the parameters become $\alpha^0 + F(u)$; correspondingly, the solution u^f of the system with feedback will satisfy the equations

$$N[u^f, \alpha^0 + F(u^f)] = Q[\alpha^0 + F(u^f)], \quad x \in \Omega, \quad (V.A.6)$$

$$B[u^f, \alpha^0 + F(u^f)] = A[\alpha^0 + F(u^f)], \quad x \in \partial\Omega. \quad (V.A.7)$$

The system response with feedback becomes $R[u^f, \alpha^0 + F(u^f)]$. The difference

$$R[u^f, \alpha^0 + F(u^f)] - R(u^0, \alpha^0), \quad x \in \mathscr{D}_R, \quad (V.A.8)$$

gives the actual effect of the feedback F on the response $R(u^0, \alpha^0)$, and can be calculated exactly only by solving Eqs. (V.A.6) and (V.A.7) anew for each F, and then using the respective solution u^f to evaluate $R[u^f, \alpha^0 + F(u^f)]$.

The sensitivity, DR_F, of the response with feedback, $R[u^f, \alpha^0 + F(u^f)]$, to parameters and feedback variations around the nominal response value $R(u^0, \alpha^0)$ is given by the Gâteaux-differential

$$DR_F \equiv \left\{ \frac{d}{d\varepsilon} R[u^0 + \varepsilon h, \alpha^0 + \varepsilon F(u^0 + \varepsilon h)] \right\}_{\varepsilon=0}, \quad (V.A.9)$$

where $h \equiv u^f - u$. For most practical applications, performing the operations indicated in Eq. (V.A.9) gives

$$DR_F = R_1'(u^0, \alpha^0) h + R_2'(u^0, \alpha^0) F(u^0), \quad x \in \Omega_R, \quad (V.A.10)$$

where R_1' and R_2' denote, respectively, the Gâteaux-derivatives of $R(u, \alpha)$ with respect to its first and second arguments.

V.A.1. The Forward Sensitivity Analysis Procedure (FSAP)

The vector of variations, $h \equiv u^f - u$, is needed in order to compute the sensitivity DR_F obtained in Eq. (V.A.10) above. To first order variations in the feedback and/or parameters, the variations $h \equiv u^f - u$ are obtained by solving the equations obtained by taking the Gâteaux-differentials of Eqs. (V.A.6) and (V.A.7) at (u^0, α^0), namely

$$N_1'(u^0, \alpha^0)h = [Q'(\alpha^0) - N_2'(u^0, \alpha^0)] F(u^0), \quad x \in \Omega, \quad (V.A.11)$$

$$B_1'(u^0, \alpha^0)h = [A'(\alpha^0) - B_2'(u^0, \alpha^0)] F(u^0), \quad x \in \partial\Omega. \quad (V.A.12)$$

Equations (V.A.11) and (V.A.12) are the *forward sensitivity equations* (*FSE*). In principle, the sensitivity DR_F can be evaluated once Eqs. (V.A.11) and (V.A.12) have been solved to determine h. However, these equations would have to be solved anew for each F, and this becomes prohibitively expensive if many effects of all possible feedback and parameter variations are to be analyzed.

V.A.2. Adjoint (Local) Sensitivity Analysis Procedure (ASAP)

The *Adjoint Sensitivity Analysis Procedure* (*ASAP*) is the alternative method for calculating the sensitivity DR_F, and the *ASAP* circumvents the need for repeatedly solving Eqs. (V.A.11) and (V.A.12). To begin with, we note that the second term on the right-side of Eq. (V.A.10) can be calculated directly, with little effort, since it does not depend on h (this is the so-called "direct effect" term). To calculate the first term on the right-side of Eq. (V.A.10), i.e., the "indirect-effect" term, the spaces $\mathcal{E}_u$, $\mathcal{E}_Q$, and $\mathcal{E}_R$ are henceforth considered to be Hilbert spaces and denoted as $\mathcal{H}_u(\Omega)$, $\mathcal{H}_Q(\Omega)$, and $\mathcal{H}_R(\Omega_R)$, respectively. The elements of $\mathcal{H}_u(\Omega)$ and $\mathcal{H}_Q(\Omega)$ are, as before, vector functions defined on the open set $\Omega \subset \mathbb{R}^J$, with a smooth boundary $\partial\Omega$. The elements of $\mathcal{H}_R(\Omega_R)$ are vector- or scalar-valued functions defined on the open set $\Omega_R \subset \mathbb{R}^m, 1 \leq m \leq J$, with a smooth boundary $\partial\Omega_R$. Of course, if $J=1$, then $\partial\Omega$ merely consists of two endpoints; similarly, if $m=1$, then $\partial\Omega_R$ consists of two endpoints only. The inner products on $\mathcal{H}_u(\Omega)$, $\mathcal{H}_Q(\Omega)$, and $\mathcal{H}_R(\Omega_R)$ are denoted by $\langle \bullet, \bullet \rangle_u$, $\langle \bullet, \bullet \rangle_Q$, and $\langle \bullet, \bullet \rangle_R$, respectively.

Since $R_1'(u^0, \alpha^0)h \in \mathcal{H}_R(\Omega_R)$, it follows that

$$R_1'(u^0,\alpha^0)h = \sum_{s\in\mathcal{S}} \langle p_s, R_1' h\rangle_R \, p_s,\qquad\text{(V.A.13)}$$

where p_s is an orthonormal basis for $\mathcal{H}_R(\Omega_R)$, $\langle\,,\,\rangle_R$ denotes the inner product in $\mathcal{H}_R$, and the series $\sum_{s\in\mathcal{S}}$ converges unconditionally over the nonzero elements in $\mathcal{S}$ (which could be infinitely many).

The functionals $\langle p_s, R_1' h\rangle_R$ are the Fourier coefficients of $R_1' h$ with respect to the basis p_s. In practice, the basis p_s is often chosen to be a set of orthogonal polynomials, particularly Chebyshev polynomials. Note that since the operator $R_1' h$ is linear in h, the functionals $\langle p_s, R_1' h\rangle_R$ are also linear in h. Furthermore, since $R_1' h \in L(\mathcal{H}_u(\Omega), \mathcal{H}_R(\Omega_R))$ and since Hilbert spaces are self-dual, it follows that $R_1'(u^0,\alpha^0)$ admits a unique adjoint operator, denoted here as the linear operator $M(u^0,\alpha^0)\in L(\mathcal{H}_R(\Omega_R), \mathcal{H}_u(\Omega))$, and defined, as customary, in terms of inner products by means of the relationship:

$$\langle p_s, R_1'(u^0,\alpha^0)h\rangle_R = \langle M(u^0,\alpha^0)p_s, h\rangle_u,\quad\text{for every } s\in\mathcal{S}.\quad\text{(V.A.14)}$$

Equations (V.A.11) and (V.A.12) reveal that the Gâteaux-derivatives $N_1' h$ and $B_1' h$ are linear in h. It is therefore possible to introduce, for every vector $\psi_s \in \mathcal{H}_Q(\Omega)$, the operator adjoint to $N_1' h$, by means of the relation

$$\langle \psi_s, N_1'(u^0,\alpha^0)h\rangle_Q = \langle L^+(u^0,\alpha^0)\psi_s, h\rangle_u + \{P(h,\psi_s)\}_{\partial\Omega},\quad s\in\mathcal{S},\quad\text{(V.A.15)}$$

where the $K\times K$ matrix $L^+ = [L_{ji}^+]$, $(i,j=1,\dots,K)$, obtained by transposing the formal adjoints L_{ij}^+ of the operators $(N_1')_{ij}$, is the formal adjoint of $N_1'(u^0,\alpha^0)$, while $\{P(h,\psi_s)\}_{\partial\Omega}$ represents the associated bilinear form evaluated on $\partial\Omega$. The domain of L^+ is determined by selecting adjoint boundary conditions, represented here in operator form as

$$\{B^+(u^0,\alpha^0;\psi_s) - A^+(\alpha^0)\}_{\partial\Omega} = 0,\quad s\in\mathcal{S}.\qquad\text{(V.A.16)}$$

These adjoint boundary conditions are determined by requiring that: (a) Eq. (V.A.16) must be independent of h and F, and (b) substitution of Eqs.

(V.A.16) and (V.A.12) in $\{P(h,\psi_s)\}_{\partial\Omega}$ must cause all terms containing unknown values of h to vanish. This selection of the adjoint boundary conditions reduces $\{P(h,\psi_s)\}_{\partial\Omega}$, to a quantity, denoted here as $\hat{P}(F,\psi_s,u^0,\alpha^0)$, that contains only known values of F,ψ_s,u^0, and α^0. In particular, $\hat{P}$ may vanish as a result of this selection of B^+ and A^+. Furthermore, this selection of B^+ and A^+, and the subsequent reduction of $\{P(h,\psi_s)\}_{\partial\Omega}$, to $\hat{P}(F,\psi_s,u^0,\alpha^0)$ also reduces Eq. (V.A.15) to

$$\langle \psi_s, N_1'(u^0,\alpha^0)h \rangle_Q - \hat{P}(F,\psi_s,u^0,\alpha^0) = \langle L^+(u^0,\alpha^0)\psi_s, h \rangle_u, \quad s \in \mathcal{S}. \quad \text{(V.A.17)}$$

Requiring now that the first terms on the right-sides of Eqs. (V.A.17) and (V.A.14), respectively, represent the same functional of h yields the relation

$$L^+(u^0,\alpha^0)\psi_s = M(u^0,\alpha^0)p_s, \quad s \in \mathcal{S}, x \in \Omega, \quad \text{(V.A.18)}$$

which holds uniquely by virtue of the Riesz representation theorem. Furthermore, using Eqs. (V.A.18) and (V.A.17) in Eq. (V.A.14) reduces the latter to

$$\langle p_s, R_1' h \rangle_R = \langle \psi_s, N_1' h \rangle_Q - \hat{P}(F,\psi_s,u^0,\alpha^0), \quad s \in \mathcal{S}. \quad \text{(V.A.19)}$$

Replacing now $N_1' h$ in Eq. (V.A.19) by the expression on the right-side of Eq. (V.A.11), substituting the resulting expression into the right-side of Eq. (V.A.13), and using Eq. (V.A.10) yields the following expression for the sensitivity $DR_F(u^0,\alpha^0,F)$ of the response R to the feedback F:

$$DR_F = R_2'(u^0,\alpha^0)F(u^0)$$
$$+ \sum_{s \in \mathcal{S}} \left\{ \langle \psi_s, [Q'(\alpha^0) - N_2'(u^0,\alpha^0)]F(u^0) \rangle_Q - \hat{P}(F,\psi_s,u^0,\alpha^0) \right\} p_s. \quad \text{(V.A.20)}$$

It is important to note that the unknown (forward) functions h do not appear in the above expression for the local sensitivities DR_F, which means that the need to solve Eqs. (V.A.11) and (V.A.12) has been eliminated. Instead, it is necessary to compute as many adjoint functions ψ_s by solving Eqs. (V.A.18) subject to the adjoint boundary conditions represented by (V.A.16) as there are nonzero terms retained in the summation representation in Eq. (V.A.20). In practice, this number of nonzero terms (and hence the number of required adjoint functions) is dictated by accuracy considerations, and varies from application to application. Note also that for systems without feedback, namely when the parameter variations are independent of the state variables u and hence do not induce feedback via u^f, then Eq. (V.A.20) reduces to

$$DR_F = R'_\alpha(u^0,\alpha^0)h_\alpha$$
$$+ \sum_{s \in \mathscr{S}} \left\{ \left\langle \psi_s, [Q'(\alpha^0) - N'_\alpha(u^0,\alpha^0)]h_\alpha \right\rangle_Q - \hat{P}(F,\psi_s,u^0,\alpha^0) \right\} p_s, \quad \text{(V.A.21)}$$

where $h_\alpha \equiv \alpha - \alpha^0$ denotes the I-component (column) vector of parameter variations around the nominal values α^0. Of course, *when R is a functional of u and α, then the summation in Eq. (V.A.20) reduces to a single term, and only a single adjoint function would need to be calculated for obtaining DR_F*.

The customary applications of DR_F as given in Eq. (V.A.20) are for: (a) ranking the importance of feedbacks or parameter variations in affecting the response R; (b) assessing the first-order local variations in R caused by feedback and/or parameter changes F by means of the functional Taylor expansion

$$R[u^f,\alpha^0 + F(u^f)] - R(u^0,\alpha^0) = DR_F(u^0,\alpha^0,F) + H.O.T(h,F) \quad \text{(V.A.22)}$$

where $H.O.T(h,F)$ denotes higher (than first-) order terms in $\|h\|$ and $\|F\|$; (c) uncertainty analysis (deterministic, response surface, Monte Carlo, etc.); (d) system improvements and/or design; (e) inclusion of experimental information (for example, via data adjustment procedures) to obtain best estimate codes; and (f) reduction of uncertainties in models.

V.B. EXAMPLE: A PARADIGM CLIMATE MODEL WITH FEEDBACK

The sensitivity theory presented in the previous Section will now be applied to analyze a paradigm one-dimensional climate model with feedback, described by a first-order nonlinear ordinary differential equation. The material presented in this Section is largely based on the work by Cacuci and Hall, *J. Atm. Sci.*, 41, 2063, 1984 (with permission). The paradigm climate model considered for illustrative purposes in the sequel is described by the equations

$$\begin{cases} du/dt + \alpha_1 u^4 + \alpha_2 = 0 \\ u(a) - u_a = 0 \end{cases} \quad \text{(V.B.1)}$$

For this model, the independent variable is the time t, which varies from an initial time $t = a$ to a final time $t = b$ (which may be infinite), the dependent variable is the temperature $u(t)$, and the two system parameters $\boldsymbol{\alpha} = (\alpha_1, \alpha_2)$ are constants that depend on the physical properties of the system, i.e., heat capacity, incident radiation, albedo, and emissivity. The initial value of u at $t = a$ is u_a, which, for simplicity, is not allowed to vary in this model.

As an illustrative response for the model described by Eq. (V.B.1), we consider the average longwave radiation, which is proportional to the functional

$$R(u, \boldsymbol{\alpha}) = \int_a^b dt \, (\alpha_1 u^4). \tag{V.B.2}$$

In the illustrative climate model, feedback can be introduced by allowing the emissivity to depend on temperature. For example, when the value of α_1 is allowed to be $\alpha_1^0 + \lambda(u - u_a)$, where λ is a constant specifying the strength of the feedback, then the feedback operator $\boldsymbol{F}(u)$ is the two-component vector

$$\boldsymbol{F}(u) = [\lambda(u - u_a), \, 0]. \tag{V.B.3}$$

The solution u^f with feedback will satisfy the equations

$$\frac{du^f}{dt} + [\alpha_1^0 + \lambda(u^f - u_a)](u^f)^4 + \alpha_2^0 = 0$$
$$u^f(a) - u_a = 0$$

while the response, expressed by Eq. (V.B.2), with feedback becomes

$$R[u^f, \boldsymbol{\alpha}^0 + \boldsymbol{F}(u^f)] = \int_a^b dt \, [\alpha_1^0 + \lambda(u^f - u_a)](u^f)^4.$$

The difference

$$R[u^f, \boldsymbol{\alpha}^0 + \boldsymbol{F}(u^f)] - R(u^0, \boldsymbol{\alpha}^0) \tag{V.B.4}$$

gives the actual effect of the feedback $\boldsymbol{F}$ on the response $R(u^0, \boldsymbol{\alpha}^0)$. In practice, this difference can be evaluated exactly only by introducing feedback into the model explicitly, calculating u^f, and reevaluating the result. For more complex climate models (e.g., atmospheric general circulation models), rerunning the model repeatedly becomes prohibitively expensive.

The sensitivity of the response R defined by Eq. (V.B.2) to the feedback F defined by Eq. (V.B.3) is given by the DR_F, which is obtained by applying the definition given in Eq. (V.A.9) to Eq. (V.B.2). This yields:

$$DR_F = \left\{ (d/d\varepsilon) \int_a^b dt \left[\alpha_1^0 + \varepsilon\lambda \left(u^0 + \varepsilon h - u_a\right)\right] \left(u^0 + \varepsilon h\right)^4 \right\}_{\varepsilon=0}$$
$$= \int_a^b dt \left[\alpha_1^0 \, 4(u^0)^3 h \right] + \int_a^b dt \left[(u^0)^4 \lambda(u^0 - u_a) \right]. \quad \text{(V.B.5)}$$

Note that, for this model, R_1' is the operator

$$R_1'(*) = \int_a^b dt \left[\alpha_1^0 \, 4(u^0)^3 (*) \right], \quad \text{(V.B.6)}$$

while R_2' is the operator

$$R_2'(*) = \left[\int_a^b dt \left[(u^0)^4 \right], 0 \right] \bullet (*). \quad \text{(V.B.7)}$$

The *Forward Sensitivity Equations*, namely the counterparts of Eqs. (V.A.11) and (V.A.12), are obtained by calculating the G-differentials of Eq. (V.B.1); this gives

$$\left((d/d\varepsilon) \left\{ (d/dt)(u^0 + \varepsilon h) + \left[\alpha_1^0 + \varepsilon\lambda\left(u^0 + \varepsilon h - u_a\right)\right]\left(u^0 + \varepsilon h\right)^4 + \alpha_2^0 \right\} \right)_{\varepsilon=0}$$
$$= \left[d/dt + \alpha_1^0 \, 4(u^0)^3 \right] h + (u^0)^4 \lambda(u^0 - u_a) = 0,$$

and

$$\{(d/d\varepsilon)[u(a) + \varepsilon h(a) - u_a]\}_{\varepsilon=0} = h(a) = 0.$$

Note that, for this model, N_1' is the operator

$$N_1'(*) = \left[d/dt + 4\alpha_1^0 \left(u^0\right)^3\right](*), \tag{V.B.8}$$

while N_2' is the operator

$$N_2'(*) = \left[\left(u^0\right)^4, 1\right] \cdot (*), \tag{V.B.9}$$

and the boundary condition becomes

$$h(a) = 0. \tag{V.B.10}$$

The *ASAP* starts by obtaining the operator L^+, formally adjoint to N_1', by means of Eq. (V.A.15). For this illustrative model, Eq. (V.A.15) takes on the form

$$\int_a^b dt \left\{ \psi \left[d/dt + \alpha_1^0 4\left(u^0\right)^3 h \right] \right\} = \int_a^b dt \left\{ h \left[-d/dt + \alpha_1^0 4\left(u^0\right)^3 \right] \psi \right\} + [\psi h]_a^b,$$

where ψ will become the adjoint function. As the above expression shows, L^+ is the operator

$$L^* = \left[-d/dt + \alpha_1^0 4\left(u^0\right)^3\right], \tag{V.B.11}$$

while $P(\psi, h)$ represents the expression

$$P(\psi, h) = [\psi h]_a^b. \tag{V.B.12}$$

Taking Eqs. (V.B.11), (V.B.12), and (V.B.6) into account shows that, for the illustrative model, the adjoint equation becomes

$$\left[-d/dt + \alpha_1^0 4\left(u^0\right)^3\right] \psi = \alpha_1^0 4\left(u^0\right)^3. \tag{V.B.13}$$

The adjoint boundary conditions are chosen to eliminate the unknown values of h from $P(\psi, h)$ in Eq. (V.B.12). The value of $h(a)$ in this equation is known from the initial conditions given in Eq. (V.B.10). Thus, the unknown value of h, namely $h(b)$, can be eliminated from Eq. (V.B.12) by choosing the adjoint boundary condition

$$\psi(b) = 0. \qquad (V.B.14)$$

Thus, the sensitivity DR_F of the response R to the feedback F becomes

$$DR_F = -\int_a^b dt \left[\psi(u^0)^4 \lambda(u^0 - u_a) \right]. \qquad (V.B.15)$$

The advantage of the adjoint method is that the adjoint solution is independent of the feedback being considered. Thus, once the adjoint solution ψ has been calculated, it is possible to estimate the effect of many different feedbacks without solving any additional differential equations.

V.C. SENSITIVITY ANALYSIS OF NONLINEAR ALGEBRAIC MODELS: THE DISCRETE ADJOINT SENSITIVITY ANALYSIS PROCEDURE (DASAP)

Most problems of practical interest are not amenable to analytical solutions, so the quantities of interest must be computed numerically. Therefore, the operator equations given in Eqs. (V.A.1) and (V.A.2) must be discretized, thereby yielding a system of coupled nonlinear algebraic equations. Similarly, the response represented by Eq. (V.A.3) must be discretized in order to calculate it numerically. In many practical problems, therefore, sensitivity and uncertainty analysis needs to be performed directly on systems of coupled nonlinear algebraic equations.

This Section, therefore, presents the mathematical formalism underlying the application *FSAP* and *ASAP* for *nonlinear algebraic systems*, such as would be obtained by *discretizing* the nonlinear system given in Eqs. (V.A.1) and (V.A.2). As has been illustrated by the application presented in Section IV.E (for the Euler's one-step method), and as will be illustrated by the applications to be presented in Sections V.D through V.F, the operations of discretization and Gâteaux-differentiation do not commute. In principle, there are two paths that can be pursued in practice for computing sensitivities, namely:

path (a): *first apply the FSAP and ASAP to differential and/or integral equations, then discretize the resulting Forward and/or Adjoint Sensitivity Equations, and finally solve the resulting algebraic equations numerically*; or

path (b): *first discretize the original (linear and/or nonlinear) differential and/or integral equations, then apply the FSAP and ASAP to the resulting linear* and/or nonlinear algebraic equations, and finally solve the resulting algebraic equations numerically.

As has been seen in Section IV.E, and as will be illustrated in Sections V.D through V.F, the algebraic equations resulting from following path (a) will be different from the algebraic equations resulting from path (b), because *the two paths will, in general, produce different truncation errors.* Of course, *it is paramount to ensure that both paths yield consistent discretizations of the same Adjoint Sensitivity Equations.* Thus, it is very useful, both theoretically and in practice, to distinguish between the procedures underlying path (a) and those underlying path (b), respectively; henceforth, therefore, the procedures underlying path (a) will continue to be called *FSAP* and *ASAP*, while the procedures underlying path (b) will be called: the *Discrete Forward Sensitivity Analysis Procedure* (*DFSAP*) and the *Discrete Adjoint Sensitivity Analysis Procedure* (*DASAP*), respectively.

The general mathematical formalisms underlying the *DFSAP* and the *DASAP*, respectively, will be presented in the remainder of this Section. The mathematical notation is based on the work by Cacuci et al., *Nucl. Sci. Eng.*, 75, 88, 1980 (with permission). Consider, therefore, that the phase-space vector x is represented at discrete points by x_j^n, where n is a time-step index taking on values $n=0$ through some final value n_f, and j is the index for a discrete point in phase-space. In this context, $n=0$ can be thought of as referring to a steady-state (i.e., time-independent) discrete representation of Eqs. (V.A.1) and (V.A.2). Note that the components denoted by the single index j of x_j^n include not only spatial but also energy, angle, etc., discretizations. The time index n is separated from the others, because of the special way of advancing time steps in (discrete) numerical methods.

Each operator N_k acting on the function $u_r(x)$ in Eq. (V.A.1) is represented discretely at time step n by a (nonlinear) matrix A_{kr}^n of dimension $M_k \times M_r$ acting on the vector

$$U_r^n = \left({}^n u_1^r, \ldots, {}^n u_{M_r}^r \right), \quad {}^n u_j^r = u_r\left(x_j^n\right). \tag{V.C.1}$$

Note that the *time-step index* will henceforth be indicated on the *left-side* of each component, but *on the right-side of each vector or matrix*, to distinguish it from the phase-space indices. A discretization of Eqs. (V.A.1) and (V.A.2) can be generally represented as a $K \times K$ system of coupled nonlinear matrix equations of the form

$$\sum_{r=1}^{K} A_{kr}^n U_r^n = B_k^n, \quad \left(k=1,2,\ldots,K; \; n=1,2,\ldots,n_f\right). \tag{V.C.2}$$

Let α denote any discrete parameter that is considered for sensitivity analysis. It may represent a discrete component of the vector of parameters $\alpha(x)$, or a parameter of purely numerical methods origin. Each component ${}^n a_{ij}^{kr}$, $(i=1,2,\ldots,M_k;\ j=1,2,\ldots,M_r)$, of the matrix A_{kr}^n is considered to be a Gâteaux differentiable function of the parameter α and of the state vectors at the same and previous time levels, i.e.,

$$ {}^n a_{ij}^{kr} = {}^n a_{ij}^{kr}\left(U_1^0,\ldots,U_1^n;U_2^0,\ldots,U_2^n;\ldots;U_K^0,\ldots,U_K^n;\alpha\right). \tag{V.C.3}$$

The vector $\boldsymbol{B}_k^n$ appearing on the right side of Eq. (V.C.2) is of the form

$$\boldsymbol{B}_k^n = \left({}^n b_1^k,\ldots,{}^n b_{M_k}^k\right), \tag{V.C.4}$$

where, for the sake of generality, each component ${}^n b_i^k$ also is considered to be a Gateaux-differentiable function of α and the vectors U_r^l, $(r=1,2,\ldots,K;\ l=0,1,\ldots,n)$.

With the definitions introduced thus far, the system of nonlinear matrix equations given in Eq. (V.C.2) is sufficiently general to represent a wide variety of practical discretizations of Eqs. (V.A.1) and (V.A.2). Note that boundary conditions are automatically incorporated in Eq. (V.C.2) as a result of the discretization process. This is in contrast to the (phase-space-dependent) operator formulation of local sensitivity analysis presented in Sec. V.A, where the boundary terms need to be treated explicitly.

Consider next the following discrete representation of the nonlinear response R given in Eq. (V.A.3):

$$R\left(U_1^0,\ldots,U_K^{n_f};\alpha\right) = W \cdot F \tag{V.C.5}$$

where W and F are the partitioned vectors

$$W = \left(W_1^0,\ldots,W_K^0;W_1^1,\ldots,W_K^1;\ldots;W_1^{n_f},\ldots,W_K^{n_f}\right), \tag{V.C.6}$$

and

$$F = \left(F_1^0,\ldots,F_K^0;F_1^1,\ldots,F_K^1;\ldots;F_1^{n_f},\ldots,F_K^{n_f}\right). \tag{V.C.7}$$

The vectors W_k^n and F_k^n have, respectively, the forms

$$W_k^n = \left({}^n w_1^k, {}^n w_2^k, \ldots, {}^n w_{M_k}^k\right),$$ (V.C.8)

and

$$F_k^n = \left({}^n f_1^k, {}^n f_2^k, \ldots, {}^n f_{M_k}^k\right),$$ (V.C.9)

where the components ${}^n w_i^k$ and ${}^n f_i^k$ are also considered to be differentiable functions of the vector α of parameters, and the state vectors U_r^l, $(r = 1, 2, \ldots, K; l = 0, 1, \ldots, n)$.

The structure of the response R given above in Eq. (V.C.5) is sufficiently general to represent most practical discretizations of Eq. (V.A.3). For example, the vector W might represent a numerical quadrature for the integral over some or all of the components of x, while F could represent a discretization of a differential or integral operator. Examples of such responses will be discussed in conjunction with the illustrative examples to be presented later in Section V.D.

V.C.1. The Discrete Forward Sensitivity Analysis Procedure (DFSAP)

The sensitivity DR is obtained by defining the vector of parameter variations h_α around the nominal parameter value α^0, defining the vector of increments

$$H_r^n = \left({}^n h_1^r, \ldots, {}^n h_{M_r}^r\right), \quad (r = 1, \ldots, K; n = 0, \ldots, n_f).$$ (V.C.10)

around the nominal value $\left(U_r^n\right)^0$, and taking the Gâteaux-(G)-derivative of the response R defined in Eq. (V.C.5), namely

$$DR \equiv \frac{d}{d\varepsilon} \left\{ R\left[\left(U_1^0\right)^0 + \varepsilon H_1^0, \ldots, \left(U_K^{n_f}\right)^0 + \varepsilon H_K^{n_f}; \alpha^0 + \varepsilon h_\alpha\right] \right\}_{\varepsilon=0}.$$ (V.C.11)

Performing the above sequence of operations leads to the following expression for the sensitivity DR of R to parameter variations h_α:

$$DR = R'_\alpha h_\alpha +$$

$$\sum_{n=0}^{n_f} \sum_{k=1}^{K} \sum_{i=1}^{M_k} \left\{ \sum_{l=0}^{n} \sum_{r=1}^{K} \sum_{j=1}^{M_r} \left({}^l h_j^r\right) \times \left[\left({}^n f_i^k\right) \partial\left({}^n w_i^k\right) / \partial\left({}^l u_j^r\right) + \left({}^n w_i^k\right) \partial\left({}^n f_i^k\right) / \partial\left({}^l u_j^r\right)\right] \right\},$$

(V.C.12)

where the "direct effect" term is defined as

$$R'_\alpha(\boldsymbol{h}_\alpha) \equiv \sum_{l=1}^{I}\left[\sum_{n=0}^{n_f}\sum_{k=1}^{K}\sum_{i=1}^{M_k}\left({}^nf_i^k\right)\frac{\partial\left({}^nw_i^k\right)}{\partial\alpha_l}+\left({}^nw_i^k\right)\frac{\partial\left({}^nf_i^k\right)}{\partial\alpha_l}\right]h_{\alpha_l}. \quad (V.C.13)$$

Note that even though all quantities in Eqs. (V.C.12) and (V.C.13) are evaluated for nominal values of the discrete state-variables $\left(U_r^n\right)^0$ and parameter(s) $\boldsymbol{\alpha}^0$, this fact will be omitted in the sequel in order to simplify the notation. Equation (V.C.12) can be simplified by introducing the partitioned vector

$$\boldsymbol{Q}^{nl} \equiv \left(\boldsymbol{Q}_1^{nl}, \boldsymbol{Q}_2^{nl}, \ldots, \boldsymbol{Q}_K^{nl}\right), \quad (V.C.14)$$

with components

$$\boldsymbol{Q}_r^{nl} \equiv \left({}_l^n q_1^r, {}_l^n q_2^r, \ldots, {}_l^n q_{M_r}^r\right), \quad (V.C.15)$$

while, in turn, the components ${}_l^n q_j^r$ are defined as

$${}_l^n q_j^r \equiv \sum_{k=1}^{K}\sum_{i=1}^{M_k}\left[\left({}^nf_i^k\right)\frac{\partial\left({}^nw_i^k\right)}{\partial\left({}^l u_j^r\right)}+\left({}^nw_i^k\right)\frac{\partial\left({}^nf_i^k\right)}{\partial\left({}^l u_j^r\right)}\right]. \quad (V.C.16)$$

It is also convenient to introduce the partitioned vectors

$$\boldsymbol{H}^n = \left(\boldsymbol{H}_1^n, \ldots, \boldsymbol{H}_K^n\right), \quad \left(n = 0, 1, \ldots, n_f\right), \quad (V.C.17)$$

with components as defined in Eq. (V.C.10). Using the definitions given in Eqs. (V.C.14) through (V.C.17), the sums in Eq. (V.C.12) can be rewritten in compact form as

$$DR = R'_\alpha \boldsymbol{h}_\alpha + \sum_{n=0}^{n_f}\sum_{l=n}^{n_f}\boldsymbol{Q}^{nl}\bullet\boldsymbol{H}^l. \quad (V.C.18)$$

The sensitivity DR can be evaluated once $\boldsymbol{H}^n$ is known. In turn, the vectors $\boldsymbol{H}^n$ are obtained by taking the G-differential of Eq. (V.C.2), i.e.,

$$\frac{d}{d\varepsilon}\left\{\sum_{r=1}^{K}A_{kr}^{n}\left[(U_{1}^{0})^{0}+\varepsilon h_{1}^{0},\ldots,(U_{K}^{n})^{0}+\varepsilon h_{K}^{n};\alpha^{0}+\varepsilon h_{\alpha}\right]\left[(U_{r}^{n})^{0}+\varepsilon h_{r}^{n}\right]\right.$$
$$\left.-B_{k}^{n}\left[(U_{1}^{0})^{0}+\varepsilon h_{1}^{0},\ldots,(U_{K}^{n})^{0}+\varepsilon h_{K}^{n};\alpha^{0}+\varepsilon h_{\alpha}\right]\right\}_{\varepsilon=0}=0.$$

Performing the above Gâteaux-differential yields

$$\sum_{r=1}^{K}\sum_{j=1}^{M_{r}}\left\{({}^{n}a_{ij}^{kr})({}^{n}h_{j}^{r})+({}^{n}u_{j}^{r})\left[\sum_{l=1}^{I}\frac{\partial({}^{n}a_{ij}^{kr})}{\partial\alpha_{l}}h_{\alpha_{l}}+\sum_{l=0}^{n}\sum_{p=1}^{K}\sum_{s=1}^{M_{p}}({}^{l}h_{s}^{p})\frac{\partial({}^{n}a_{ij}^{kr})}{\partial({}^{l}u_{s}^{p})}\right]\right\}$$
$$=\sum_{l=1}^{I}\frac{\partial({}^{n}b_{i}^{k})}{\partial\alpha_{l}}h_{\alpha_{l}}+\sum_{l=0}^{n}\sum_{p=1}^{K}\sum_{s=1}^{M_{p}}({}^{l}h_{s}^{p})\frac{\partial({}^{n}b_{i}^{k})}{\partial({}^{l}u_{s}^{p})}.$$

(V.C.19)

After considerable rearrangement of the multiple sums, Eq. (V.C.19) can be rewritten as

$$\sum_{r=1}^{K}\sum_{j=1}^{M_{r}}\left[({}^{n}c_{ij}^{kr})({}^{n}h_{j}^{r})-\sum_{l=0}^{n-1}({}^{n}_{l}d_{ij}^{kr})({}^{l}h_{j}^{r})\right]={}^{n}e_{i}^{k},\qquad(V.C.20)$$

where

$${}^{n}_{l}d_{ij}^{kr}\equiv\frac{\partial({}^{n}b_{i}^{k})}{\partial({}^{l}u_{j}^{r})}-\sum_{p=1}^{K}\sum_{s=1}^{M_{p}}({}^{n}u_{s}^{p})\frac{\partial({}^{n}a_{is}^{kp})}{\partial({}^{l}u_{j}^{r})},\qquad(V.C.21)$$

$${}^{n}c_{ij}^{kr}\equiv{}^{n}a_{ij}^{kr}-{}^{n}_{n}d_{ij}^{kr},\qquad(V.C.22)$$

$${}^{n}e_{i}^{k}\equiv\sum_{l=1}^{I}\left[\frac{\partial({}^{n}b_{i}^{k})}{\partial\alpha_{l}}-\sum_{p=1}^{K}\sum_{s=1}^{M_{p}}({}^{n}u_{s}^{p})\frac{\partial({}^{n}a_{is}^{kp})}{\partial\alpha_{l}}\right]h_{\alpha_{l}}.\qquad(V.C.23)$$

The system of equations given in Eq. (V.C.20) can be written in matrix form as

$$[C^{n}]H^{n}=\sum_{l=0}^{n-1}[D^{nl}]H^{l}+E^{n},\quad(n=1,2,\ldots,n_{f}),\qquad(V.C.24)$$

where the partitioned matrices $[C^n] \equiv (C_{kr}^n)$ and $[D^{nl}] \equiv (D_{kr}^{nl})$, $(k, r = 1, 2, \ldots, K)$, are constructed from the submatrices

$$C_{kr}^n \equiv \begin{bmatrix} {}^n c_{11}^{kr} & \cdots & {}^n c_{1M_r}^{kr} \\ \vdots & & \vdots \\ {}^n c_{M_k 1}^{kr} & \cdots & {}^n c_{M_k M_r}^{kr} \end{bmatrix}, \quad \text{(V.C.25)}$$

and

$$D_{kr}^{nl} \equiv \begin{bmatrix} {}^n_l d_{11}^{kr} & \cdots & {}^n_l d_{1M_r}^{kr} \\ \vdots & & \vdots \\ {}^n_l d_{M_k 1}^{kr} & \cdots & {}^n_l d_{M_k M_r}^{kr} \end{bmatrix}, \quad \text{(V.C.26)}$$

while the partitioned vector

$$E^n \equiv (E_1^n, \ldots, E_K^n) \quad \text{(V.C.27)}$$

is constructed from vectors

$$E_k^n \equiv ({}^n e_1^k, \ldots, {}^n e_{M_k}^k). \quad \text{(V.C.28)}$$

Equation (V.C.24) represents the *Discretized Forward Sensitivity Equations* (*DFSE*), and is the discrete counterpart of the *Forward Sensitivity Equations* (*FSE*) represented by Eqs. (V.A.11) and (V.A.12) of Section V.A. As expected, Eq. (V.C.24) is linear in H^n, and its solution can be used directly in Eq. (V.C.18) to evaluate DR. However, as has already been discussed, it is impractical to compute DR this way if many system parameter variations h_α are to be considered. In such cases, it is advantageous to compute DR by using the solutions (independent of any quantities involving h_α) of the system that is adjoint to Eq. (V.C.24). This alternative method will be referred to as the *Discrete Adjoint Sensitivity Analysis Procedure* (*DASAP*), in analogy to the *ASAP* for phase-space-dependent operators previously presented in Section V.A.

V.C.2. The Discrete Adjoint Sensitivity Analysis Procedure (DASAP)

To construct the adjoint system, we introduce the partitioned vectors Ψ^n, $(n = 1,\ldots,n_f)$, defined as

$$\Psi^n \equiv (\Psi_1^n, \Psi_2^n, \ldots, \Psi_K^n), \qquad \text{(V.C.29)}$$

where

$$\Psi_k^n \equiv ({}^n\psi_1^k, \ldots, {}^n\psi_{M_k}^k). \qquad \text{(V.C.30)}$$

Multiplying Eq. (V.C.24) on the left by $(\Psi^n)^T$ (where T denotes transposition), and summing the resulting equation over n gives

$$\sum_{n=1}^{n_f} (\Psi^n)^T [C^n] H^n = \sum_{n=1}^{n_f}\sum_{l=0}^{n-1} (\Psi^n)^T [D^{nl}] H^l + \sum_{n=1}^{n_f} (\Psi^n)^T E^n. \qquad \text{(V.C.31)}$$

Equation (V.C.31) is now rearranged by performing the following operations:
  (a)   Transpose the terms involving matrices;
  (b)   Interchange the order of summation in the double sum;
  (c)   Separate the contribution from $n = 0$.
The above sequence of operations transform Eq. (V.C.31) into the equivalent form

$$\sum_{l=1}^{n_f} (H^n)^T \left\{ [C^n]^T \Psi^n - \sum_{l=n+1}^{n_f} [D^{nl}]^T \Psi^l \right\} = \sum_{l=1}^{n_f} (H^0)^T [D^{l0}]^T \Psi^l + \sum_{n=1}^{n_f} (\Psi^n)^T E^n.$$

$$\text{(V.C.32)}$$

Comparison of Eq. (V.C.32) with Eq. (V.C.18) reveals that by requiring that the vectors Ψ^n satisfy the adjoint system

$$[C^n]^T \Psi^n = \sum_{l=n+1}^{n_f} [D^{nl}]^T \Psi^l + \sum_{l=n}^{n_f} Q^{nl}, \quad (n = n_f, n_f - 1,\ldots,1), \qquad \text{(V.C.33)}$$

the sensitivity DR can be expressed in terms of the adjoint vectors Ψ^n in the form

$$DR = R'_\alpha h_\alpha + \sum_{n=1}^{n_f} (\boldsymbol{\Psi}^n)^T \boldsymbol{E}^n$$
$$+ \left[\sum_{n=1}^{n_f} (\boldsymbol{\Psi}^n)^T [\boldsymbol{D}^{n0}]^T + \sum_{n=0}^{n_f} (\boldsymbol{Q}^{n0})^T \right] \boldsymbol{H}^0. \tag{V.C.34}$$

The adjoint system given in Eq. (V.C.33) is the discrete counterpart of the adjoint system given in Eqs. (V.A.18) and (V.A.16). Note that the adjoint boundary conditions corresponding to Eq. (V.A.16) are automatically incorporated into the discrete adjoint system represented by Eq. (V.C.33) for $\boldsymbol{\Psi}^n$. Furthermore, the same correspondence is noted between Eqs. (V.C.34) and (V.A.20) or (V.A.21). The source term $\sum_{l=n}^{n_f} \boldsymbol{Q}^{nl}$ for the adjoint system is recognized to correspond to the Gâteaux-derivative of R with respect to $\boldsymbol{U}^n$. Finally, note that if sensitivity to initial conditions is not desired, then $\boldsymbol{H}^0 = 0$ and Eq. (V.C.34) simplifies considerably.

The methodology leading to Eqs. (V.C.33) and (V.C.34) can be employed in a straightforward manner to perform sensitivity studies on many responses of interest. *An important advantage of this procedure is that sensitivity analysis can also be performed for certain parameters introduced by the discretization process. This is not possible within the framework of either the FSAP or ASAP for phase-space-dependent operators.* This additional capability of the *DASAP* will be illustrated with an example presented in the next Section.

V.D. ILLUSTRATIVE EXAMPLE: A TRANSIENT NONLINEAR REACTOR THERMAL HYDRAULICS PROBLEM

This Section presents an application of the general sensitivity analysis theory developed in Sections V.A and V.C to a typical heat transfer problem in a fuel rod of a nuclear reactor, cooled on the outside by a fluid. In this application, the main emphasis is placed on illustrating that the *ASAP* (applied to differential equations) and the *Discrete ASAP (DASAP)*, applied to the discretized form of the original nonlinear differential equation, produce *consistent numerical approximations of the adjoint differential equation and responses, but produce different truncation errors*. This is because the *ASAP* and *DASAP* formalisms generally lead to difference equations that are distinct from each other, even though they correspond to the same adjoint differential equation. The material presented in this Section is based largely on the work by Cacuci et al., *Nucl. Sci. Eng.*, 75, 88, 1980 (with permission).

Consider, therefore, a cylindrically symmetric configuration of a typical reactor fuel rod, which includes the fuel, gap, and cladding regions; the cladding is surrounded by a coolant to remove the heat. Assuming (a) radial conduction in the fuel rod, (b) axial convection and single phase flow in the coolant, and (c) energy conservation in the gap, the temperatures $T(r,z,t)$ and $T_c(z,t)$ in the fuel rod and coolant, respectively, satisfy the following system of coupled nonlinear partial differential equations:

$$\rho C_p(T)\frac{\partial T}{\partial t} - \frac{1}{r}\frac{\partial}{\partial r}\left[k(T)\frac{\partial T}{\partial r}\right] = Q(r,z,t),$$

$$0 < r < R_f,\ R_g < r < R_c,\ 0 \leq z \leq L,\ 0 < t < t_f,$$

$$T(r,z,0) = T_0(r,z),$$

$$\left.\frac{\partial T}{\partial r}\right|_{r=0} = 0,$$

$$\left(rk\frac{\partial T}{\partial r}\right)_{R_f} = \left(rk\frac{\partial T}{\partial r}\right)_{R_g},$$

$$\left(k\frac{\partial T}{\partial r} + h_g T\right)_{R_f} = \left(h_g T\right)_{R_g},$$

$$\left(k\frac{\partial T}{\partial r} + h_c T\right)_{R_c} = h_c T_c,$$

$$\frac{\partial}{\partial t}(A\rho C_p T_c) + \frac{\partial}{\partial z}(wC_p T_c) = 2\pi R_c h_c [T - T_c]_{r=R_c},\quad 0 \leq z \leq L, 0 < t < t_f,$$

$$T_c(z,0) = T_{c0}(z),$$

$$T_c(0,t) = T_{c1}(t),$$

(V.D.1)

where the following notations have been used:

R_f, R_g, R_c = outer radii of fuel, gap, and cladding
L = fuel rod height
ρ = density
C_p = heat capacity
k = thermal conductivity
Q = nuclear heat source density
h_g = gap conductance
h_c = coolant-cladding heat transfer coefficient
A = coolant-channel cross-sectional flow area
w = coolant mass flow rate.

Note that in Eq. (V.D.1) the fuel, cladding, and coolant have distinct material properties, even though the same symbols $(k, \rho, C_p, etc.)$ have been used for notational simplicity.

Introducing the discrete mesh (r_i, z_k, t_n), $(i = 1, 2, \ldots, I;\ k = 1, 2, \ldots, K;\ n = 1, 2, \ldots, n_f)$, the system given in Eq. (V.D.1) is solved numerically by using a Crank-Nicolson-type procedure to compute each radial profile of discrete fuel rod temperatures, $T_{i,k+1/2}^n \equiv T(r_i, z_{k+1/2}, t_n)$, and by employing weighting factors for both derivative terms to obtain the discrete coolant temperatures, $T_{c,k}^n \equiv T_c(z_k, t_n)$. The discrete mesh points (r_i, z_k, t_n) involve uniform spacing in the axial and time dimensions and nonuniform radial spacing. Thus, the step-sizes are defined by $\Delta t = t_n - t_{n-1}, (n = 1, 2, \ldots, n_f);\ \Delta z = z_{k+1} - z_k$, $(k = 1, 2, \ldots, K-1)$; and $\Delta r_i = r_{i+1} - r_i, (i = 1, 2, \ldots, I-1)$. The parameter I_f represents the number of radial points in the fuel; the symbol V_i represents the volume of the mesh cell surrounding the radial point r_i; the quantities $\hat{r}_i$ represent radial intermediate points, $r_i < \hat{r}_i < r_{i+1}$. The resulting system of nonlinear difference equations is

$$\lambda_{i,k}^n T_{i-1,k+1/2}^n + \beta_{i,k}^n T_{i,k+1/2}^n + \gamma_{i,k}^n T_{i+1,k+1/2}^n = \mu_{i,k}^n,$$
$$\lambda_{I,k}^n T_{I-1,k+1/2}^n + \beta_{I,k}^n T_{I,k+1/2}^n + \gamma_{I,k}^n \left(T_{c,k+1}^n + T_{c,k}^n\right) = \mu_{I,k}^n, \quad \text{(V.D.2)}$$
$$\lambda_{I+1,k}^n T_{I,k+1/2}^n + \beta_{I+1,k}^n T_{c,k+1}^n + \nu_k^n T_{c,k}^n = \mu_{I+1,k}^n,$$
$$\left(i = 1, \ldots, I-1, k = 1, \ldots, K-1, n = 1, \ldots, n_f\right).$$

The definitions of the coefficients $\lambda_{i,k}^n, \beta_{i,k}^n, \gamma_{i,k}^n, \mu_{i,k}^n$, and ν_k^n are listed below.

$$\gamma_{i,k}^n \equiv \begin{cases} -\dfrac{\Delta t r_i}{\Delta r_i} k_{i,k+1/2}^n, & (i = 1, \ldots, I_f - 1, I_f + 1, \ldots, I-1), \\ -R_f \Delta t h_{g,k+1/2}^n, & i = I_f, \\ -\dfrac{1}{2} R_c \Delta t h_{c,k+1/2}^n, & i = I, \end{cases}$$

$$\beta_{i,k}^n \equiv \begin{cases} V_i(\rho C_p)_{i,k+1/2}^{n-1/2} - \alpha_{i,k}^n - \gamma_{i,k}^n, & (i = 1,2,\ldots,I-1), \\ V_I(\rho C_p)_{I,k+1/2}^{n-1/2} - \alpha_{I,k}^n - 2\gamma_{I,k}^n, & i = I, \\ \theta_1(A\rho C_p)_{k+1}^n + \theta_2 \dfrac{\Delta t}{\Delta z}(wC_p)_{k+1}^n + 2\pi R_c \Delta t \theta_1 \theta_2 h_{c,k+1}^n, & i = I+1, \end{cases}$$

$$\mu_{i,k}^n \equiv \begin{cases} \Delta t V_i Q_{i,k+1/2}^{n-1/2} + V_i(\rho C_p)_{i,k+1/2}^{n-1/2} T_{i,k+1/2}^{n-1} + \lambda_{i,k}^{n-1}\left(T_{i,k+1/2}^{n-1} - T_{i-1,k+1/2}^{n-1}\right) \\ \quad -\gamma_{i,k}^{n-1}\left(T_{i+1,k+1}^{n-1} - T_{i,k+1/2}^{n-1}\right), \qquad (i = 1,2,\ldots,I-1), \\ \Delta t V_I Q_{I,k+1/2}^{n-1/2} + V_I(\rho C_p)_{I,k+1/2}^{n-1/2} T_{I,k+1/2}^{n-1} + \lambda_{I,k}^{n-1}\left(T_{I,k+1/2}^{n-1} - T_{I-1,k+1/2}^{n-1}\right) \\ \quad + 2\gamma_{I,k}^{n-1}\left(T_{I,k+1/2}^{n-1} - T_{c,k+1/2}^{n-1}\right), \qquad i = I, \\ T_{c,k}^{n-1}\left[(1-\theta_1)(A\rho C_p)_k^{n-1} + (1-\theta_2)\dfrac{\Delta t}{\Delta z}(wC_p)_k^{n-1}\right. \\ \qquad\left. - 2\pi R_c \Delta t (1-\theta_1)(1-\theta_2) h_{c,k}^{n-1}\right] \\ \quad + T_{c,k+1}^{n-1}\left[\theta_1(A\rho C_p)_{k+1}^{n-1} - (1-\theta_2)\dfrac{\Delta t}{\Delta z}(wC_p)_{k+1}^{n-1} - 2\pi R_c \Delta t \theta_1 (1-\theta_2) h_{c,k+1}^{n-1}\right] \\ \quad + 2\pi R_c \Delta t (1-\theta_2) T_{I,k+1/2}^{n-1}\left[\theta_1 h_{c,k+1}^{n-1} + (1-\theta_1) h_{c,k}^{n-1}\right], \quad i = I+1, \end{cases}$$

$$v_k^n \equiv (1-\theta_1)(A\rho C_p)_k^n - \theta_2 \dfrac{\Delta t}{\Delta z}(wC_p)_k^n + 2\pi R_c \Delta t (1-\theta_1)\theta_2 h_{c,k}^n.$$

The quantities θ_1 and θ_2 $(0 \leq \theta_1 \leq 1, 0 \leq \theta_2 \leq 1)$ appearing above are weighting factors used in the discretization of the coolant equation in the system of equations shown in Eq. (V.D.1). For instance, the time derivative is approximated using the axial weighting parameter θ_1 as follows:

$$\left[\dfrac{\partial}{\partial t}(A\rho C_p)\right]_{k+1/2}^{n-1/2} = \dfrac{\theta_1}{\Delta t}\left[(A\rho C_p T_c)_{k+1}^n - (A\rho C_p T_c)_{k+1}^{n-1}\right]$$

$$+ \dfrac{(1-\theta_1)}{\Delta t}\left[(A\rho C_p T_c)_k^n - (A\rho C_p T_c)_k^{n-1}\right]$$

$$+ \dfrac{\Delta z}{2}(2\theta_1 - 1)\dfrac{\partial^2}{\partial z \partial t}(A\rho C_p T_c) + O\left[(\Delta z)^2 + (\Delta t)^2\right].$$

The parameter θ_2 is used similarly in the time variable and, as such, describes the "degree of implicitness" of the difference approximation. For example,

$\theta_2 = 0$, $\theta_2 = 1$, and $\theta_2 = 1/2$ correspond, respectively, to the explicit, fully implicit, and Crank-Nicholson schemes.

It can be shown that Eq. (V.D.2) is a *consistent and stable* numerical approximation to Eq. (V.D.1). Since Eq. (V.D.2) is nonlinear, it must be solved by iteration; three to five iterations were required to converge the values of the dependent variables to within four significant figures at all points in space and time.

Two different types of responses will be considered in this example. The first type of response, denoted in the sequel as R_1, is a linear functional of the fuel temperature, $T(r,z,t)$, and coolant temperature, $T_c(z,t)$, of the form

$$R_1 = \iiint f_1(r,z,t) T(r,z,t) \, 2\pi r \, dr \, dz \, dt \\ + \iint g_1(z,t) T_c(z,t) \, dz \, dt, \qquad \text{(V.D.3)}$$

where f_1 and g_1 are weight functions or distributions. The above form can be used to represent either a point or an average value of the respective temperatures. For instance, if the response is the temperature in the fuel rod at the particular point $(\bar{r}, \bar{z}, \bar{t})$, then

$$f_1(r,z,t) = \frac{1}{2\pi r} \delta(r - \bar{r}) \delta(z - \bar{z}) \delta(t - \bar{t}), \\ g_1(z,t) = 0. \qquad \text{(V.D.4)}$$

The second type of response considered in this example is representative of a nonlinear operator acting on a dependent variable. Such a response, of considerable practical interest, is the linear heat rate (i.e., power added to coolant per unit axial length) at a point $(R_c, \bar{z}, \bar{t})$; this response will be denoted in this Section by R_2, and is defined as

$$R_2 = 2\pi R_c \left[-k(T) \frac{\partial T}{\partial r} \right]_{(R_c, \bar{z}, \bar{t})}. \qquad \text{(V.D.5)}$$

V.D.1. Adjoint Sensitivity Analysis Procedure (ASAP)

Applying the *ASAP* presented in Section V.A to the model described by Eq. (V.C.1) and the response R_1 defined in Eq. (V.D.3) leads to the following adjoint system:

$$-\rho C_p \frac{\partial \psi}{\partial t} - \frac{1}{r} k(T) \frac{\partial}{\partial r}\left(r \frac{\partial \psi}{\partial r}\right) = f_1(r,z,t),$$

$$-(A\rho C_p)' \frac{\partial \psi_c}{\partial t} - (wC_p)' \frac{\partial \psi_c}{\partial z} - 2\pi R_c h_c'[\psi - \psi_c]_{r=R_c} = g_1(z,t),$$

$$\psi(r,z,t_f) = 0,$$

$$\left.\frac{\partial \psi}{\partial r}\right|_{r=0} = 0,$$

$$\left(k\frac{\partial \psi}{\partial r} + h_g' \psi\right)_{R_f} = (h_g \psi)_{R_g}, \qquad \text{(V.D.6)}$$

$$(R_f h_g'' \psi)_{R_f} = \left(R_g k \frac{\partial \psi}{\partial r} + R_f h_g'' \psi\right)_{R_g},$$

$$\left(k\frac{\partial \psi}{\partial r} + h_c \psi\right)_{R_c} = h_c \psi_c,$$

$$\psi_c(z,t_f) = 0,$$

$$\psi_c(L,t) = 0$$

where the following definitions have been used in the above equation:

$$(A\rho C_p)' \equiv A\rho C_p + T_c \frac{\partial}{\partial T_c}(A\rho C_p); \quad (wC_p)' \equiv wC_p + T_c \frac{\partial}{\partial T_c}(wC_p);$$

$$h_g' \equiv h_g(T) + \frac{\partial h_g}{\partial \tau_1}(\tau_1 - \tau_2); \quad \tau_1 \equiv T(R_f,z,t); \quad \tau_2 \equiv T(R_g,z,t);$$

$$h_g'' \equiv h_g + \frac{\partial h_g}{\partial \tau_2}(\tau_2 - \tau_1); \quad h_c' \equiv h_c - \left[(T - T_c)\frac{\partial h_c}{\partial T_c}\right]_{R_c}.$$

In Eq. (V.D.6), ψ is the adjoint function corresponding to the increment $h_T(r,z,t)$ around the nominal value, $T^0(r,z,t)$, of the fuel temperature, while ψ_c is the adjoint function corresponding to the increment $h_{T_c}(z,t)$ around the nominal value $T_c^0(z,t)$ of the coolant temperature. Carrying through the procedure underlying the *ASAP* leads ultimately to the following expression for the sensitivity DR_1 of R_1 with respect to any generic parameter variations h_{α_l} around the nominal value α_l^0:

$$\frac{DR_1}{h_{\alpha_l}} = \int_t \int_z \int_r \psi \left[\frac{\partial Q}{\partial \alpha_l} - \frac{\partial(\rho C_p)}{\partial \alpha_l} \frac{\partial T}{\partial t} + \frac{1}{r} \frac{\partial}{\partial r}\left(r \frac{\partial k}{\partial \alpha_l} \frac{\partial T}{\partial r} \right) \right] 2\pi r dr dz dt$$

$$- 2\pi R_f \int_t \int_z [\psi(R_f, z, t) - \psi(R_g, z, t)] \; [T(R_f, z, t) - T(R_g, z, t)] \frac{\partial h_g}{\partial \alpha_l} dz dt$$

$$- 2\pi R_c \int_t \int_z [(T - T_c)(\psi - \psi_c)]_{r = R_c} \frac{\partial h_c}{\partial \alpha_l} dz dt + 2\pi \int_z \int_r [\rho C_p \psi]_{t=0} \frac{\partial T_0}{\partial \alpha_l} r dr dz$$

$$- \int_t \int_z \psi_c \left\{ \frac{\partial}{\partial t}\left[T_c \frac{\partial}{\partial \alpha_l}(A\rho C_p) \right] + \frac{\partial}{\partial z}\left[T_c \frac{\partial}{\partial \alpha_l}(\rho C_p) \right] \right\} dz dt$$

$$+ \int_z \left[\psi_c (A\rho C_p)' \right]_{t=0} \times \frac{\partial T_{c0}}{\partial \alpha_l} dz + \int_t \left[\psi_c (wC_p)' \right]_{z=0} \frac{\partial T_{c1}}{\partial \alpha_l} dt.$$

(V.D.7)

Note that the right side of Eq. (V.D.7) has been divided through h_{α_l} (which thus appears on the left side), for notational simplicity. Note also that the direct effect term is identically zero, for the response R_1, since f_1 and g_1 in Eq. (V.D.3) are independent of α_l.

The adjoint system given in Eq. (V.D.6) must be solved numerically. For this purpose, Eq. (V.D.6) is finite-differenced using the same discrete mesh in phase space as for Eq. (V.D.1). The tridiagonal structure of the resulting difference equations closely resembles the structure of Eq. (V.D.2). However, the adjoint system is linear and its solution is therefore obtained without iteration.

Consider next the nonlinear response R_2 defined by Eq. (V.D.5). The adjoint system corresponding to this response can be shown to be the same as Eq. (V.D.6), but with the source terms f_1 and g_1 replaced, respectively, by the functions f_2 and g_2, defined as follows:

$$f_2 \equiv \frac{R_c}{r} k(T) \delta'(r - R_c) \delta(z - \bar{z}) \delta(t - \bar{t}), \quad g_2 \equiv 0. \qquad \text{(V.D.8)}$$

The expression for the sensitivity DR_2 of R_2 with respect to any parameter variation h_{α_l} around the nominal values α_l^0 has the same form as Eq. (V.D.7), except for the addition of the *nonzero direct effect term*

$$\frac{\partial R_2^0 / \partial \alpha_l}{h_{\alpha_l}} = -2\pi \left[r \frac{\partial k}{\partial \alpha_l} \frac{\partial T}{\partial r} \right]_{(R_c, \bar{z}, \bar{t})} - 2\pi \frac{\partial R_c}{\partial \alpha_l} \left[k \frac{\partial T}{\partial r} \right]_{(R_c, \bar{z}, \bar{t})}. \qquad \text{(V.D.9)}$$

As an aside, note here that, in view of the boundary condition at $r = R_c$ in Eq. (V.D.1), the response R_2 could have been expressed in the equivalent form

$$R_2 = 2\pi R_c [h_c(T - T_c)]_{(R_c, \bar{z}, \bar{t})}.$$

Nevertheless, the expression of R_2 given in Eq. (V.D.5) was used, rather than the expression above, to illustrate the application of the *ASAP* to a response involving both nonlinearities and a derivative of the state-function $T(r, z, t)$.

V.D.2. Discrete Adjoint Sensitivity Analysis Procedure (DASAP)

We now apply the *Discrete Adjoint Sensitivity Analysis Procedure* (*DASAP*) presented in Section V.C to the discretized (nonlinear) equations that model this illustrative example. Thus, comparing Eq. (V.D.2) above with the general form represented by Eq. (V.C.2), we note the following correspondences: the (column) vector U_k^n defined in Eq. (V.C.1) corresponds, in the illustrative example, to the vector of discrete temperatures,

$$U_k^n = \left(T_{1,k+1/2}^n, \ldots, T_{I,k+1/2}^n, T_{c,k+1}^n \right). \qquad \text{(V.D.10)}$$

Furthermore, the components $^n b_i^k$ of the vector $\boldsymbol{B}_k^n$ in Eq. (V.C.2) correspond to the quantities $\mu_{i,k}^n$, namely:

$$^n b_i^k = \mu_{i,k}^n, \quad (i = 1, 2, \ldots, I+1), \qquad \text{(V.D.11)}$$

while the components $^n a_{ij}^{kr}$ of the matrices A_{kr}^n in Eq. (V.C.2) are identified as follows:

$$^n a_{ij}^{kr} = 0, \ r \neq k-1, k, \qquad \text{(V.D.12a)}$$

$$^n a_{ij}^{kk} = \begin{cases} \beta_{j,k}^n, & i = j \\ \lambda_{j,k}^n, & i = j+1 \\ \gamma_{j,k}^n, & i = j-1 \\ 0, & \text{otherwise.} \end{cases} \qquad \text{(V.D.12b)}$$

$$^n a_{i,j}^{k,k-1} = \begin{cases} \gamma_{i,k}^n, & i=I, j=I+1 \\ v_k^n, & i=j=I+1 \\ 0, & \text{otherwise.} \end{cases} \qquad (\text{V.D.12c})$$

The responses of interest are discretized to fit the general form of Eq. (V.C.5). Thus, if the response R_1^D (the superscript "D" denoting the discrete, DASAP, formalism) is a fuel rod temperature at the phase-space point $(r_{\bar{i}}, z_{\bar{k}}, t_{\bar{n}})$, then the respective components of the response are identified as

$$^n f_i^k = T_{i,k+1/2}^n, \quad ^n w_i^k = \begin{cases} 1, & i=\bar{i}, k=\bar{k}, n=\bar{n} \\ 0, & \text{otherwise.} \end{cases} \qquad (\text{V.D.13})$$

In like manner, the discretization of the nonlinear response R_2, defined in Eq. (V.D.5), at $(r_I, z_{\bar{k}}, t_{\bar{n}})$ is

$$R_2^D = -2\pi R_c k_{I,\bar{k}}^{\bar{n}} \frac{1}{\Delta r_{I-1}} \left[T_{I,\bar{k}}^{\bar{n}} - T_{I-1,\bar{k}}^{\bar{n}} \right], \qquad (\text{V.D.14})$$

which implies the correspondences

$$^n f_i^j = k_{i,j}^n \left(T_{i,j}^n - T_{i-1,j}^n \right),$$
$$^n w_i^j = \begin{cases} -2\pi R_c / \Delta r_{i-1}, & i=I, j=\bar{k}, n=\bar{n} \\ 0, & \text{otherwise.} \end{cases} \qquad (\text{V.D.15})$$

Furthermore, we need to determine the components $^n_l q_i^j$ of the source vector Q_n^l for the adjoint equations. Comparing the general form given in Eq. (V.C.16) to Eqs. (V.D.13) and (V.D.14), respectively, shows that the fuel rod temperature response R_1^D generates the components

$$^n_l q_i^j = \begin{cases} 1, & i=\bar{i}, j=\bar{k}, n=l=\bar{n} \\ 0, & \text{otherwise,} \end{cases} \qquad (\text{V.D.16})$$

while the nonlinear response R_2^D generates the components

$$_l^n q_i^j = \begin{cases} -\dfrac{2\pi R_c}{\Delta r_{i-1}} \left[\left(\dfrac{\partial k}{\partial T}\right)_{i,j}^n \left(T_{i,j}^n - T_{i-1,j}^n\right) + k_{i,j}^n \right], \\ \qquad \text{for } j = \bar{k},\ i = I,\ n = l = \bar{n} \\ \dfrac{2\pi R_c}{\Delta r_i} k_{i+1,j}^n, \qquad \text{for } j = \bar{k},\ i = I-1,\ n = l = \bar{n} \\ 0, \qquad \text{otherwise}. \end{cases}$$
(V.D.17)

The components $^n d_{ij}^{kr}$ and $^n c_{ij}^{kr}$, needed to construct the matrices $\mathbf{D}^{ne}$ and $\mathbf{C}^n$, respectively, for the discrete adjoint system given by Eq. (V.C.33), are determined by applying the definitions given in Eqs. (V.C.21) and (V.C.22) to the nonlinear difference equations Eq. (V.D.2), underlying our illustrative example. The respective computations involve lengthy but straightforward algebra; so they will be omitted here.

Note from Eqs. (V.D.16) and (V.D.17) that the components $_l^n q_i^j$ vanish whenever $l \neq n$. Moreover, the quantities $_l^n d_{ij}^{kr}$ also vanish whenever $l \neq n-1$. Consequently, the sums over l in the general form of the discrete adjoint system presented in Eq. (V.A.33) reduce to a single nonzero term, yielding the following system of linear matrix equations to be solved for the adjoint vectors $\mathbf{\Psi}^n$:

$$\mathbf{\Psi}^{n_f + 1} \equiv 0,$$
$$\left[\mathbf{C}^n\right]^T \mathbf{\Psi}^n = \left[\mathbf{D}^{n+1,n}\right]^T \mathbf{\Psi}^{n+1} + \mathbf{Q}^{nn}, \quad (n = n_f, n_f - 1, \ldots, 1).$$
(V.D.18)

As indicated above, Eq. (V.D.18) is solved backwards in time, beginning at the final time-step index n_f. The adjoint solutions $\mathbf{\Psi}^n$ are to be folded with the vectors $\mathbf{E}^n$ to compute the various sensitivities. Since the quantities $\left(^n e_i^k\right) \alpha_l$ actually prove more convenient to compute in practice, the relative sensitivities $(DR/R)/(h_{\alpha_l}/\alpha_l)$ are obtained directly by slightly rearranging Eq. (V.C.34). For example, if α_l represents the coolant mass flow rate scale factor, then

$$\left(^n e_i^k\right)\alpha_l = \begin{cases} 0, \qquad (i = 1, 2, \ldots, I), \\ (1 - \theta_2) \dfrac{\Delta t}{\Delta z} \left[(wC_p T_c)_k^{n-1} - (wC_p T_c)_{k+1}^{n-1} \right] \\ + \theta_2 \dfrac{\Delta t}{\Delta z} \left[(wC_p T_c)_k^n - (wC_p T_c)_{k+1}^n \right], \ i = I+1. \end{cases}$$
(V.D.19)

As has been previously mentioned, the direct effect term $\left[\partial R_2^0 / \partial \alpha_l\right]/h_{\alpha_l}$ for the response R_2 is not identically zero. Its contribution to $\left(DR_2^D / R_2^D\right)/\left(h_{\alpha_l} / \alpha_l\right)$ is computed using Eq. (V.C.13), to obtain

$$\frac{\alpha_l}{R_2^D} \frac{\partial R_2^D / \partial \alpha_l}{h_{\alpha_l}} = \begin{cases} \dfrac{1}{R_2^D} \dfrac{2\pi R_c R_g}{\Delta r_{I-1}(R_c - R_g)} k_{I,\bar{k}}^{\bar{n}}\left(T_{I-1,\bar{k}}^{\bar{n}} - T_{I,\bar{k}}^{\bar{n}}\right), & \alpha_l = R_g \\[2mm] -\dfrac{1}{R_2^D} \dfrac{2\pi R_c R_g}{\Delta r_{I-1}(R_c - R_g)} k_{I,\bar{k}}^{\bar{n}}\left(T_{I-1,\bar{k}}^{\bar{n}} - T_{I,\bar{k}}^{\bar{n}}\right), & \alpha_l = R_c \\[2mm] \dfrac{1}{R_2^D} \dfrac{2\pi R_c}{\Delta r_{I-1}} k_{I,\bar{k}}^{\bar{n}}\left(T_{I-1,\bar{k}}^{\bar{n}} - T_{I,\bar{k}}^{\bar{n}}\right), & \alpha_l = k \\[2mm] 0, & \text{otherwise.} \end{cases} \quad \text{(V.D.20)}$$

Comparing Eqs. (V.D.20) and (V.D.9) reveals that the *DASAP* produces a nonzero term for $\left[\partial R_2^0 / \partial \alpha_l\right]/h_{\alpha_l}$ when $\alpha_l = R_g$, while the *ASAP* does not. This situation arises because the parameter R_g does not appear in the formulation of the original response R_2 defined in Eq. (V.D.5), but is introduced into the discrete form that represents R_2, i.e., into Eq. (V.D.14) through the outermost cladding stepsize Δr_{I-1}.

V.D.3. Comparative Discussion of the ASAP and DASAP Formalisms

As has been shown, the same nonlinear system of difference equations, namely Eq. (V.D.2), must be solved within both the ASAP and the DASAP to compute the base-case forward solution, namely the distribution of the fuel rod and coolant temperatures. Note, however, that the adjoint system of difference equations derived by employing the DASAP formalism is not identical to the difference approximation for the differential adjoint system shown in Eq. (V.D.6), produced by the ASAP formalism.

For the purposes of comparison, consider the adjoint difference equations for a typical interior node $(r_i, z_{k+1/2}, t_n)$ in the fuel rod. The difference equation solved when using the ASAP formalism is

$$p_{i-1}k_{i,k+1/2}^{n-1}\left(\psi_{i,k+1/2}^{n-1} - \psi_{i-1,k+1/2}^{n-1}\right) - p_i k_{i,k+1/2}^{n-1}\left(\psi_{i+1,k+1/2}^{n-1} - \psi_{i,k+1/2}^{n-1}\right)$$
$$+ V_i\left(\rho C_p\right)_{i,k+1/2}^{n-1/2}\left(\psi_{i,k+1/2}^{n-1} - \psi_{i,k+1/2}^{n}\right) + p_{i-1}k_{i,k+1/2}^{n}\left(\psi_{i,k+1/2}^{n} - \psi_{i-1,k+1/2}^{n}\right) \quad \text{(V.D.21)}$$
$$- p_i k_{i,k+1/2}^{n}\left(\psi_{i+1,k+1/2}^{n} - \psi_{i,k+1/2}^{n}\right) - \frac{2}{\Delta z} F_{i,k+1/2}^{n-1/2} = 0,$$

while the corresponding adjoint equation employed in the *DASAP* formalism is

$$\left[\frac{1}{2}p_{i-1}\left(\frac{\partial k}{\partial T}\right)_{\hat{i}-1,k+1/2}^{n}\left(T_{i,k+1/2}^{n} - T_{i-1,k+1/2}^{n}\right) + p_{i-1}k_{i-1,k+1/2}^{n}\right]\left({}^n\psi_i^k - {}^n\psi_{i-1}^k\right)$$
$$+ \left[\frac{1}{2}p_i\left(\frac{\partial k}{\partial T}\right)_{\hat{i},k+1/2}^{n}\left(T_{i+1,k+1/2}^{n} - T_{i,k+1/2}^{n}\right) - p_i k_{i,k+1/2}^{n}\right]\left({}^n\psi_{i+1}^k - {}^n\psi_i^k\right)$$
$$+ \left\{\frac{1}{2}V_i\left[\frac{\partial(\rho C_p)}{\partial T}\right]_{i,k+1/2}^{n-1/2}\left(T_{i,k+1/2}^{n} - T_{i,k+1/2}^{n-1}\right) + V_i\left(\rho C_p\right)_{i,k+1/2}^{n-1/2}\right\}\left({}^n\psi_i^k\right)$$
$$+ \left[\frac{1}{2}p_{i-1}\left(\frac{\partial k}{\partial T}\right)_{\hat{i}-1,k+1/2}^{n}\left(T_{i,k+1/2}^{n} - T_{i-1,k+1/2}^{n}\right) + p_{i-1}k_{i-1,k+1/2}^{n}\right]\left({}^{n+1}\psi_i^k - {}^{n+1}\psi_{i-1}^k\right)$$
$$+ \left[\frac{1}{2}p_i\left(\frac{\partial k}{\partial T}\right)_{\hat{i},k+1/2}^{n}\left(T_{i+1,k+1/2}^{n} - T_{i,k+1/2}^{n}\right) - p_i k_{i,k+1/2}^{n}\right]\left({}^{n+1}\psi_{i+1}^k - {}^{n+1}\psi_i^k\right)$$
$$+ \left\{\frac{1}{2}V_i\left[\frac{\partial(\rho C_p)}{\partial T}\right]_{i,k+1/2}^{n+1/2}\left(T_{i,k+1/2}^{n+1} - T_{i,k+1/2}^{n}\right) - V_i\left(\rho C_p\right)_{i,k+1/2}^{n+1/2}\right\}\left({}^{n+1}\psi_i^k\right) - {}_n q_i^k = 0.$$

(V.D.22)

As a point in common, both Eqs. (V.D.21) and (V.D.22) involve adjoint functions at two time levels, with the unknown time level determined by the solution of a tridiagonal matrix equation at each axial level. In addition, the solution procedure for both the *ASAP* and the *DASAP* formalisms begins at time index $n = n_f$ and continues backwards in time to $n = 1$; within each time step, calculations start at axial level $k = K - 1$, and progress downstream to $k = 1$.

The most obvious difference between Eqs. (V.D.21) and (V.D.22) is the presence of the terms $\partial k / \partial T$ and $\partial(\rho C_p)/\partial T$ in the adjoint equation given by Eq. (V.D.22), which results from the *DASAP* formalism. Furthermore, Eq. (V.D.22) contains temperature terms evaluated at three time levels $(t_{n-1}, t_n, \text{ and } t_{n+1})$, but the temperature values at t_n are clearly dominant. On the other hand, Eq. (V.D.21) produced by the *ASAP* formalism contains temperatures evaluated at only two time steps $(t_n$ and $t_{n+1})$, of equal importance.

To understand the origin of these distinctions, consider again the procedures used to obtain Eqs. (V.D.21) and (V.D.22). Equation (V.D.21) is a difference equation obtained by integrating the respective (continuous) partial differential equation over each discrete mesh interval. The Crank-Nicolson procedure then produces equal weighting of terms at time levels t_n and t_{n+1}. This fact implies that

$$[\text{Eq. (V.D.21)}] = \int_{t_{n-1}}^{t_n} \int_{z_k}^{z_{k+1}} \int_{\hat{r}_{i-1}}^{\hat{r}_i} \left[-\rho C_p \frac{\partial \psi}{\partial t} - \frac{1}{r} k \frac{\partial}{\partial r}\left(r \frac{\partial \psi}{\partial r} \right) - f_1(r,z,t) \right] r \, dr \, dz \, dt$$
$$+ O\left[(\Delta t)^2 + \Delta t (\Delta r_i + \Delta r_{i-1}) + \Delta t (\hat{r}_i - \hat{r}_{i-1})^2 + \Delta t (\Delta z)^2 \right] ,$$
(V.D.23)

which indicates that Eq. (V.D.21) is numerically consistent with the differential equation, since the higher order terms vanish as the time- and step-sizes approach zero. On the other hand, the adjoint equation (V.D.22) produced by the *DASAP* formalism is the result of algebraic manipulation rather than direct differencing. Nevertheless, using the relationship

$$^n\psi_i^k = \begin{cases} \pi \Delta z \, \psi(r_i, z_{k+1/2}, t_{n-1/2}), & (i = 1, 2, \ldots, I), \\ \Delta z \, \psi_c(z_{k+1/2}, t_{n-1/2}), & i = I+1, \end{cases}$$
(V.D.24)

it can be shown that Eq. (V.D.22) is *also* a consistent numerical approximation to the fuel rod conduction equation of the system given in Eq. (V.D.6). Thus, *the differences between the ASAP and DASAP occur because Eq. (V.D.22) corresponds to a **different** form of the same partial differential equation, namely*

$$[\text{Eq. (V.D.22)}] = \int_{t_{n-1}}^{t_n} \int_{z_k}^{z_{k+1}} \int_{\hat{r}_{i-1}}^{\hat{r}_i} \left[-\frac{\partial}{\partial t}(\rho C_p \psi) + \psi \frac{\partial T}{\partial t} \frac{\partial}{\partial T}(\rho C_p) \right.$$
$$\left. - \frac{1}{r} \frac{\partial}{\partial r}\left(rk \frac{\partial \psi}{\partial r} \right) + \frac{\partial k}{\partial T} \frac{\partial T}{\partial r} \frac{\partial \psi}{\partial r} - f_1(r,z,t) \right] r \, dr \, dz \, dt$$
(V.D.25)
$$+ O\left[(\Delta t)^2 + \Delta t (\Delta r_i + \Delta r_{i-1}) + \Delta t (\hat{r}_i - \hat{r}_{i-1})^2 + \Delta t (\Delta z)^2 \right].$$

Thus, both the ASAP and DASAP formalisms solve consistent numerical approximations of the adjoint differential equation. However, the two approaches do have different truncation errors since the coefficients of the higher order terms are not the same in Eqs. (V.D.24) and (V.D.25).

Note that Eqs. (V.D.24) and (V.D.25) only apply to radial mesh elements in the interior of each material in the fuel rod. However, similar consistency results can also be obtained for the boundary and interface regions in the fuel rod. Furthermore, it can be shown that both the *ASAP* and *DASAP* formalisms produce consistent approximations to the same form of the integrated coolant equation of the adjoint differential system given in Eq. (V.D.6), even though the *ASAP* and *DASAP* formalisms may lead to adjoint difference equations that are not identical.

Finally, consider the relative sensitivity coefficient $(DR/R)/(h_{\alpha_l}/\alpha_l)$ for the parameter $\alpha_l = \overline{w}$ (coolant mass flow rate scale factor). For the temperature response defined by Eq. (V.D.3), the *ASAP* formalism yields through Eq. (V.D.7) the integral

$$\frac{DR_1^0/R_1^0}{h_w/\overline{w}} = -\frac{1}{R_1^0}\int_0^{t_f}\int_0^L \psi_c \frac{\partial}{\partial z}(wC_pT_c)dzdt, \qquad (V.D.26)$$

which can be discretized as follows:

$$\frac{DR_1^0/R_1^0}{h_w/\overline{w}} = \frac{\Delta t}{2R_1^0}\sum_{n=1}^{n_f}\sum_{k=1}^{K-1}\left\{\psi_{c,k+1/2}^{n-1}\left[(wC_pT_c)_k^{n-1}-(wC_pT_c)_{k+1}^{n-1}\right]\right.$$
$$\left. + \psi_{c,k+1/2}^{n}\left[(wC_pT_c)_k^{n}-(wC_pT_c)_{k+1}^{n}\right]\right\}, \qquad (V.D.27)$$

where $\psi_{c,k+1/2}^{n} \equiv \psi_c(z_{k+1/2},t_n)$. On the other hand, when substituting Eq. (V.D.19) into Eq. (V.C.34), the *DASAP* formalism yields the sensitivity

$$\frac{DR_1^D/R_1^D}{h_w/\overline{w}} = \frac{\Delta t}{\Delta z R_1^D}\sum_{n=1}^{n_f}\sum_{k=1}^{K-1}\left({}^n\psi_{I+1}^k\right)\left\{(1-\theta_2)\left[(wC_pT_c)_k^{n-1}-(wC_pT_c)_{k+1}^{n-1}\right]\right.$$
$$\left. +\theta_2\left[(wC_pT_c)_k^{n}-(wC_pT_c)_{k+1}^{n}\right]\right\}. \qquad (V.D.28)$$

Using Eq. (V.D.24), the sensitivity produced by *DASAP* can be expressed in terms of the function ψ_c as

$$\frac{DR_1^D/R_1^D}{h_w/\overline{w}} = \frac{\Delta t}{R_1^D}\sum_{n=1}^{n_f}\sum_{k=1}^{K-1}\left(\psi_{c,k+1/2}^{n-1/2}\right)\times\left\{(1-\theta_2)\left[(wC_pT_c)_k^{n-1}-(wC_pT_c)_{k+1}^{n-1}\right]\right.$$
$$\left. +\theta_2\left[(wC_pT_c)_k^{n}-(wC_pT_c)_{k+1}^{n}\right]\right\}, \qquad (V.D.29)$$

which is, at the same time, a consistent approximation to Eq. (V.D.26). The close similarity between Eqs. (V.D.29) and (V.D.27) is now evident. On the other hand, the distinctions between these two expressions are that in Eq.

(V.D.29) the adjoint function ψ_c is evaluated at midpoint rather than endpoint values, and the user-specified input value of θ_2 is used. By contrast, Eq. (V.D.27) uses $\theta_2 = 1/2$.

In conclusion, both the *ASAP* and the *DASAP* formalisms yield consistent approximations to both the adjoint differential system and the response integrals of the operator formalism. As has been shown, however, the two formalisms lead in general to different truncation errors in the final results. For the illustrative problem presented in this Section, the effects of these differences are relatively minimal, but this may not necessarily be true in general.

V.E. ILLUSTRATIVE EXAMPLE: SENSITIVITY ANALYSIS OF THE TWO-STEP RUNGE-KUTTA METHOD FOR NONLINEAR DIFFERENTIAL EQUATIONS

This Section presents the application of the *Discrete Forward Sensitivity Analysis Procedure* (*DFSAP*), followed by the *Discrete Adjoint Sensitivity Analysis Procedure* (*DASAP*) applied to the nonlinear algebraic equations obtained after discretizing a nonlinear ordinary differential equation by using the Runge-Kutta method. This discretization method, known also under the name "Modified Euler" or "Improved Polygon" method, is one of the most popular methods for solving numerically differential equations, and is particularly useful for discretizing the time-dependency in such equations.

Thus, consider the first order nonlinear ordinary differential

$$\frac{du}{dt} = f(u,t), \quad u^0 = u(0). \tag{V.E.1}$$

Discretizing the above equation by means of the two-step Runge-Kutta method yields the following system of nonlinear algebraic equations:

$$\begin{cases} u^0 = u(0); \\ v^n = u^n + \frac{\Delta t_n}{2} f(u^n, t^n); \quad (n = 0, \ldots, N-1); \\ u^{n+1} = u^n + \Delta t_n f\left(v^n, t^n + \frac{\Delta t_n}{2}\right); \quad (n = 0, \ldots, N-1). \end{cases} \tag{V.E.2}$$

Consider, along with Eq. (V.E.1), a response of the same form as in Eq. (IV.E.3), namely

$$R \equiv \int_0^{t_f} \gamma(t)\,d(t)\,u(t)\,dt \;, \tag{V.E.3}$$

where, as discussed in Section IV.E for the one-step Euler method, $\gamma(t)$ is a weight function to be used for numerical quadrature and/or time-step sensitivity analysis, while $d(t)$ allows flexibility in modeling several types of responses. Thus, the expression of the discretized response is of the same form as given by Eq. (IV.E.10), namely:

$$R = \sum_{s=0}^{N} \Delta t_s^q \gamma^s d^s u^s \;. \tag{V.E.4}$$

The response sensitivity is obtained by calculating the Gâteaux-differential of Eq. (V.E.4); the resulting expression has the same form as given in Eq. (IV.E.11), namely:

$$DR = \sum_{s=0}^{N} \gamma^s \Delta t_s^q d^s h^n + \sum_{s=0}^{N} \delta\!\left(\Delta t_s^q\right) \gamma^s d^s u^s \;. \tag{V.E.5}$$

The variation h^n, which is needed in order to evaluate the response sensitivity DR above, is determined now by calculating the Gâteaux-differential of Eq. (V.E.2). The resulting algebraic equations for h^n are the *Discrete Forward Sensitivity Equations (DFSE)*

$$\begin{cases} h^0 = \delta u^0 \;; \\ g^n = h^n + \dfrac{\Delta t_n}{2}(f'_u)^n h^n + \dfrac{1}{2}\delta(\Delta t_n) f(u^n) + \dfrac{\Delta t_n}{2}\sum_i (f'_{\alpha_i})^n \delta\alpha_i \;; (n=0,\ldots,N-1); \\ h^{n+1} = h^n + \Delta t_n (f'_v)^n_{t^{n+1/2}} g^n + \delta(\Delta t_n) f(v^n) + \Delta t_n \sum_i (f'_{\alpha_i}(v^n)) \delta\alpha_i \;; \\ \hspace{4cm} (n=0,\ldots,N-1). \end{cases} \tag{V.E.6}$$

Note that the above *DFSE* are linear in h^n and, except for the terms containing $\delta(\Delta t_n)$, it is identical to the Runge-Kutta discretization of the differential *FSE*

$$\frac{dh}{dt} = f'_u(u)h + \sum_i f'_{\alpha_i}(u)\delta\alpha_i \;;\quad h(0) = \delta u^0 \tag{V.E.7}$$

that would have been obtained by applying the *FSAP* to Eq. (V.E.1). The *DFSE* represented by Eq. (V.E.6) can be written more compactly by defining the quantities

$$a^n \equiv 1 + \frac{\Delta t_n}{2}(f'_u)^n \, ; \, q_2^n \equiv \frac{\Delta t_n}{2}\sum_i (f'_{\alpha_i})^n \delta\alpha_i + \frac{1}{2}\delta(\Delta t_n)f(u^n); (n = 0,\ldots,N-1);$$

$$b^n \equiv \Delta t_n (f'_v)^n_{t^{n+1/2}} \, ; \, \begin{cases} q_1^{n+1} \equiv \Delta t_n \sum_i (f'_{\alpha_i}(v^n))^n \delta\alpha_i + \delta(\Delta t_n)f(v^n); \\ \qquad\qquad\qquad\qquad (n = 0,\ldots,N-1); \\ q_1^0 = \delta u^0 \end{cases}$$

and replacing them in Eq. (V.E.6), which becomes, in turn

$$\begin{cases} h^0 = \delta u^0 \equiv q_1^0 \, ; \\ h^n a^n + g^n = q_2^n \, , \quad (n = 0,\ldots,N-1); \\ -h^n + h^{n+1} + b^n g^n = q_1^{n+1} \, , \quad (n = 0,\ldots,N-1). \end{cases} \qquad\text{(V.E.8)}$$

The above system of algebraic equations can now be written in matrix form by defining the vectors

$$\mathbf{h} = (h^0,\ldots,h^N)^T_{1\times(N+1)}, \quad \mathbf{g} = (g^0,\ldots,g^{N-1})^T_{1\times N}, \qquad\text{(V.E.9)}$$

$$\mathbf{q}_1 \equiv (\delta u^0, q_1^1, q_1^2, \ldots, q_1^N), \quad \mathbf{q}_2 \equiv (q_2^0, \ldots, q_2^{N-1}) \qquad\text{(V.E.10)}$$

and the matrices

$$\begin{pmatrix} 1 & 0 & 0 & \cdots & 0 & 0 \\ -1 & 1 & 0 & \cdots & 0 & 0 \\ 0 & -1 & 1 & \cdots & 0 & 0 \\ \cdots & \cdots & \cdots & \cdots & \cdots \\ 0 & 0 & 0 & \cdots & -1 & 1 \end{pmatrix} \equiv [A_{11}]_{(N+1)\times(N+1)} \, ;$$

$$\begin{pmatrix} 0 & 0 & \cdots & 0 \\ b^0 & 0 & \cdots & 0 \\ 0 & b^1 & \cdots & 0 \\ \cdots & \cdots & \cdots & \cdots \\ 0 & 0 & \cdots & b^{N-1} \end{pmatrix} \equiv [A_{12}]_{(N+1) \times N} ; \qquad \text{(V.E.11)}$$

$$\begin{pmatrix} a^0 & 0 & \cdots & 0 & 0 \\ 0 & a^1 & \cdots & 0 & 0 \\ \cdots & \cdots & \cdots & \cdots & \cdots \\ 0 & 0 & \cdots & a^{N-1} & 0 \end{pmatrix} \equiv [A_{21}]_{N \times (N+1)} ;$$

$$\begin{pmatrix} 1 & 0 & \cdots & 0 \\ 0 & 1 & \cdots & 0 \\ \cdots & \cdots & \cdots & \cdots \\ 0 & 0 & \cdots & 1 \end{pmatrix} \equiv [A_{22}]_{N \times N} .$$

Introducing the definitions from Eqs. (V.E.9) through (V.E.11) into Eq. (V.E.8) yields the matrix equation

$$\begin{pmatrix} A_{11} & A_{12} \\ A_{21} & A_{22} \end{pmatrix} \begin{pmatrix} h \\ g \end{pmatrix} = \begin{pmatrix} q_1 \\ q_2 \end{pmatrix} . \qquad \text{(V.E.12)}$$

To apply the *Discrete Adjoint Sensitivity Analysis Procedure* (*DASAP*), Eq. (V.E.12) is multiplied on the left by the (partitioned) row-vector (ϕ, ψ), which has the same structure as the vector (h, g), namely

$$\begin{cases} \phi = \left(\phi^0, \ldots, \phi^N \right)^T_{1 \times (N+1)} \\ \psi = \left(\psi^0, \ldots, \psi^{N-1} \right)^T_{1 \times N} \end{cases} \qquad \text{(V.E.13)}$$

but is otherwise unspecified yet, to obtain

$$\begin{pmatrix} \phi & \psi \end{pmatrix} \begin{pmatrix} A_{11} & A_{12} \\ A_{21} & A_{22} \end{pmatrix} \begin{pmatrix} h \\ g \end{pmatrix} = \begin{pmatrix} \phi & \psi \end{pmatrix} \begin{pmatrix} q_1 \\ q_2 \end{pmatrix} . \qquad \text{(V.E.14)}$$

Transposing Eq. (V.E.14) yields

$$(h \quad g)\begin{pmatrix} A_{11}^T & A_{21}^T \\ A_{12}^T & A_{22}^T \end{pmatrix}\begin{pmatrix} \phi \\ \psi \end{pmatrix} = (q_1 \quad q_2)\begin{pmatrix} \phi \\ \psi \end{pmatrix} \quad \text{(V.E.15)}$$
$$= q_1\phi + q_2\psi.$$

The (partitioned) row-vector (ϕ,ψ) is now specified to be the solution of the block-matrix equation

$$\begin{pmatrix} A_{11}^T & A_{21}^T \\ A_{12}^T & A_{22}^T \end{pmatrix}\begin{pmatrix} \phi \\ \psi \end{pmatrix} = \begin{pmatrix} r_1 \\ r_2 \end{pmatrix}. \quad \text{(V.E.16)}$$

The above equation is the *Discrete Adjoint Sensitivity Equation* (*DASE*) for a general response of the form (r_1, r_2). For the specific response sensitivity DR given by Eq. (V.E.5), the expressions of (r_1, r_2) are obtained by replacing Eq. (V.E.16) on the left side of Eq. (V.E.15) to obtain

$$hr_1 + gr_2 = q_1\phi + q_2\psi, \quad \text{(V.E.17)}$$

and by setting

$$\begin{cases} r_2 = 0 \\ r_1 = \left(\gamma^0 \Delta t_0^q d^0, \ldots, \gamma^N \Delta t_N^q d^N\right)^T \end{cases} \quad \text{(V.E.18)}$$

in Eq. (V.E.17). Comparing now the resulting equation to the expression of the sensitivity DR given by Eq. (V.E.5) yields the result

$$DR = q_1\phi + q_2\psi + \sum_{s=0}^{N} \delta\left(\Delta t_s^q\right)\gamma^s d^s u^s$$
$$- \delta u^0 \phi^0 + \sum_{s=1}^{N} q_1^s \phi^s + \sum_{s=0}^{N-1} q_2^s \psi^s \quad \text{(V.E.19)}$$
$$+ \sum_{s=0}^{N} \delta\left(\Delta t_s^q\right)\gamma^s d^s u^s.$$

In component form, the *DASE*, namely Eq. (V.E.16), takes on the form

$$\begin{cases} A_{11}^T \phi + A_{21}^T \psi = r_1 \\ A_{12}^T \phi + A_{22}^T \psi = 0 \end{cases}, \quad \text{(V.E.20)}$$

or, equivalently,

$$\begin{pmatrix} 1 & -1 & 0 & \cdots & 0 \\ 0 & 1 & -1 & \cdots & 0 \\ 0 & 0 & 1 & \cdots & 0 \\ \cdots & \cdots & \cdots & \cdots & \cdots \\ 0 & 0 & 0 & \cdots & -1 \\ 0 & 0 & 0 & \cdots & 1 \end{pmatrix} \begin{pmatrix} \phi^0 \\ \phi^1 \\ \phi^2 \\ \vdots \\ \phi^{N-1} \\ \phi^N \end{pmatrix} + \begin{pmatrix} a^0 & 0 & 0 & \cdots & 0 \\ 0 & a^1 & 0 & \cdots & 0 \\ 0 & 0 & a^2 & \cdots & 0 \\ \cdots & \cdots & \cdots & \cdots & \cdots \\ 0 & 0 & 0 & \cdots & a^{N-1} \\ 0 & 0 & 0 & \cdots & 0 \end{pmatrix} \begin{pmatrix} \psi^0 \\ \psi^1 \\ \psi^2 \\ \vdots \\ \psi^{N-2} \\ \psi^{N-1} \end{pmatrix} = \begin{pmatrix} r_1^0 \\ r_1^1 \\ r_1^2 \\ \vdots \\ r_1^{N-1} \\ r_1^N \end{pmatrix}$$

and

$$\begin{pmatrix} 0 & b^0 & 0 & \cdots & 0 \\ 0 & 0 & b^1 & \cdots & 0 \\ 0 & 0 & 0 & \cdots & 0 \\ \cdots & \cdots & \cdots & \cdots & \cdots \\ 0 & 0 & 0 & \cdots & 0 \\ 0 & 0 & 0 & \cdots & b^{N-1} \end{pmatrix} \begin{pmatrix} \phi^0 \\ \phi^1 \\ \phi^2 \\ \vdots \\ \phi^{N-1} \\ \phi^N \end{pmatrix} + \begin{pmatrix} 1 & 0 & 0 & \cdots & 0 \\ 0 & 1 & 0 & \cdots & 0 \\ 0 & 0 & 1 & \cdots & 0 \\ \cdots & \cdots & \cdots & \cdots & \cdots \\ 0 & 0 & 0 & \cdots & 0 \\ 0 & 0 & 0 & \cdots & 1 \end{pmatrix} \begin{pmatrix} \psi^0 \\ \psi^1 \\ \psi^2 \\ \vdots \\ \psi^{N-2} \\ \psi^{N-1} \end{pmatrix} = \begin{pmatrix} 0 \\ 0 \\ 0 \\ \vdots \\ 0 \\ 0 \end{pmatrix}.$$

Performing the vector-matrix operations in the above equations explicitly gives the following system of linear algebraic equations for the *DASE*:

$$\begin{cases} \phi^N = r_1^N = \gamma^N \Delta t_N^q d^N \\ \phi^n - \phi^{n+1} - \left[1 + \dfrac{\Delta t_n}{2} (f_u')^n \right] \psi^n = \gamma^n \Delta t_n^q d^n; (n = 0, \ldots, N-1); \quad \text{(V.E.21)} \\ \left[-\Delta t_n (f_v')_{t^{n+1/2}}^n \right] \phi^{n+1} + \psi^n = 0, (n = 0, \ldots, N-1). \end{cases}$$

Eliminating ψ^n from Eq. (V.E.21) leads to

$$\phi^n - \phi^{n+1} \left[1 + \Delta t_n (f_v')_{t^{n+1/2}}^n + \dfrac{(\Delta t_n)^2}{2} (f_v')_{t^{n+1/2}}^n (f_u')^n \right] = \gamma^n \Delta t_n^q d^n; \quad \text{(V.E.22)}$$
$$(n = 0, \ldots, N-1).$$

On the other hand, eliminating the intermediate function g^n from Eq. (V.E.6) yields

$$\begin{cases} h^o = \delta u^o \\ -h^n c^n + h^{n+1} = q_1^{n+1} - b^n q_2^n; \quad (n = 0,...,N-1), \text{ where } c \equiv 1 + a^n b^n. \end{cases} \quad \text{(V.E.23)}$$

In matrix form, the above equation becomes

$$\begin{pmatrix} 1 & 0 & 0 & \cdots & 0 & 0 \\ -c^o & 1 & 0 & \cdots & 0 & 0 \\ 0 & -c^1 & 0 & \cdots & 0 & 0 \\ \cdots & \cdots & \cdots & \cdots & \cdots & \cdots \\ 0 & 0 & 0 & \cdots & 1 & 0 \\ 0 & 0 & 0 & \cdots & -c^{N-1} & 1 \end{pmatrix} \begin{pmatrix} h^o \\ h^1 \\ h^2 \\ \vdots \\ h^{N-1} \\ h^N \end{pmatrix} = \begin{pmatrix} \delta u^o \\ q^1 - b^o q_2^o \\ q^2 - b^1 q_2^1 \\ \vdots \\ q^{N-1} - b^{N-2} q^{N-2} \\ q^N - b^{N-1} q^{N-1} \end{pmatrix}. \quad \text{(V.E.24)}$$

The adjoint equation corresponding to Eq. (V.E.26) is

$$\begin{pmatrix} 1 & -c^o & 0 & \cdots & 0 & 0 \\ 0 & 1 & -c^1 & \cdots & 0 & 0 \\ 0 & 0 & 0 & \cdots & 0 & 0 \\ \cdots & \cdots & \cdots & \cdots & \cdots & \cdots \\ 0 & 0 & 0 & \cdots & 1 & -c^{N-1} \\ 0 & 0 & 0 & \cdots & 0 & 1 \end{pmatrix} \begin{pmatrix} \phi^o \\ \phi^1 \\ \phi^2 \\ \vdots \\ \phi^{N-1} \\ \phi^N \end{pmatrix} = \begin{pmatrix} \gamma^o \Delta t_o^q d^o \\ \vdots \\ \vdots \\ \vdots \\ \vdots \\ \gamma^N \Delta t_N^q d^N \end{pmatrix} \quad \text{(V.E.25)}$$

or, in component form:

$$\begin{cases} \phi^n - \phi^{n+1} c^n = \gamma^n \Delta t_n^q d^n, \quad (n = 0,...,N-1); \\ \phi^N = \gamma^N \Delta t_N^q d^N. \end{cases} \quad \text{(V.E.26)}$$

Replacing $c^n = 1 + a^n b^n$ in the above equation gives

$$\phi^n - \phi^{n+1} \left[1 + \Delta t_n (f_v')_{t^{n+1/2}}^n + \frac{(\Delta t_n)^2}{2} (f_v')_{t^{n+1/2}}^n (f_u')^n \right] = \gamma^n \Delta t_n^q d^n, \quad (n = 0,....,N-1)$$

which is identical to (V.E.22). This underscores the fact that the *DASE* can be derived either by using *explicitly* the "intermediate-time" functions g^n (and correspondingly, the adjoint functions ψ^n) or by eliminating them at the outset from the *DFSE*; the two alternative approaches are completely equivalent to one another.

V.F. ILLUSTRATIVE EXAMPLE: SENSITIVITY ANALYSIS OF A SCALAR NONLINEAR HYPERBOLIC PROBLEM

This Section presents yet another application of the *Discrete Forward Sensitivity Analysis Procedure* (*DFSAP*) and the *Discrete Adjoint Sensitivity Analysis Procedure* (*DASAP*). The paradigm problem considered in the following comprises the nonlinear algebraic equations obtained by using the Lax scheme to discretize a typical nonlinear partial differential equation underlying practical applications involving conservation laws. Thus, consider the partial differential equation

$$\begin{cases} \dfrac{\partial u}{\partial t} + \dfrac{\partial}{\partial x} f(u) = 0 \\ \\ u(x,0) = u^{init}(x); \quad x_{in} < x < x_{out}; \text{ at } t = 0 \\ u(x_{in},t) = u_{inlet}(t); \quad 0 < t < t_f; \text{ at } x = 0. \end{cases} \qquad \text{(V.F.1)}$$

Particular cases of importance to practical problems are

(a) $f(u) = u$, leading to a linear parabolic problem,

(b) $f(u) = \dfrac{1}{2} u^2$, leading to a nonlinear parabolic problem (i.e., a nonviscous Burgers' equation).

The response considered for the above problem is a linear functional of $u(x,t)$ of the form

$$R \equiv \int_0^{t_f} \int_{x_{in}}^{x_{out}} \gamma u(x,t) d(x,t) dx\, dt. \qquad \text{(V.F.2)}$$

The sensitivity DR of R is obtained by taking the Gâteaux-differential of Eq. (V.F.2) to obtain

$$DR = \int_0^{t_f} \int_{x_{in}}^{x_{out}} h(x,t) d(x,t) dx\, dt + \int_0^{t_f} \int_{x_{in}}^{x_{out}} u(x,t) \left(\sum_i \frac{\partial [d(x,t)]}{\partial \alpha_i} \delta\alpha_i \right) dx\, dt. \qquad \text{(V.F.3)}$$

Note that the customary superscript "0," denoting "base-case" value, has been omitted in Eq. (V.F.3) above, and will also be omitted throughout the section, to keep the notation as simple as possible.

The function $h(x,t)$ needed to evaluate the sensitivity DR could, in principle, be obtained by taking the Gâteaux-differential of Eq. (V.F.1) to obtain the *Forward Sensitivity Equations (FSE)*:

$$\begin{cases} \dfrac{\partial h}{\partial t} + \dfrac{\partial}{\partial x}(f'_u h) = -\dfrac{\partial}{\partial x}\sum_i \dfrac{\partial f}{\partial \alpha_i}\delta\alpha_i \\ h(x,0) = \delta u_0(x); \text{ at } t = 0 \\ h(x_{in},t) = \delta u_{in}(t); \text{ at } x = x_{in}. \end{cases} \quad (V.F.4)$$

The (differential) form of the adjoint system corresponding to Eq. (V.F.4) is obtained by applying the *ASAP*, as described in Section V.A. Omitting, for brevity, the intermediate *ASAP* steps, we give directly the end-form of the (differential) *Adjoint Sensitivity Equations*:

$$\begin{cases} -\dfrac{\partial \phi}{\partial t} - f'_u(x,t)\dfrac{\partial \phi}{\partial x} = d(x,t), \\ \phi(x,t_f) = 0, \text{ at } t = t_f, \\ \phi(x_{out},t) = 0, \text{ at } x = x_{out}. \end{cases} \quad (V.F.5)$$

We now apply the *DFSAP* and the *DASAP* to analyze the discretized versions of Eqs. (V.F.1) and (V.F.2), and thereby derive the *Discrete Adjoint Sensitivity Equations (DASE)*. Thus, the response R defined in Eq. (V.F.2) can be generally represented in discretized form as

$$R = \sum_{n=0}^{N}\sum_{j=0}^{J} \Delta t_n^q \, \Delta x_j^q \, \gamma_j^n \, d_j^n \, u_j^n. \quad (V.F.6)$$

Taking the Gâteaux-differential of Eq. (V.F.6) leads to the following expression of the discrete response sensitivity, denoted in the sequel as DDR:

$$DDR = \sum_{n=0}^{N}\sum_{j=0}^{J} \Delta t_n^q \, \Delta x_j^q \, \gamma_j^n \, d_j^n \, h_j^n + \sum_{n=0}^{N}\sum_{j=0}^{J} u_j^n \left[\sum_i \dfrac{\partial}{\partial \alpha_i}\left(\Delta t_n^q \, \Delta x_j^q \, \gamma_j^n \, d_j^n\right)\right]. \quad (V.F.7)$$

The first term on the right side of Eq. (V.F.7) is the indirect effect, which can be written in compact form as:

$$(DDR_i) \equiv \sum_{n=0}^{N}\sum_{j=0}^{J} q_j^n\, h_j^n, \text{ where } q_j^n \equiv \Delta t_n^q\, \Delta x_j^q\, \gamma_j^n\, d_j^n. \tag{V.F.8}$$

The procedure that will be now used to discretize Eq. (V.F.1) is the well known Lax scheme, which leads to the following system of nonlinear coupled algebraic equations:

$$u_j^o = u_j^{init}, \;(j=0,...,J), \; n=0; \; \text{(initial condition)}$$
$$u_0^n = u_{inlet}^n, \;(j=0,\, n=0,...,N); \text{ (boundary condition)}$$
$$u_j^n - \frac{1}{2}\left(u_{j+1}^{n-1} + u_{j-1}^{n-1}\right) + \frac{\Delta t_n}{2\Delta x_j}\left[f\!\left(u_{j+1}^{n-1}\right) - f\!\left(u_{j-1}^{n-1}\right)\right] = 0;\; (j=1,...,J-1;\, n=1,...,N);$$
$$u_J^n - u_J^{n-1} + \frac{\Delta t_n}{\Delta x_j}\left[f\!\left(u_J^{n-1}\right) - f\!\left(u_{J-1}^{n-1}\right)\right] = 0,\; j=J;\,(n=1,...,N).$$

$$\tag{V.F.9}$$

Taking the G-differential of the above system yields the following form for the *Discrete Forward Sensitivity Equations* (DFSE):

$$h_j^0 = \delta u_j^{init}, \;(j=0,...,J);\; n=0;$$
$$h_0^n = \delta u_{inlet}^n, \; j=0;\,(n=0,...,N); \tag{V.F.10}$$
$$h_j^n + a_{j-1}^{n-1} h_{j-1}^{n-1} + b_j^{n-1} h_j^{n-1} + c_{j+1}^{n-1} h_{j+1}^{n-1} = s_j^{n-1};\;(n=1,...,N;\, j=1,...,J),$$

where the various coefficients are defined below:

$$a_{j-1}^{n-1} \equiv -\left[\frac{1}{2} + \frac{\Delta t_n}{2\Delta x_j} f_u'\!\left(u_{j-1}^{n-1}\right)\right],\;(j=1,...,J-1);\; a_{J-1}^{n-1} \equiv -\frac{\Delta t_n}{\Delta x_J} f_u'\!\left(u_{J-1}^{n-1}\right);$$

$$b_j^{n-1} \equiv 0,\;(j=0,...,J-1);$$

$$b_J^{n-1} \equiv -1 + \frac{\Delta t_n}{\Delta x_J} f_u'\!\left(u_J^{n-1}\right);$$

$$c_{j+1}^{n-1} \equiv -\frac{1}{2} + \frac{\Delta t_n}{2\Delta x_j} f_u'\!\left(u_{j+1}^{n-1}\right),\;(j=1,...,J-1);$$

$$c_{J+1}^{n-1} \equiv 0;\; c_0^{n-1} \equiv c_1^{n-1} \equiv 0;$$

$$s_j^{n-1} \equiv \sum_{i=1}^{I} \frac{\partial}{\partial \alpha_i}\left\{\frac{\Delta t_n}{2\Delta x_j}\left[f\!\left(u_{j+1}^{n-1}\right) - f\!\left(u_{j-1}^{n-1}\right)\right]\right\}\delta\alpha_i,\;(j=1,...,J-1);$$

$$s_J^{n-1} \equiv \sum_{i=1}^{I} \frac{\partial}{\partial \alpha_i} \left\{ \frac{\Delta t_n}{\Delta x_J} \left[f\left(u_J^{n-1}\right) - f\left(u_{J-1}^{n-1}\right) \right] \right\}, \quad j = J. \tag{V.F.11}$$

As expected, Eq. (V.F.10) is linear in h_j^n, and can be written in matrix form as

$$H_{(J+1)}^0 \equiv \left(\delta u_0^{init}, \ldots, \delta u_J^{init}\right)^T \equiv \delta U^0$$
$$I_{(J+1)\times(J+1)} H_{(J+1)}^n + A_{(J+1)\times(J+1)}^{n-1} H_{(J+1)}^{n-1} = S_{(J+1)}^{n-1}; \quad (n = 1,\ldots,N), \tag{V.F.12}$$

where the dimensions of the various matrices have been explicitly shown, and where $A_{(J+1)\times(J+1)}^{n-1}$ is the tri-diagonal matrix

$$\begin{pmatrix} 0 & 0 & 0 & 0 & \cdots & 0 \\ a_0^{n-1} & b_1^{n-1} & c_2^{n-1} & 0 & \cdots & 0 \\ 0 & a_1^{n-1} & b_2^{n-1} & c_3^{n-1} & \cdots & 0 \\ 0 & 0 & a_2^{n-1} & b_3^{n-1} & \cdots & 0 \\ \cdots & \cdots & \cdots & \cdots & \cdots & \cdots \\ 0 & 0 & 0 & 0 & \cdots & b_J^{n-1} \end{pmatrix} \equiv A_{(J+1)\times(J+1)}^{n-1}.$$

The above system of matrix equation can now be written as a single matrix equation, in the form

$$AV = S, \tag{V.F.13}$$

where the block matrices A, V, and S are defined as

$$A \equiv \begin{pmatrix} I & 0 & 0 & \cdots & 0 & 0 \\ A^0 & I & 0 & \cdots & 0 & 0 \\ 0 & A^1 & I & \cdots & 0 & 0 \\ \cdots & \cdots & \cdots & \cdots & \cdots & \cdots \\ 0 & 0 & 0 & \cdots & I & 0 \\ 0 & 0 & 0 & \cdots & A^{N-1} & I \end{pmatrix}; V \equiv \begin{pmatrix} V^0 \\ V^1 \\ V^2 \\ \vdots \\ V^{N-1} \\ V^N \end{pmatrix}; S \equiv \begin{pmatrix} \delta U^0 \\ S^0 \\ S^1 \\ \vdots \\ S^{N-2} \\ S^{N-1} \end{pmatrix}.$$

The corresponding *Discrete Adjoint Sensitivity System (DASE)* corresponding to Eq. (V.F.13) is obtained by following the procedure detailed in Section IV.C, which leads to the (adjoint) system

$$A^T \Phi = Q, \qquad (V.F.14)$$

where the adjoint function is the block-vector

$$\Phi \equiv \left(\Phi^0, \ldots, \Phi^N\right)^T;$$
$$\Phi^n \equiv \left(\varphi_0^n, \ldots, \varphi_J^n\right)^T, \ (n=0,\ldots,N).$$

The components of the block-vector Q are defined as

$$Q \equiv \left(Q^0, \ldots, Q^N\right)^T; \ Q^n \equiv \left(q_0^n, \ldots, q_J^n\right)^T, (n=0,\ldots,N), \qquad (V.F.15)$$

where the quantities q_j^n, $(j=0,\ldots,J)$ are as defined in Eq. (V.F.8). The block-matrix equation (V.F.14) can be simplified by first writing it in component form, namely

$$\begin{pmatrix} I & (A^0)^T & 0 & \cdots & 0 & 0 \\ 0 & I & (A^1)^T & \cdots & 0 & 0 \\ 0 & 0 & I & \cdots & 0 & 0 \\ \cdots & \cdots & \cdots & \cdots & \cdots \\ 0 & 0 & 0 & \cdots & I & (A^{N-1})^T \\ 0 & 0 & 0 & \cdots & 0 & I \end{pmatrix} \begin{pmatrix} \Phi^0 \\ \Phi^1 \\ \Phi^2 \\ \vdots \\ \Phi^{N-1} \\ \Phi^N \end{pmatrix} = \begin{pmatrix} Q^0 \\ Q^1 \\ Q^2 \\ \vdots \\ Q^{N-1} \\ Q^N \end{pmatrix}$$

and performing the respective (block) matrix-vector multiplication to obtain

$$\begin{cases} \Phi^n + (A^n)^T \Phi^{n+1} = Q^n, \ (n=0,\ldots,N-1); \\ \Phi^N = Q^N, \ n = N. \end{cases}$$

For each n, the above matrix equations can be written in component form as:

$$\begin{pmatrix} 1 & 0 & 0 & \cdots & 0 & 0 \\ 0 & 1 & 0 & \cdots & 0 & 0 \\ 0 & 0 & 1 & \cdots & 0 & 0 \\ \cdots & \cdots & \cdots & \cdots & \cdots \\ 0 & 0 & 0 & \cdots & 1 & 0 \\ 0 & 0 & 0 & \cdots & 0 & 1 \end{pmatrix} \begin{pmatrix} \varphi_0^n \\ \varphi_1^n \\ \varphi_2^n \\ \vdots \\ \varphi_{J-1}^n \\ \varphi_J^n \end{pmatrix} +$$

Local Sensitivity and Uncertainty Analysis of Nonlinear Systems

$$\begin{pmatrix} 0 & a_0^n & 0 & \cdots & 0 & 0 \\ 0 & b_1^n & a_1^n & \cdots & 0 & 0 \\ 0 & c_2^n & b_2^n & \cdots & 0 & 0 \\ \cdots & \cdots & \cdots & \cdots & \cdots & \\ 0 & 0 & 0 & \cdots & b_{J-1}^n & a_{J-1}^n \\ 0 & 0 & 0 & \cdots & c_J^n & b_J^n \end{pmatrix} \begin{pmatrix} \varphi_0^{n+1} \\ \varphi_1^{n+1} \\ \varphi_2^{n+1} \\ \vdots \\ \varphi_{J-1}^{n+1} \\ \varphi_J^{n+1} \end{pmatrix} = \begin{pmatrix} q_0^n \\ q_1^n \\ q_2^n \\ \vdots \\ q_{J-1}^n \\ q_J^n \end{pmatrix}.$$

Performing now the matrix-vector multiplication in the above expression reduces it to

$$\begin{cases} \varphi_j^N = q_j^N, \ (j = 0,\ldots,J); \ n = N; \\ \varphi_j^n + c_j^n \varphi_{j-1}^{n+1} + b_j^n \varphi_j^{n+1} + a_j^n \varphi_{j+1}^{n+1} = q_j^n, \\ \qquad (j = 0,\ldots,J; \ n = 0,\ldots,N-1). \end{cases} \qquad \text{(V.F.16)}$$

Introducing the explicit expressions for a_j^n, b_j^n, and c_j^n in Eq. (V.F.16) yields the component form of the *DASE* [i.e., Eq. (V.F.17)], which is consistent with the differential *Adjoint Sensitivity Equations* given in Eq. (V.F.5) but differs from the discrete form that would be obtained by applying the Lax scheme to it.

$$\begin{cases} \varphi_0^n - \left[\dfrac{1}{2} + \dfrac{\Delta t_n}{2x_0} f_u'(u_0^n)\right] \varphi_1^{n+1} = q_0^n \\[4pt] \varphi_1^n - \left[\dfrac{1}{2} + \dfrac{\Delta t_n}{2x_1} f_u'(u_1^n)\right] \varphi_2^{n+1} = q_1^n \\[4pt] \varphi_j^n - \left[\dfrac{1}{2} - \dfrac{\Delta t_n}{2x_j} f_u'(u_j^n)\right] \varphi_{j-1}^{n+1} \\[4pt] \quad - \left[\dfrac{1}{2} + \dfrac{\Delta t_n}{2x_j} f_u'(u_j^n)\right] \varphi_{j+1}^{n+1} = q_j^n; \ (j = 2,\ldots,J-2); \\[4pt] \varphi_{J-1}^n - \left[\dfrac{1}{2} - \dfrac{\Delta t_n}{2x_{J-1}} f_u'(u_{J-1}^n)\right] \varphi_{J-2}^{n+1} \\[4pt] \quad - \dfrac{\Delta t_n}{x_{J-1}} f_u'(u_{J-1}^n) \varphi_J^{n+1} = q_{J-1}^n; \ j = J-1 \\[4pt] \varphi_J^n - \left[\dfrac{1}{2} - \dfrac{\Delta t_n}{2x_J} f_u'(u_J^n)\right] \varphi_{J-1}^{n+1} + \left[-1 + \dfrac{\Delta t_n}{x_J} f_u'(u_J^n)\right] \varphi_J^{n+1} = q_J^n. \end{cases} \qquad \text{(V.F.17)}$$

Chapter VI

GLOBAL OPTIMIZATION AND SENSITIVITY ANALYSIS

As has been discussed in Chapters IV and V, the scope of local sensitivity analysis is to calculate exactly and efficiently the sensitivities of the system's response to variations in the system's parameters, around their nominal values. As has also been shown in those chapters, the sensitivities are given by the first Gâteaux-differential of the system's response, calculated at the nominal value of the system's dependent variables (i.e., state functions) and parameters. Two procedures were developed for calculating the sensitivities, namely the *Forward Sensitivity Analysis Procedure* (*FSAP*) and the *Adjoint Sensitivity Analysis Procedure* (*ASAP*). Once they became available, the sensitivities could be used for various purposes, such as for ranking the respective parameters in order of their relative importance to the response, for assessing changes in the response due to parameter variations, or for performing uncertainty analysis by using the propagation of errors (moments) procedure presented in Section III.F. In particular, the changes in the response due to parameter variations can be calculated by using the multivariate Taylor series expansion given in Eq. (III.F.3), which is reproduced, for convenience, below:

$$R(\alpha_1,\ldots,\alpha_k) \equiv R\left(\alpha_1^0 + \delta\alpha_1,\ldots,\alpha_k^0 + \delta\alpha_k\right)$$

$$= R(\boldsymbol{\alpha}^0) + \sum_{i_1=1}^{k}\left(\frac{\partial R}{\partial \alpha_{i_1}}\right)_{\boldsymbol{\alpha}_0}\delta\alpha_{i_1} + \frac{1}{2}\sum_{i_1,i_2=1}^{k}\left(\frac{\partial^2 R}{\partial \alpha_{i_1}\partial \alpha_{i_2}}\right)_{\boldsymbol{\alpha}_0}\delta\alpha_{i_1}\delta\alpha_{i_2}$$

$$+\frac{1}{3!}\sum_{i_1,i_2,i_3=1}^{k}\left(\frac{\partial^3 R}{\partial \alpha_{i_1}\partial \alpha_{i_2}\partial \alpha_{i_3}}\right)_{\boldsymbol{\alpha}_0}\delta\alpha_{i_1}\delta\alpha_{i_2}\delta\alpha_{i_3} + \cdots \qquad \text{(III.F.3)}$$

$$+\frac{1}{n!}\sum_{i_1,i_2,\ldots,i_n=1}^{k}\left(\frac{\partial^n R}{\partial \alpha_{i_1}\partial \alpha_{i_2}\ldots\partial \alpha_{i_n}}\right)_{\boldsymbol{\alpha}_0}\delta\alpha_{i_1}\ldots\delta\alpha_{i_n} + \cdots.$$

As has been discussed in Chapters IV and V, the *FSAP* is conceptually easier to develop and implement than the *ASAP*; however, the *FSAP* is advantageous to employ only if, in the problem under consideration, the number of different responses of interest exceeds the number of system parameters and/or parameter variations to be considered. This situation, though, does not often occur in practice, since most problems of practical interest comprise a large number of parameters and comparatively few responses. In such situations, it is by far more advantageous to employ the *ASAP*. As has been shown in Chapter IV, *the adjoint sensitivity equations underlying the ASAP can be formulated and solved independently of the original (forward) equations* if *the physical systems* under consideration *are linear in the dependent (state) variables*. On the other hand, when the *physical systems under consideration involve operators that act nonlinearly on the dependent variables, the adjoint sensitivity equations underlying the ASAP depend on the base-case values of the dependent (state) variables and cannot, therefore, be solved independently of the original (forward) equations.*

As has been highlighted in Chapters IV and V, the concept of local sensitivity was defined in terms of the first Gâteaux-differential of the system's response, calculated at the nominal value of the system's dependent variables (i.e., state functions) and parameters. Thus, the local sensitivities would give only the first order contributions to the response variation $R\left(\alpha_1^0 + \delta\alpha_1, \ldots, \alpha_k^0 + \delta\alpha_k\right)$ in Eq. (III.F.3) and/or to the response statistics in Eqs. (III.F.17) through (III.F.20). As these equations show, the contributions from the higher order terms in $\|\delta\alpha\|$ would require knowledge of the higher order Gâteaux-differentials of the response and, consequently, of the higher order Gâteaux-differentials of the operator equations underlying the mathematical description of the physical system under investigation. Although several techniques have been proposed in the literature (see the References Section) for calculating the higher order Gâteaux-differentials of the response and system's operator equations, none of these techniques has proved routinely practicable for realistic problems. This is because *the systems of equations that need to be solved for obtaining the second (and higher) order Gâteaux-differentials of the response and system's operator equations are very large and depend on the perturbation* $\delta\alpha$. Thus, even the calculation of the second-order Gâteaux-differentials of the response and system's operator equations is just as difficult as undertaking the complete task of computing the exact value of perturbed response $R\left(\alpha_1^0 + \delta\alpha_1, \ldots, \alpha_k^0 + \delta\alpha_k\right)$. Moreover, the (operator) Taylor series defined in Eq. (I.D.13), namely

$$F(x_0 + h) = F(x_0) + \delta F(x_0; h) + \frac{1}{2}\delta^2 F(x_0; h) + \ldots$$
$$+ \frac{1}{(n-1)!}\delta^{n-1} F(x_0; h) + \int_0^1 \frac{(1-t)^{n-1}}{(n-1)!}\delta^n F(x_0 + th; h)\,dt,$$

(I.D.13)

clearly shows that the *Taylor series is a local concept, valid within some radius of convergence of the series around the base-case value* x_0. This means that even if the Gâteaux-differentials of the response $R(\alpha)$ around α^0 were available to all orders, they would still merely provide local, but not global, information. They would yield little, if any, information about the important global features of the physical system, namely the critical points of $R(\alpha)$ and the bifurcation branches and/or turning points of the system's state variables. Furthermore, it would be very difficult, if not impossible, to calculate the requisite higher-order Gâteaux-differentials.

As has been shown in Chapters IV and V, the objective of *local sensitivity analysis* is to analyze the behavior of the system responses locally around a chosen point or trajectory in the combined phase space of parameters and state variables. On the other hand, the objective of *global sensitivity analysis* is to determine all of the system's critical points (bifurcations, turning points, response extrema) in the combined phase space formed by the parameters, state variables, and adjoint variables, and subsequently analyze these critical points by local sensitivity analysis. To devise a framework for attaining the objective of global sensitivity analysis, we consider, as in Chapters IV and V, a general representation of a physical system, modeled mathematically in terms of: (a) linear and/or nonlinear equations that relate the system's independent variables and parameters to the system's state (i.e., dependent) variables, (b) inequality and/or equality constraints that delimit the ranges of the system's parameters, and (c) one or several system responses (or objective functions, or indices of performance) that are to be analyzed as the parameters vary over their respective ranges. The material presented in this Chapter is based largely on the work by Cacuci, *Nucl. Sci. Eng.*, 104, 78, 1990 (with permission).

VI.A. MATHEMATICAL FRAMEWORK

Throughout this section, the mathematical model of the physical system and associated responses are represented by nonlinear algebraic relationships as would be required prior to numerical computations. This description in terms of algebraic (as opposed to differential and/or integral) operators in finite dimensional vector spaces simplifies considerably the mathematical manipulations to follow without detracting from the conceptual generality and applicability of the underlying methodology. In the sequel, all vectors are considered to be column vectors, while the superscript T will be used to denote transposition.

The canonical discretized mathematical representation of the physical system under consideration is as follows:

1. m linear and/or nonlinear equations:

$$N(\varphi,\alpha)=0, \quad N:\mathscr{D}_N \subset \mathscr{D}_\varphi \times \mathbb{R}^i \to \mathbb{R}^m, \qquad \text{(VI.A.1)}$$

where

$$N \equiv [N_1(\varphi,\alpha),\ldots,N_m(\varphi,\alpha)] \qquad \text{(VI.A.2)}$$

is an m-component column vector whose components are linear and/or nonlinear operators; N is here defined on a domain $\mathscr{D}_N$ and takes values in the Euclidean space $\mathbb{R}^m$. Each component of N is considered to operate on the vectors α and φ, where

$$\alpha \equiv (\alpha_1,\ldots,\alpha_i), \quad \alpha \in \mathbb{R}^i, \qquad \text{(VI.A.3)}$$

is an i-component column vector, comprising the system parameters, and

$$\varphi \equiv (\varphi_1,\ldots,\varphi_m), \quad \varphi: \mathscr{D}_\varphi \subset \mathbb{R}^m, \qquad \text{(VI.A.4)}$$

is an m-component column vector, comprising the system's dependent variables, defined on a domain $\mathscr{D}_\varphi \subset \mathbb{R}^m$. Note that the components of both α and φ are considered here to be scalar quantities, taking on values in the real Euclidean spaces $\mathbb{R}^i$ and $\mathbb{R}^m$, respectively. Since Eq. (VI.A.1) is a canonical representation for problems that have been fully discretized in preparation for a numerical solution, it automatically comprises all initial and/or boundary conditions that may have appeared in the originally continuous-variable description of the physical problem.

2. k inequality and/or equality constraints:

$$g(\alpha) \leq 0, \quad g: \mathscr{D}_g \subset \mathbb{R}^i \to \mathbb{R}^k, \qquad \text{(VI.A.5)}$$

where $g(\alpha) \equiv [g_1(\alpha),\ldots,g_k(\alpha)]$ is a k-component column vector, defined on a domain $\mathscr{D}_g$ that delimits, directly or indirectly, the range of the parameters α_i.

3. a system response:

$$R(\varphi,\alpha), \quad R: \mathscr{D}_P \subset \mathscr{D}_\varphi \times \mathbb{R}^i \to \mathbb{R}^1. \qquad \text{(VI.A.6)}$$

In the discrete formulation framework considered here, R is a real-valued functional defined on a domain $\mathscr{D}_P$, having its range in $\mathbb{R}^1$.

VI.B. CRITICAL POINTS AND GLOBAL OPTIMIZATION

This Section presents the general framework for analyzing the global optimization of the physical system defined above in Section VI.A. This framework provides a natural setting for subsequently formulating (in Section VI.C) a global framework for sensitivity and uncertainty analysis. Consider, therefore, the fundamental problem of global optimization, namely to find the points that minimize or maximize the system response $R(\varphi,\alpha)$ subject to the equality and inequality constraints represented by Eqs. (VI.A.1) and (VI.A.5). This problem is typically handled by introducing the Lagrange functional $L(\varphi, y, \alpha, z)$, defined as

$$L(\varphi, y, \alpha, z) \equiv R(\varphi,\alpha) + \langle y, N(\varphi,\alpha) \rangle_m + \langle z, g(\alpha) \rangle_k, \qquad \text{(VI.B.1)}$$

where the angular brackets denote inner products in $\mathbb{R}^m$ and $\mathbb{R}^k$, respectively; i.e.,

$$\langle a, b \rangle_n = a \bullet b = \sum_{i=1}^{n} a_i b_i; \quad a, b \in \mathbb{R}^n; \quad (n = m \text{ or } k), \qquad \text{(VI.B.2)}$$

while

$$y = (y_1, \ldots, y_m) \qquad \text{(VI.B.3)}$$

and

$$z = (z_1, \ldots, z_k) \qquad \text{(VI.B.4)}$$

are column vectors of Lagrange multipliers. The critical points (i.e., extrema) of R are found among the points that cause the first Gâteaux-variation δL of L to vanish for arbitrary variations $\delta\varphi, \delta y, \delta z$, and $\delta\alpha$. From Eq. (VI.B.1), δL is obtained as

$$\delta L(\varphi, y, \alpha, z) = \delta\varphi \bullet N^+(\varphi,y,\alpha) + \delta y \bullet N(\varphi,\alpha) + \delta z \bullet g(\alpha) + \delta\alpha \bullet S(\varphi, y, \alpha, z),$$
$$\text{(VI.B.5)}$$

where the column vectors N^+ and S are defined as

$$N^+(\varphi,y,\alpha) \equiv \nabla_\varphi R + (\nabla_\varphi N) y \qquad \text{(VI.B.6)}$$

and

$$S(\varphi, y, \alpha, z) \equiv \nabla_\alpha R + (\nabla_\alpha N)y + (\nabla_\alpha g)z, \qquad \text{(VI.B.7)}$$

respectively. The gradient vectors and matrices appearing above in Eqs. (VI.B.6) and (VI.B.7) are defined as follows:

$$\nabla_\varphi R \equiv (\partial R/\partial \varphi_p)_{m \times 1}, \quad \nabla_\alpha R \equiv (\partial R/\partial \alpha_q)_{i \times 1}, \qquad \text{(VI.B.8)}$$

$$\nabla_\varphi N \equiv (\partial N_r/\partial \varphi_p)_{m \times m}, \quad \nabla_\alpha N \equiv (\partial N_r/\partial \alpha_q)_{i \times m}, \qquad \text{(VI.B.9)}$$

$$\nabla g \equiv (\partial g_s/\partial \alpha_q)_{i \times k}. \qquad \text{(VI.B.10)}$$

Note that the derivatives appearing in Eqs. (VI.B.8-10) above are first-order partial Gâteaux-derivatives.

The requirements that the first Gâteaux-variation δL of L vanish, for arbitrary $\delta\varphi, \delta y, \delta z$, and $\delta\alpha$, together with the constraints $g \leq 0$, lead to the following (Karush-Kuhn-Tucker) necessary conditions:

$$N^+(\varphi, y, \alpha) = 0, \; S(\varphi, y, \alpha, z) = 0, \; N(\varphi, \alpha) = 0, \; z \bullet g = 0, \; g \leq 0, \; z \geq 0,$$
$$\text{(VI.B.11)}$$

for the minima of $R(\varphi, \alpha)$ and similar conditions (except that $z \leq 0$) for the maxima of $R(\varphi, \alpha)$. The inequalities in Eq. (VI.B.11) imply a lack of global differentiability so a direct solution is usually hampered by computational difficulties. Such computational difficulties can be mitigated by recasting the last three conditions in Eq. (VI.B.11) into the following equivalent form which involves equalities only:

$$\boldsymbol{K} \equiv (K_1, \ldots, K_k) = \boldsymbol{0}, \qquad \text{(VI.B.12)}$$

where the component K_i of the column vector $\boldsymbol{K}$ is defined as

$$K_i \equiv (g_i + z_i)^2 + g_i|g_i| - z_i|z_i|. \qquad \text{(VI.B.13)}$$

Using Eq. (VI.B.12) in Eq. (VI.B.11) makes it possible to recast the latter into the equivalent form

$$F(u) \equiv [N^+(u), N(u), S(u), K(u)] = \boldsymbol{0}, \qquad \text{(VI.B.14)}$$

where the components of the column vector

$$u \equiv (\varphi, y, \alpha, z), \quad u \in \mathbb{R}^{2m+i+k} \qquad (VI.B.15)$$

are, respectively: the dependent variables, φ; their corresponding Lagrange multipliers, y; the parameters, α; and the Lagrange multipliers, z, corresponding to the inequalities constraints, g. As will be shown in the sequel, choosing the above structure of F allows considerable simplifications in the global numerical procedure to be devised for finding the roots and critical points of Eq. (VI.B.14). It is important to note that F is globally differentiable if R, N, and g are differentiable twice globally. The Jacobian matrix $F'(u)$ of $F(u)$ has the block matrix structure

$$F'(u) = \begin{bmatrix} \nabla_\varphi N^+ & \nabla_\varphi N & \nabla_\varphi S & 0 & 0 \\ (\nabla_\varphi N)^T & 0 & (\nabla_\alpha N)^T & 0 & 0 \\ (\nabla_\varphi S)^T & \nabla_\alpha N & \nabla_\alpha S & \nabla g_{\mathcal{A}} & \nabla g_{\mathcal{I}} \\ 0 & 0 & 2Z_{\mathcal{A}}(\nabla g_{\mathcal{A}})^T & 0 & 0 \\ 0 & 0 & 0 & 0 & 2C_{\mathcal{I}} \end{bmatrix}, \qquad (VI.B.16)$$

where $\mathcal{A} \equiv \{j \mid g_j = 0\}$ and $\mathcal{I} \equiv \{j \mid g_j < 0\}$ denote the set of indices corresponding to the active and inactive constraints, respectively, while the remaining quantities are defined as follows:

$$\nabla_\varphi N^+ \equiv \left[\partial^2 R / \partial \varphi_p \partial \varphi_r + \left\langle y, \partial^2 N / \partial \varphi_p \partial \varphi_r \right\rangle_m \right]_{m \times m}, \qquad (VI.B.17)$$

$$\nabla_\varphi S \equiv \left[\partial^2 R / \partial \varphi_p \partial \alpha_q + \left\langle y, \partial^2 N / \partial \varphi_p \partial \alpha_q \right\rangle_m \right]_{m \times i}, \qquad (VI.B.18)$$

$$\nabla_\alpha S \equiv \left[\partial^2 R / \partial \alpha_q \partial \alpha_r + \left\langle y, \partial^2 N / \partial \alpha_q \partial \alpha_r \right\rangle_m + \left\langle z, \partial^2 g / \partial \alpha_q \partial \alpha_r \right\rangle_k \right]_{i \times i},$$

$$\qquad (VI.B.19)$$
$$Z_{\mathcal{A}} \equiv diag(z_j)_{j \in \mathcal{A}}, \qquad (VI.B.20)$$
$$C_{\mathcal{I}} \equiv diag(g_j)_{j \in \mathcal{I}}, \qquad (VI.B.21)$$
$$\nabla g_{\mathcal{A}} \equiv (\partial g_j / \partial \alpha_q)_{(j \in \mathcal{A}) \times i}, \qquad (VI.B.22)$$
$$\nabla g_{\mathcal{I}} \equiv (\partial g_j / \partial \alpha_q)_{(j \in \mathcal{I}) \times i}. \qquad (VI.B.23)$$

All derivatives appearing in Eqs. (VI.B.17-23) are partial (first- and second-order, respectively) Gâteaux-derivatives.

Note that in the two extreme situations when the constraints are either all inactive or all active, the matrices $\nabla g_{\mathcal{A}}$ and $2Z_{\mathcal{A}}(\nabla g_{\mathcal{A}})^T$ or the matrices $\nabla g_{\mathcal{I}}$ and $C_{\mathcal{I}}$ disappear, respectively, from the structure of $F'(u)$ in Eq. (VI.B.16).

By introducing the equivalence (VI.B.11)$\Leftrightarrow$(VI.B.14), all inequalities have disappeared from $F(u) = 0$. Furthermore, the Jacobian $F'(u)$ is nonsingular at the zeros of $F(u)$ so efficient numerical methods, such as locally superlinearly convergent quasi-Newton methods, can be used to find these zeros. The Jacobian $F'(u)$ vanishes, though, at the bifurcation and limit and/or turning points present in our system. Such critical points need to be located by using global methods that are capable of avoiding local nonconvergence problems (a class of such methods is presented in Sec. VI.D, below).

VI.C. SENSITIVITY ANALYSIS

The fundamental problem in sensitivity analysis is to determine the effects caused by variations $\delta\alpha$ around nominal parameter values α^0 in the response $R(\varphi,\alpha)$, subject to the constraints represented by Eqs. (VI.A.1) and (VI.A.5). Within the framework of sensitivity analysis, therefore, both α^0 and $\delta\alpha$ are known, at the outset. As shown in Chapters IV and V, the preliminary steps in sensitivity analysis are (a) to use the known values of α^0 for solving Eq. (VI.A.1), namely $N(\varphi^0,\alpha^0) = 0$, to obtain the nominal values φ^0 of the state (i.e., dependent) variables φ; (b) to use the nominal values α^0 and φ^0 subsequently, to obtain the nominal value $R^0 \equiv R(\varphi^0,\alpha^0)$ of the response R (while verifying that the constraints $g(\alpha^0) \le 0$ are satisfied at α^0). Then, since both α^0 and $\delta\alpha$ are known, Eq. (VI.A.1) can, in principle, be solved anew for $\alpha = \alpha^0 + \delta\alpha$ to obtain the corresponding dependent (state) variable function $\varphi = \varphi(\alpha)$. The new value of the dependent variable, $\varphi(\alpha)$, thus obtained is used together with α to calculate the new response value $R(\varphi(\alpha),\alpha)$. Variations $\delta\alpha$ that would violate the constraints $g(\alpha) \le 0$ could be handled *a priori*, since the constraints $g(\alpha)$ do not depend on $\varphi(\alpha)$.

At this stage, it is important to highlight the fundamentally distinct roles played by the system parameters α in optimization (as discussed in Section VI.B) versus sensitivity analysis. In optimization, although the nominal parameter values α^0 may be used as starting points for the optimization procedure, the parameters α are allowed to vary freely over their respective ranges while the optimization procedure determines their final values α^* that optimize the system response R. Thus, the α's are (part of the vector of) unknowns together with the state variables φ, and Lagrange multipliers y and z. Thus, the final values, α^*, at the optimal point in phase-space are not known *a priori*, but are obtained after solving Eq. (VI.B.14). This is in contradistinction with the framework of sensitivity analysis, where the α's are not unknown, but are given by the known relation $\alpha = \alpha^0 + \delta\alpha$.

Consequently, the framework of sensitivity analysis will differ from the framework of global optimization described in Section VI.B; these differences will become apparent in the sequel.

In sensitivity analysis, variations $\delta\alpha$ around α^0 induce variations

$$\Delta R(\varphi^0, \alpha^0) \equiv R(\varphi, \alpha) - R(\varphi^0, \alpha^0)$$
$$= \delta\alpha \bullet \{\nabla_\alpha R\}_{(\varphi^0, \alpha^0)} + \delta\varphi \bullet \{\nabla_\varphi R\}_{(\varphi^0, \alpha^0)} + O(\|\Delta\alpha\|^2, \|\Delta\varphi\|^2).$$
(VI.C.1)

The above expression explicitly indicates the important fact that the respective gradients are to be evaluated at the nominal values α^0 and $\varphi^0 = \varphi(\alpha^0)$, obtained from the solution of $N(\varphi^0, \alpha^0) = 0$. As has been shown in Chapters IV and V, the variations $\delta\varphi$ and $\delta\alpha$ in Eq. (VI.C.1) are not independent, but are related to each other through the relationship

$$\begin{bmatrix} \{\nabla_\varphi N\}^T_{(\varphi^0, \alpha^0)} & \{\nabla_\alpha N\}^T_{(\varphi^0, \alpha^0)} \\ 0 & \{\nabla_\alpha g\}^T_{(\varphi^0, \alpha^0)} \end{bmatrix} \begin{pmatrix} \delta\varphi \\ \delta\alpha \end{pmatrix} = O(\|\Delta\alpha\|^2, \|\Delta\varphi\|^2),$$
(VI.C.2)

which are obtained by taking Gâteaux-differentials of Eqs. (VI.A.1) and (VI.A.5), respectively. Forming the inner product of Eq. (VI.C.2) with the partitioned vector (y, z), and transposing the resulting expression (which corresponds to calculating the respective adjoints in the continuous-variable case) yields

$$\delta\varphi \bullet \{\nabla_\varphi N\}_{(\varphi^0, \alpha^0)} y + \delta\alpha \bullet [\{\nabla_\alpha N\}_{(\varphi^0, \alpha^0)} y + \{\nabla_\alpha g\}_{(\varphi^0, \alpha^0)} z] = O(\|\Delta\alpha\|^2, \|\Delta\varphi\|^2).$$
(VI.C.3)

Requiring next that the Lagrange multipliers (which are actually the adjoint functions) y^0 be the solutions of the adjoint equations,

$$\{\nabla_\varphi N\}_{(\varphi^0,\alpha^0)} y^0 + \{\nabla_\varphi R\}_{(\varphi^0,\alpha^0)} = \mathbf{0}, \tag{VI.C.4}$$

and using their resulting expressions in Eq. (VI.C.3) gives

$$\delta\varphi \bullet \{\nabla_\varphi R\}_{(\varphi^0,\alpha^0)} = \delta\alpha \bullet \left[\{\nabla_\alpha N\}_{(\varphi^0,\alpha^0)} y^0 + \{\nabla_\alpha g\}_{(\varphi^0,\alpha^0)} z^0 \right] + O\!\left(\|\Delta\alpha\|^2, \|\Delta\varphi\|^2\right). \tag{VI.C.5}$$

Finally, replacing Eq. (VI.C.5) in Eq. (VI.C.1) gives

$$\Delta R(\varphi^0, \alpha^0) = \delta\alpha \bullet S(\varphi^0, y^0, \alpha^0, z^0) + O\!\left(\|\Delta\alpha\|^2, \|\Delta\varphi\|^2\right), \tag{VI.C.6}$$

where the components of the vector $S(\varphi^0, y^0, \alpha^0, z^0)$, defined as

$$S(\varphi^0, y^0, \alpha^0, z^0) \equiv \{\nabla_\alpha R\}_{(\varphi^0,\alpha^0)} + \{\nabla_\alpha N\}_{(\varphi^0,\alpha^0)} y^0 + \{\nabla_\alpha g\}_{(\varphi^0,\alpha^0)} z^0, \tag{VI.C.7}$$

contain the first-order sensitivities at α^0 of $R(\varphi^0, \alpha^0)$ to variations $\delta\alpha$, and where the adjoint variables y^0 are the solutions of Eq. (VI.C.4) while the Lagrange multipliers z are chosen such as to enforce the linear independence of the constraints, i.e., $z \bullet g = 0$; note that the adjoint system is occasionally defined with $(-y^0)$ replacing y^0 in Eq. (VI.C.4).

Adopting a variational point of view, the sensitivity expression given by Eq. (VI.C.7) can also be obtained by using the same Lagrange functional as introduced used in (VI.B.1), in the previous Section for the optimization procedure, namely:

$$L(\varphi, y, \alpha, z) \equiv R(\varphi,\alpha) + \langle y, N(\varphi,\alpha)\rangle_m + \langle z, g(\alpha)\rangle_k. \tag{VI.B.1}$$

The first Gâteaux-variation δL of L has, of course, the same form as given in Eq. (VI.B.5), namely

$$\begin{aligned}\delta L(\varphi, y, \alpha, z) = \delta\varphi \bullet [\nabla_\varphi R + (\nabla_\varphi N) y] + \delta y \bullet N(\varphi,\alpha) + \delta z \bullet g(\alpha) \\ + \delta\alpha \bullet [\nabla_\alpha R + (\nabla_\alpha N) y + (\nabla_\alpha g) z].\end{aligned} \tag{VI.B.5}$$

Imposing the requirement that δL be stationary at (φ^0, α^0) with respect to arbitrary variations $\delta\varphi, \delta y, \delta z$ but not $\delta\alpha$ in the above equation yields the expression

$$\delta L(\varphi^0, y^0, \alpha^0, z^0) = \delta\alpha \cdot \left[\{\nabla_\alpha R\}_{(\varphi^0,\alpha^0)} + \{\nabla_\alpha N\}_{(\varphi^0,\alpha^0)} y^0 + \{\nabla_\alpha g\}_{(\varphi^0,\alpha^0)} z^0\right]$$
$$= \Delta R(\varphi^0, \alpha^0) + O(\|\Delta\alpha\|^2, \|\Delta\varphi\|^2),$$
(VI.C.8)

which is the same expression for the sensitivity ΔR of R at (φ^0, α^0) as provided by Eq. (VI.C.6); of course, the quantities φ^0, y^0, α^0, and z^0 in Eq. (VI.C.8) are the same as in Eq. (VI.C.6).

Comparing Eqs. (VI.C.5) and (VI.B.5) reveals additional fundamental conceptual differences between the frameworks of optimization and sensitivity analysis. It is thus apparent that although both frameworks use the first Gâteaux-variation $\delta L(\varphi, y, \alpha, z)$ of *the same Lagrangian functional $L(\varphi, y, \alpha, z)$, but the respective uses are conceptually quite distinct*. On the one hand, the sensitivity analysis framework revolves fundamentally around the *a priori known* nominal point α^0, and imposes on $\delta L(\varphi, y, \alpha, z)$ the requirements that (a) δL be stationary at α^0 with respect to δy, which implies that $N(\varphi^0, \alpha^0) = 0$, thereby fixing the nominal values φ^0; (b) δL be stationary at (φ^0, α^0) with respect to $\delta\varphi$, which implies that the adjoint variables y^0 must satisfy the adjoint equation (VI.C.4) at (φ^0, α^0); and (c) δL be stationary at (φ^0, α^0) with respect to δz, which ensures that the constraints $g(\alpha^0)$ are satisfied while fixing the values z^0 of the Lagrange multipliers z.

On the other hand, the optimization theory framework for finding the critical points of $R(\varphi, \alpha)$ leads to Eq. (VI.B.5), which imposes the requirements that (a) δL be stationary with respect to δy, implying that $N(\varphi^*, \alpha^*) = 0$ at the *yet unknown critical point* (φ^*, α^*); (b) δL be stationary with respect to $\delta\varphi$, implying that y^* must satisfy at (φ^*, α^*) the adjoint equation $N^+(\varphi^*, y^*, \alpha^*) \equiv \{\nabla_\varphi N\}_{(\varphi^*,\alpha^*)} y^* + \{\nabla_\varphi R\}_{(\varphi^*,\alpha^*)} = 0$; (c) δL be stationary with respect to δz, implying that the respective constraints must be satisfied at the yet unknown critical point (φ^*, α^*); and, *in contradistinction to the sensitivity analysis framework*, the requirement that (d) δL be stationary with respect to $\delta\alpha$, implying that

$$S(\varphi^*, y^*, \alpha^*, z^*) \equiv \{\nabla_\alpha P\}_{(\varphi^*,\alpha^*)} + \{\nabla_\alpha N\}_{(\varphi^*,\alpha^*)} y^* + \{\nabla_\alpha g\}_{(\varphi^*,\alpha^*)} z^* = 0 \quad \text{(VI.C.9)}$$

at the critical point (φ^*, α^*). Note that the critical point (φ^*, α^*) is not *a priori* known, but is to be determined from the solution of Eq. (VI.B.14). In other words, conditions (a) through (d) constitute a system of $2m+i+k$ equations whose simultaneous solution yields the regular critical points (φ^*,α^*) of $R(\varphi^*,\alpha^*)$ together with the respective values y^* and z^* of the Lagrange multipliers. *Note the important fact that condition (d) above cannot be imposed within the framework of sensitivity analysis, since δL cannot be required to vanish for known variations $\delta\alpha$ around a fixed, a priori known point (φ^0,α^0); δL will vanish for variations $\delta\alpha$ only if the respective point (φ^0,α^0) happens to coincide with a regular critical point (φ^*,α^*).*

VI.D. GLOBAL COMPUTATION OF FIXED POINTS

It is important to note that the vector $F(u)$ defined in Eq. (VI.B.14) contains all the features necessary for unifying the scopes of global optimization with those of sensitivity analysis. This is because, one the one hand, by setting to zero all *four* components of $F(u)$, namely: $N(u), N^+(u), K(u)$, and $S(u)$, yields the respective critical points u^*, as has been shown in Section VI.B. On the other hand, as has been shown in Section VI.C, setting only *three* components of $F(u)$, i.e., $N(u), N^+(u)$, and $K(u)$, to zero at *any* point u^0 and calculating its fourth component $S(u^0)$ yields the first-order local sensitivities at u^0. Of course, $S(u^0)$ would vanish if u^0 happened to coincide with a critical point u^* of $R(\varphi,\alpha)$. Devising a computational algorithm that could determine the features mentioned in the foregoing, over the global space of allowed parameter variations, would extend the scopes of both optimization and sensitivity analysis while unifying their underlying computational methodologies. A computational algorithm for achieving all of these goals will now be described in the remainder of this Section.

The initial information available in practical problems comprises the nominal design values α^0 of the problem's parameters α, and the ranges over which the respective parameters can vary, as expressed mathematically by the constraints in Eq. (VI.A.5). As the parameters α vary over their respective ranges, the state variables φ and the response R vary in phase space. The objective of the computational algorithm, therefore, is to determine all of the critical points where the solution path φ bifurcates (i.e., splits in two or more branches) and where the response R attains maxima, minima, and/or saddle points. As discussed in Sec. VI.B, the bifurcation points occur at the zeros of the determinant of the Jacobian matrix $F'(u)$, denoted as $Det[F'(u)]$, while the maxima, minima and/or saddle points of R occur at the zeros of $F(u)$, respectively.

To determine all the zeros of both $F(u)$ and $Det[F'(u)]$, we must use an algorithm that embodies the following properties: (a) efficiently avoids non-convergence problems at the singularities of $F'(u)$, (b) avoids getting bogged down (as many local methods do) at the first zero of $F(u)$ that it may happen to find, and (c) finds all the fixed points $u_i{}^*$ of $F(u)$ regardless of the starting point u_0. It is apparent that local computational methods would not satisfy all these requirements; a global method would need to be devised for this purpose.

Perhaps the most powerful mathematical techniques for obtaining global results are the homotopy theory-based continuation methods. Typically, these methods compute the solution u^* of a fixed-point equation such as Eq. (VI.B.14) by embedding it into a one-parameter family of equations of the form $G(u,\lambda) = 0$, where $\lambda \in \mathbb{R}^1$ is a real scalar. Most of the direct procedures to follow the path $u(\lambda)$ typically encounter difficulties (slow convergence, small steps, or even failure) at points where the Fréchet-derivative G'_u does not have a bounded inverse. Such difficulties can be efficiently circumvented by the "pseudo-arc-length" (i.e., distance along a local tangent) continuation methods, which employ a scalar constraint in addition to the homotopy $G(u,\lambda) = 0$, thus "inflating" the original problem into one of higher dimension.

To determine the roots of $F(u) = 0$ and $F'(u) = 0$ by means of the pseudo-arc-length continuation method, the homotopy $G(u, \lambda) = 0$ is specified to be of the form

$$G(u, \lambda) \equiv F[u(s)] - \lambda(s) F(u_0) = 0 . \qquad (\text{VI.D.1})$$

In addition to the above homotopy, we impose the condition that

$$\|w(s)\|^2 + \mu(s) = 1, \quad w(s) \equiv \frac{d}{ds} u(s), \quad \mu(s) \equiv \frac{d}{ds} \lambda(s) \qquad (\text{VI.D.2})$$

where $s \in \mathbb{R}^1$ is a real parameter, $\lambda(s) \in \mathbb{R}^1$ is a function of s, $F[u(s)]$ is as defined in Eq. (VI.B.14), while u_0 denotes a fixed (e.g., starting) value of u. Note that Eq. (VI.D.1) implies that $u(s)$ must satisfy the differential equation obtained by setting to zero its first Gâteaux-derivative with respect to s, namely

$$F'(u) w(s) + \mu(s) F(u_0) = 0 . \qquad (\text{VI.D.3})$$

Furthermore, the imposition of Eq. (VI.D.2) makes s the arclength parameter along the path $[u(s), \lambda(s)]$ in the inflated space $\mathbb{R}^{2m+i+k+1}$. As Eqs. (VI.D.1) and

(VI.D.3) imply, the values of s for which $\lambda(s)$ vanishes determine the zeros of F (and, consequently, the extrema, of R), while the values of s for which $\mu(s)$ vanishes determine the zeros of F' (and, consequently, the bifurcation, limit, and turning points in the problem under consideration). The Jacobian matrix $F'(u)$ appearing in Eq. (VI.D.3) has, of course, the same block matrix structure as given in Eq. (VI.B.16).

The numerical solution of Eqs. (VI.D.1) through (VI.D.3) is computed as follows. Suppose that (u_0, λ_0) is a known solution point of Eq. (VI.D.1). Then, corresponding to this point, there is the pseudo-arc-length parameter point $s_0 \in \mathbb{R}^1$. The starting point is thus set to be $[u(s_0), \lambda(s_0)] \equiv (u_0, \lambda_0)$. The direction of the tangent along the solution path at (u_0, λ_0) is given by $[w(s_0), \mu(s_0)]$, which is obtained by solving Eqs. (VI.D.2) and (VI.D.3) at $s = s_0$. The next point $[u(s), \lambda(s)]$ on the solution path, for $s \ne s_0$ (but s near s_0), is obtained by solving the following system of equations:

$$G[u(s), \lambda(s)] \equiv F[u(s)] - \lambda(s) F[u(s_0)] = 0 \qquad \text{(VI.D.4)}$$

and

$$M[u(s), \lambda(s), s] \equiv w(s_0) \bullet [u(s) - u(s_0)] + \mu(s_0) \bullet [\lambda(s) - \lambda(s_0)] - (s - s_0) = 0. \qquad \text{(VI.D.5)}$$

When (u_0, λ_0) is a regular or a simple limit point of $G(u, \lambda)$, the Jacobian

$$J \equiv \begin{bmatrix} F'(u) & -F(u_0) \\ (w_0)^T & \mu_0 \end{bmatrix}$$

of Eqs. (VI.D.4) and (VI.D.5) is nonsingular. Furthermore, while the bifurcations in Eqs. (VI.D.4) and (VI.D.5) are identical to those of $F(u)$, the structure of Eqs. (VI.D.4) and (VI.D.5) at a bifurcation point possesses distinctly advantageous convergence properties over that of $F(u)$.

To compute $[u(s), \lambda(s)]$ from Eqs. (VI.D.4) and (VI.D.5), it is advantageous to use Newton's method because of its simplicity and superior convergence properties (quadratic or, at worst, superlinear at bifurcation points). Applying Newton's method to Eqs. (VI.D.4) and (VI.D.5) and rearranging the resulting expressions yield the iterations

$$\Delta u^\nu \equiv u^{\nu+1} - u^\nu, \quad \Delta \lambda^\nu \equiv \lambda^{\nu+1} - \lambda^\nu; \quad (\nu = 0, 1, \ldots) \qquad \text{(VI.D.6)}$$

where

$$\Delta \lambda^\nu = [u(s_0) \bullet b - M(u^\nu, \lambda^\nu, s)]/[w(s_0) \bullet a + \mu(s_0)] \qquad \text{(VI.D.7)}$$

and

$$\Delta u^\nu = (\Delta \lambda^\nu) a - b. \qquad \text{(VI.D.8)}$$

In the above equations, the vectors a and b denote, respectively, the solutions of

$$F'(u^\nu) \bullet a = F(u_0) \qquad \text{(VI.D.9)}$$

and

$$F'(u^\nu) \bullet b = G(u^\nu, \lambda^\nu). \qquad \text{(VI.D.10)}$$

The initial point $u_0 = (\varphi_0, y_0, \alpha_0, z_0)$ for the Newton iteration is chosen as follows: initially, only the nominal values α_0 of the system parameters α are specified, usually as part of the definition of the physical system under consideration. If any of the constraints $g(\alpha)$ are to remain strict equalities over the entire range of variations of the parameters α, then we include them in the definition of $N(\varphi, \alpha)$, thereby redefining (a) the parameters α that are to be considered as independent variables, and (b) the corresponding structure of the vector φ of dependent variables. The corresponding nominal values φ_0 of the state variables φ are then obtained by solving the equation $N(\varphi_0, \alpha_0) = 0$. The nominal values y_0 are obtained as the solutions of the adjoint equations

$$N^+(\varphi_0, \alpha_0, y) = \nabla_\varphi R(\varphi_0, \alpha_0) + [\nabla_\varphi N(\varphi_0, \alpha_0)] y_0 = 0. \qquad \text{(VI.D.11)}$$

Furthermore, the inequality constraints $g_j(\alpha)$ are arranged to be inactive at α_0, so that the inequalities $g_j(\alpha_0) < 0$ are satisfied. This is always possible by appropriately defining the respective functions g_j. Consequently, the initial values for the respective Lagrange multipliers z are $z_0 = 0$. Finally, having obtained all the components of u_0 as $u_0 = (\varphi_0, y_0, \alpha_0, 0)$, we compute the (first-order) sensitivities at $u_0 = (\varphi_0, y_0, \alpha_0, 0)$, namely, $S(u_0) = \nabla_\alpha P(\varphi_0, \alpha_0)$

$+[\nabla_\alpha N(\varphi_0, \alpha_0)] y_0$. This also completes the calculation for the starting value $F(u_0)$, whose components are thus $F(u_0) = [0, 0, S(u_0), 0]$.

Having obtained u_0, we select the starting value $\lambda_0 = \lambda(s_0)$ by noting that the point $(u_0, \lambda_0 = 1)$ satisfies Eq. (VI.D.1); the initial directions (w_0, μ_0) are then obtained by solving Eqs. (VI.D.2) and (VI.D.3) at $(u_0, 1)$. The initial guess for the Newton method is provided by a single Euler step

$$u^0 = u_0 + (s - s_0) w_0, \quad \text{for } v = 0 \qquad (\text{VI.D.12})$$

and

$$\lambda^0 = \lambda_0 + (s - s_0) \mu_0, \quad \text{for } v = 0, \qquad (\text{VI.D.13})$$

where $s - s_0$ can be estimated from convergence theorems for Newton-like methods within a ball of radius $|s - s_0|$ around s_0.

It is advantageous to use the largest step-length $|s - s_0|$ that still assures convergence of the iterative process. For the illustrative examples to be presented in the next section, the step-length $|s - s_0|$ was estimated from the expression

$$|s - s_0| = \left\{ \| F(u_0 + \Delta w_0) - 2 F(u_0) + F(u_0 - \Delta w_0) \| \Delta^{-2} \gamma(J) \right\}^{-1/2}, \quad (\text{VI.D.14})$$

where $\gamma(J)$ is the condition number of the Jacobian J of Eqs. (VI.D.4) and (VI.D.5), and Δ is a small increment (for example, the computations described in Section VI.E were performed using the value $\Delta = 10^{-2}$). The expression for $|s - s_0|$ given in Eq. (VI.D.14) was obtained by using the Kantorovich sufficient condition for the convergence of the Newton method in a ball around s_0, by neglecting all terms of order higher than two in $|s - s_0|$, and approximating the Hessian $F''(u_0)$ by finite differences. While this procedure is useful for many practical applications, it may not always be optimal; therefore, alternative means of obtaining an optimal maximum step length $|s - s_0|$ remain of interest.

Calculating the Newton iterates $\Delta \lambda^v$ and Δu^v from Eqs. (VI.D.7) and (VI.D.8), respectively, requires solving Eqs. (VI.D.9) and (VI.D.10) to obtain the vectors a and b, respectively. For large-scale problems, $F'(u)$ is a very large matrix, so a direct solution of Eqs. (VI.D.9) and (VI.D.10) would be impractical. Note, though, that the special structure of $F'(u)$ was deliberately created by the structural arrangements in the definitions of $F(u)$ and u in order to simplify the subsequent task of solving Eqs. (VI.D.9) and (VI.D.10), by exploiting the positioning of the zero-submatrices and the easy invertibility of the diagonal matrices $Z_{\mathcal{A}}$ and $C_{\mathcal{I}}$. By partitioning the vector a in the form

$$a = (a_\varphi, a_y, a_\alpha, a_{\mathcal{A}}, a_{\mathcal{I}}) \tag{VI.D.15}$$

with components having the same dimensions as those of $\varphi, y, \alpha, \mathcal{A}$ (i.e., number of active constraints), and $\mathcal{I}$ (i.e., number of inactive constraints), respectively, the task solving Eq. (VI.D.9) is reduced to solving the following matrix equations:

1. When all constraints are inactive, the set $\mathcal{A}$ is empty; it then follows from Eq. (VI.D.9) that $a_{\mathcal{A}} = 0$ and $a_{\mathcal{I}} = 0$. Hence, Eq. (VI.D.9) is reduced to solving

$$F_1(u^v) \begin{pmatrix} a_\varphi \\ a_y \\ a_\alpha \end{pmatrix} = \begin{pmatrix} 0 \\ 0 \\ S(u_0) \end{pmatrix}, \tag{VI.D.16}$$

where $F_1(u)$ is the symmetric matrix defined as

$$F_1(u) \equiv \begin{bmatrix} \nabla_\varphi N^+ & \nabla_\varphi N & \nabla_\varphi S \\ (\nabla_\varphi N)^T & 0 & (\nabla_\alpha N)^T \\ (\nabla_\varphi S)^T & \nabla_\alpha N & \nabla_\alpha S \end{bmatrix}. \tag{VI.D.17}$$

2. When some constraints are active, then $\mathcal{A}$ is not empty but Eq. (VI.D.9) still implies that $a_{\mathcal{I}} = 0$; in this case, Eq. (VI.D.9) is reduced to solving

$$F_2(u^v) \begin{pmatrix} a_\varphi \\ a_y \\ a_\alpha \\ a_{\mathcal{A}} \end{pmatrix} = \begin{pmatrix} 0 \\ 0 \\ S(u_0) \\ 0 \end{pmatrix}, \tag{VI.D.18}$$

where $F_2(u)$ is the symmetric matrix defined as

$$F_2(u) = \begin{bmatrix} & & \vdots & 0 \\ & F_1(u) & \vdots & 0 \\ \cdots & \cdots & \vdots & \nabla g_{\mathcal{A}} \\ 0 & 0 & (\nabla g_{\mathcal{A}})^T & 0 \end{bmatrix}. \qquad \text{(VI.D.19)}$$

In preparation for solving Eq. (VI.D.10), the vectors b and G are partitioned as follows:

$$b = (b_\varphi, b_y, b_\alpha, b_{\mathcal{A}}, b_{\vartheta}) \qquad \text{(VI.D.20)}$$

and

$$G = (G_\varphi, G_y, G_\alpha, G_{\mathcal{A}}, G_{\vartheta}), \qquad \text{(VI.D.21)}$$

respectively. This way, Eq. (VI.D.10) is reduced to solving one or the other of the following matrix equations:

(a) When all constraints are inactive, it follows from Eq. (VI.D.10) that

$$b_{\mathcal{A}} = 0, \quad b_{\vartheta} = \frac{1}{2}[C_{\vartheta}(\alpha^v, z^v)]^{-1} K(\alpha^v, z^v); \qquad \text{(VI.D.22)}$$

hence, the remaining components of b are obtained by solving

$$F_1(u^v)\begin{pmatrix} b_\varphi \\ b_y \\ b_\alpha \end{pmatrix} = \begin{pmatrix} G_\varphi \\ G_y \\ G_\alpha - [\nabla g(\alpha^v, z^v)]b_{\vartheta} \end{pmatrix}. \qquad \text{(VI.D.23)}$$

(b) When some constraints are active, then b_{ϑ} is still given by Eq. (VI.D.22), while solving Eq. (VI.D.10) is reduced to solving the system

$$F_2(u^v)\begin{pmatrix} b_\varphi \\ b_y \\ b_\alpha \\ b_{\mathcal{A}} \end{pmatrix} = \begin{pmatrix} G_\varphi \\ G_y \\ G_\alpha - [\nabla g_1(\alpha^v, z^v)]b_1 \\ \frac{1}{2}[Z_{\mathcal{A}}(z^v)]^{-1} K_{\mathcal{A}}(\alpha^v, z^v) \end{pmatrix}. \qquad \text{(VI.D.24)}$$

The matrix $F'(u)$ and, equivalently, the matrix F_1 (or F_2) become singular at the bifurcation points. It is therefore essential to determine correctly the rank of

F_1 or F_2 when solving the equations involving these matrices. In actual computations, singular value decomposition procedures are recommended to determine the null vectors of F_1 (or, respectively, F_2); these null vectors play an essential role for continuing the solution through a bifurcation point.

The Newton algorithm described in the foregoing generates a sequence of points $[u(s_j),\lambda(s_j)]$ and corresponding tangent directions $[w(s_j),\mu(s_j)]$, starting at $j=0$ and continuing until the algorithm terminates at some point j_{final}. When $\lambda(s)$ changes sign between two points, $[u(s_{r-1}),\lambda(s_{r-1})]$ and $[u(s_r),\lambda(s_r)]$, on the solution path, then a root $u^* \equiv u(s^*)$ of $F(u)$ must correspond to the root $\lambda(s^*)=0$ that occurs at $s^* \in (s_{r-1},s_r)$. On the other hand, the points $u^{**} \equiv u(s^{**})$ where the determinant of $F'(u)$ vanishes correspond to the points where $\mu(s^{**})=0$. These points are found analogously to those where $\lambda(s^*)=0$, but by monitoring the sign changes in $\mu(s)$ instead of those in $\lambda(s)$. To find the precise locations of the roots $\lambda(s^*)=0$ and $\mu(s^{**})=0$, it is useful to switch, in the respective interval (s_{r-1},s_r), from the Newton marching algorithm to the secant method coupled with *regula falsi*, to ensure rapid convergence.

VI.E. A CONSTRAINED OPTIMIZATION PROBLEM FOR TESTING THE GLOBAL COMPUTATIONAL ALGORITHM

The illustrative problem presented in this Section can be used both to benchmark the optimization segment of the global algorithm described in the previous Section, and to benchmark the accuracy of the sub-algorithms dealing with inequality constraints. The illustrative problem is to minimize the response $R(\alpha) \equiv \alpha_1^2 + \alpha_2^2 + \alpha_3^2 + 40\alpha_1 + 20\alpha_2 - 3000$, subject to the constraints $g(\alpha) \leq 0$, where

$$g(\alpha) = [g_1(\alpha), g_2(\alpha), g_3(\alpha)]^T = \begin{pmatrix} \alpha_1 - 50 \\ \alpha_1 + \alpha_2 - 100 \\ \alpha_1 + \alpha_2 + \alpha_3 - 150 \end{pmatrix}.$$

For this illustrative problem, we note that $u = (\alpha, z)$, $z = (z_1, z_2, z_3)$, and $K = (K_1, K_2, K_3)$, where $K_i = (g_i + z_i)^2 + g_i|g_i| - z_i|z_i|$, for $i = 1, 2, 3$. Furthermore, ∇g is obtained as

$$\nabla g = \begin{bmatrix} 1 & 1 & 1 \\ 0 & 1 & 1 \\ 0 & 0 & 1 \end{bmatrix},$$

so that Eq. (VI.B.14) becomes, in this case, $F(u) = [S(u), K(u)]^T = 0$, with

$$S(u) = \begin{pmatrix} 2\alpha_1 + 40 + z_1 + z_2 + z_3 \\ 2\alpha_2 + 20 + z_2 + z_3 \\ 2\alpha_3 + z_3 \end{pmatrix}.$$

The starting point for the global computational algorithm was chosen, for convenience, at $\alpha_0 = (0,0,0)$, so that all the constraints were inactive [i.e., $g(\alpha_0) < 0$]; this implies that $z_0 = (0,0,0)$ and, consequently, $u_0 = (0,0,0,0,0,0)$. The global algorithm quickly converged to the exact solution $u^* = (-20, -10, 0, 0, 0, 0)$, for which the response has a minimum value of $R_{min} = -3500$; the constraints remained inactive throughout this test problem.

To test the global computational algorithm when the constraints change from inactive to active, the previous problem was modified to read: maximize $R(\alpha)$ subject to $g(\alpha) \geq 0$, using the same expressions for $R(\alpha)$ and $g(\alpha)$. The computational algorithm was started at various points α_0 such that the constraints $g(\alpha_0)$ were inactive at α_0. For any one of these starting points, the algorithm quickly converged to the exact solution $u^* = (50, 50, 50, -20, -20, -100)$, for which the response attains a maximum value of $R = 7500$. Note that all three constraints changed from inactive to active while the algorithm converged to the solution u^* for this test case.

VI.F. ILLUSTRATIVE EXAMPLE: GLOBAL ANALYSIS OF A NONLINEAR LUMPED-PARAMETER MODEL OF A BOILING WATER REACTOR

The objective of the illustrative problem presented in this Section is to determine the critical and first-order bifurcation points of a paradigm nonlinear lumped parameter model describing the dynamic behavior of a boiling water reactor (BWR). This model comprises the following nonlinear system of five coupled differential equations:

$$N(\varphi,\alpha) = \begin{pmatrix} \dfrac{d\varphi_1}{dt} - \varphi_1(\alpha_1\varphi_4 + \alpha_2\varphi_3 - \alpha_3)/\alpha_4 - \alpha_5\varphi_2 - (\alpha_1\varphi_4 + \alpha_2\varphi_3)/\alpha_4 \\ \dfrac{d\varphi_2}{dt} - \varphi_1\alpha_3/\alpha_4 + \alpha_5\varphi_2 \\ \dfrac{d\varphi_3}{dt} - \alpha_6[\varphi_1 + \alpha_7 H(t)] + \alpha_8\varphi_3 \\ \dfrac{d\varphi_4}{dt} - \varphi_5 \\ \dfrac{d\varphi_5}{dt} + \alpha_9\varphi_5 + \alpha_{10}\varphi_4 + k\varphi_3 \end{pmatrix} = 0,$$

where $H(t)$ is the Heaviside unit-step function, while the components of the (column) vectors $\boldsymbol{\varphi} = (\varphi_1, \varphi_2, \varphi_3, \varphi_4, \varphi_5)$ and $\boldsymbol{\alpha} = (\alpha_1, \ldots, \alpha_{10}, \alpha_{11} \equiv k)$ are defined as follows: $\varphi_1 \equiv$ excess neutron population; $\varphi_2 \equiv$ excess population of delayed neutron precursors; $\varphi_3 \equiv$ excess fuel temperature; $\varphi_4 \equiv$ relative excess coolant density; $\varphi_5 \equiv$ time rate of change of relative excess coolant density. The parameters $\alpha_i, (i = 1, \ldots, 11)$, which appear in the above equation, take on the following numerical values: $\alpha_1 = 0.15$, $\alpha_2 = -2.61 \times 10^{-5} (K^{-1})$, $\alpha_3 = 0.0056$, $\alpha_4 = 4 \times 10^{-5} (s)$, $\alpha_5 = 0.08 (s^{-1})$, $\alpha_6 = -25.04 (K \cdot s^{-1})$, $\alpha_7 = -0.1$, $\alpha_8 = 0.24 (s^{-1})$, $\alpha_9 = 2.25 (s^{-1})$, $\alpha_{10} = -6.82 (s^{-1})$, and $\alpha_{11} \equiv k =$ heat transfer coefficient.

Consider, for illustrative purposes, that the system response is the excess fuel temperature, so that $R(\boldsymbol{\varphi},\boldsymbol{\alpha}) = \varphi_3$. Furthermore, consider that the parameters α_1 through α_{10} are kept fixed at their respective nominal values, and only k is allowed to vary freely. Setting $\boldsymbol{u} = (\boldsymbol{\varphi}, \boldsymbol{y}, \alpha)$, $\boldsymbol{\varphi} = (\varphi_1, \ldots, \varphi_5)$, $\boldsymbol{y} = (y_1, \ldots, y_5)$, and $\alpha = k$, the components $\boldsymbol{N}^+(\boldsymbol{u})$ and $\boldsymbol{S}(\boldsymbol{u})$ of the vector $\boldsymbol{F}(\boldsymbol{u})$, as defined in Eq. (VI.B.14), take on the forms

$$\boldsymbol{N}^+(\boldsymbol{u}) = \begin{pmatrix} 0 \\ 0 \\ 1 \\ 0 \\ 0 \end{pmatrix} + \begin{pmatrix} -(\alpha_1\varphi_4 + \alpha_2\varphi_3 - \alpha_3)/\alpha_4 & -\alpha_3/\alpha_4 & -\alpha_6 & 0 & 0 \\ -\alpha_5 & \alpha_5 & 0 & 0 & 0 \\ -(\varphi_1 + 1)\alpha_2/\alpha_4 & 0 & \alpha_8 & 0 & -k \\ -(\varphi_1 + 1)\alpha_1/\alpha_4 & 0 & 0 & 0 & \alpha_{10} \\ 0 & 0 & 0 & -1 & \alpha_9 \end{pmatrix} \begin{pmatrix} y_1 \\ y_2 \\ y_3 \\ y_4 \\ y_5 \end{pmatrix},$$

and $\boldsymbol{S}(\boldsymbol{u}) = (0,0,0,0,-\varphi_3) \bullet \boldsymbol{y} = -\varphi_3 y_5$, respectively. Solving the equation $\boldsymbol{F}(\boldsymbol{u}) \equiv [\boldsymbol{N}^+(\boldsymbol{u}), \boldsymbol{N}(\boldsymbol{u}), \boldsymbol{S}(\boldsymbol{u}), \boldsymbol{K}(\boldsymbol{u})] = 0$ yields the critical points. Starting the global computational algorithm arbitrarily at the phase-space point $\boldsymbol{\varphi}_0 = (0,0,0,0,0)$ and using starting values in the range $k_0 > -0.013$, the stable critical point $\varphi^* = (0.1,175.0,0,0,0)$ was quickly located; the algorithm then located a bifurcation point at $k^* = -0.01320 \left(K^{-1} \cdot s^{-2} \right)$. Note that this paradigm BWR dynamic model has a stable critical point located at

$$\varphi^*_{exact} = \left[-\alpha_7, -\alpha_7 \alpha_3 / (\alpha_4 \alpha_5), 0, 0, 0 \right] = (0.1, 175.0, 0, 0, 0)$$

and an unstable manifold with coordinates

$$\left[-1, -\alpha_3/(\alpha_4 \alpha_5), (\alpha_7 - 1)\alpha_6/\alpha_8, k\alpha_6(\alpha_7 - 1)/(\alpha_{10}\alpha_8), 0 \right]$$
$$= (-1, -1750.0, -118.47, -17.379k, 0).$$

The additional quantities needed to construct the Jacobian matrix $F'(u)$ are obtained, respectively, as $\nabla_\alpha N^+(u) = (0, 0, -y_5, 0, 0)^T$, $\nabla_\alpha S(u) = 0$, and

$$\nabla_\varphi N^+ = \begin{bmatrix} 0 & 0 & -y_1 \alpha_2/\alpha_4 & -y_1 \alpha_1/\alpha_4 & 0 \\ 0 & 0 & 0 & 0 & 0 \\ -y_1 \alpha_2/\alpha_4 & 0 & 0 & 0 & 0 \\ -y_1 \alpha_1/\alpha_4 & 0 & 0 & 0 & 0 \\ 0 & 0 & 0 & 0 & 0 \end{bmatrix}.$$

When k was decreased below -0.020, the algorithm located the manifold of unstable critical points with coordinates $[-1, -1750.0, -118.5, \varphi_4(k), 0]$, where the $\varphi_4(k)$ coordinate is proportional to k. These numerical results were obtained using convergence criteria of 10^{-8} for the Newton algorithm and 10^{-16} for the secant method, respectively, while the roots were located with an accuracy of better than five significant decimals.

REFERENCES

Given the wide spectrum of topics covered in this volume, it is very difficult to provide an *exhaustive* list of references without inadvertently omitting somebody. Therefore, the list of references below is intended to provide a representative sample of the most recent books and review articles on the various topics covered in this volume. The interested reader can pursue individual subjects in more depth by consulting the references contained within the books and articles listed below.

1. Bauer, H., *Probability Theory and Elements of Measure Theory*, 2nd ed., Academic Press, New York, 1981.
2. Bode, H.W., *Network Analysis and Feedback Amplifier Design*, Van Nostrand-Reinhold, Princeton, New Jersey, 1945.
3. Bogaevski, V.N. and Povzner, A., *Algebraic Methods in Nonlinear Perturbation Theory*, Springer-Verlag, New York, 1991.
4. Box, G.E.P., Hunter, W.G., and Hunter, J.S., *Statistics for Experimenters*, John Wiley & Sons, New York, 1978, chap.15.
5. Boyack, B.E. et al., An overview of the code scaling, applicability and uncertainty evaluation methodology, *Nucl. Eng. Des.*, 119, 1, 1990.
6. Cacuci, D.G. et al., Sensitivity theory for general systems of nonlinear equations, *Nucl. Sci. Eng.*, 75, 88, 1980.
7. Cacuci, D.G., Sensitivity theory for nonlinear systems. I. Nonlinear functional analysis approach, *J. Math. Phys.*, 22, 2794, 1981.
8. Cacuci, D.G., Sensitivity theory for nonlinear systems. II. Extensions to additonal classes of responses, *J. Math. Phys.*, 22, 2803, 1981.
9. Cacuci, D.G. and Hall, M.C.G., Efficient Estimation of Feedback Effects with Application to Climate Models, *J. Atmos. Sci.*, 41, 2063, 1984.
10. Cacuci, D.G., The forward and the adjoint methods of sensitivity analysis, in *Uncertainty Analysis*, Ronen, Y., Ed., CRC Press, Boca Raton, FL, 1988, 71.
11. Cacuci, D.G., Global optimization and sensitivity analysis, *Nucl. Sci. Eng.*, 104, 78, 1990.
12. Committee on the Safety of Nuclear Installations (CSNI), OECD Nuclear Energy Agency, Report of a CSNI Workshop on Uncertainty Analysis Methods, Vol. 1, 1994.
13. Conover, W.J., *Practical Nonparametric Statistics*, 2nd ed., John Wiley & Sons, New York, 1980, 357 and 456.
14. Cowan, G., *Statistical Data Analysis*, Clarendon Press, Oxford, 1998.
15. Cruz, J.B., *System Sensitivity Analysis*, Dowden, Hutchinson and Ross, Stroudsburg, PA, 1973.
16. D'Auria, F. and Giannotti, W., Development of a code with the capability of internal assessment of uncertainty, *Nuclear Technology*, 131, 159, 2000.

17. Deif, A.S., *Advanced Matrix Theory for Scientists and Engineers*, Abacus Press-Halsted Press, Kent, England, 1982.
18. Deif, A.S., *Sensitivity Analysis in Linear Systems*, Springer-Verlag, New York, 1986.
19. Demiralp, M. and Rabitz, H., Chemical kinetic functional sensitivity analysis: elementary sensitivities, *J. Chem. Phys.*, 74, 3362, 1981.
20. Dickinson, R.P. and Gelinas, R.J., Sensitivity analysis of ordinary differential equation systems - a direct method, *J. Comput. Phys.*, 21, 123, 1976.
21. Dieudonne, J., *Foundations of Modern Analysis*, Vol. I-VI, Academic Press, New York, 1969.
22. Drazin, P.G., *Nonlinear Systems*, Cambridge University Press, London, 1994.
23. Eckhaus, W., *Asymptotic Analysis of Singular Perturbations*, North-Holland, Amsterdam, 1979.
24. Eslami, M., *Theory of Sensitivity in Dynamic Systems*, Springer-Verlag, Heidelberg, 1994.
25. Fiacco, A.V., Ed., *Sensitivity, Stability, and Parametric Analysis* (A publication of the Mathematical Programming Society), North-Holland, Amsterdam, 1984.
26. Frank, P.M., *Introduction to System Sensitivity Theory*, Academic Press, New York, 1978.
27. Gandini, A., Generalized Perturbation Theory (GPT) methods: a heuristic approach, in *Advances in Nuclear Science and Technology*, Vol. 19, Plenum Press, New York, 1987.
28. Gandini, A., Uncertainty analysis and experimental data transposition methods based on perturbation theory, in *Uncertainty Analysis*, Ronen, Y., Ed., CRC Press, Boca Raton, FL, 1988.
29. Greenspan, E., New developments in sensitivity theory, in *Advances in Nuclear Science and Technology*, Vol. 14, Plenum Press, New York, 1982.
30. Gourlay, A.R. and Watson, G.A., *Computational Methods for Matrix Eigenproblems*, John Wiley & Sons, New York, 1973.
31. Holmes, M.H., *Introduction to Perturbation Methods*, Springer-Verlag, New York, 1995.
32. Horn, R.A. and Johnson, C.R., *Matrix Analysis*, Cambridge University Press, 1996.
33. Jaynes, E.T., *Papers on Probability, Statistics, and Statistical Physics*, Rosencrantz, R.D., Ed., Reidel, Dordrecht, 1983.
34. Kahn, D.W., *Introduction to Global Analysis*, Academic Press, New York, 1980.
35. Kanwal, R.P., *Linear Integral Equations, Theory and Technique*, Academic Press, New York, 1971.
36. Kato, T., *Perturbation Theory for Linear Operators*, Springer-Verlag, Berlin, 1963.

37. Kendal, M. and Stuart, A., *The Advanced Theory of Statistics*, 3rd ed., Vol. 1-3, Hafner Press, New York, 1976.
38. Kevorkian, J. and Cole, J.D., *Multiple Scale and Singular Perturbation Methods*, Springer-Verlag, Heidelberg, 1996.
39. Kokotovic, P.V. et al., Singular perturbations: order reduction in control system design, *JACC*, 1972.
40. Krasnosel'skii, M.A. et al., *Approximate Solution of Operator Equations*, Wolters-Noordhoff Publishing, Groningen, 1972.
41. Lillie, R.A. et al., Sensitivity/Uncertainty analysis for free-in-air tissue kerma at Hiroshima and Nagasaki due to initial radiation, *Nucl. Sci. Eng.*, 100, 105, 1988.
42. Lions, J.L., *Some Methods in the Mathematical Analysis of Systems and Their Control*, Gordon and Breach Science Publishers, New York, 1981.
43. Liusternik, L.A. and Sobolev, V.J., *Elements of Functional Analysis*, Frederick Ungar, New York, 1961.
44. March-Leuba, J., Cacuci, D.G., and Perez, R.B., Universality and aperiodic behavior of nuclear reactors, *Nucl. Sci. Eng.*, 86, 401, 1984.
45. March-Leuba, J., Cacuci, D.G., and Perez, R.B., Nonlinear dynamics and stability of boiling water reactors. I. Qualitative analysis, *Nucl. Sci. Eng.*, 93, 111, 1986.
46. McKay, M.D., Sensitivity and uncertainty analysis using a statistical sample of input values, in *Uncertainty Analysis*, Ronen, Y., Ed., CRC Press, Boca Raton, FL, 1988.
47. Morse, M. and Cairns, S.S., *Critical Point Theory in Global Analysis and Differential Topology*, Academic Press, New York, 1969.
48. Myers, R.H., *Response Surface Methodology*, Allyn and Bacon, Boston, 1971.
49. Navon, I.M. et al., Variational data assimilation with an adiabatic version of the NMC spectral model, *Mont. Wea. Rev.*, 120, 1433, 1992.
50. Nayfeh, A.H., *Perturbation Methods*, John Wiley & Sons, New York, 1973.
51. Nowinski, J.L., *Applications of Functional Analysis in Engineering*, Plenum Press, New York, 1981.
52. O'Malley, R.E., Jr., *Singular Perturbation Methods for Ordinary Differential Equations*, Springer-Verlag, New York, 1991.
53. Ortega, J.M. and Rheinboldt, W.C., *Iterative Solution of Nonlinear Equations in Several Variables*, Academic Press, New York, 1970.
54. Ostrowski, A.M., *Solution of Equations and Systems of Equations*, Academic Press, New York, 1966.
55. Pontryagin, L.S. et al., *The Mathematical Theory of Optimal Processes*, John Wiley & Sons, New York, 1962.
56. Rabinovich, S.G., *Measurement Errors and Uncertainties: Theory and Practice*, 2nd ed., Springer-Verlag, New York, 2000.
57. Rall, L.B., Ed., *Nonlinear Functional Analysis and Applications*, Academic Press, New York, 1971.

58. Reed, M. and Simon, B., *Methods of Modern Mathematical Physics*, Vol. I-IV, Academic Press, New York, 1980.
59. Rief, H., Monte Carlo uncertainty analysis, in *Uncertainty Analysis,* Ronen, Y., Ed., CRC Press, Boca Raton, FL, 1988.
60. Ronen, Y., Uncertainty analysis based on sensitivity analysis, in *Uncertainty Analysis,* Ronen, Y., Ed., CRC Press, Boca Raton, FL, 1988.
61. Rosenwasser, E. and Yusupov, R., *Sensitivity of Automatic Control Systems*, CRC Press, Boca Raton, FL, 2000.
62. Ross, S.M., *Introduction to Probability Models*, 7th ed., Academic Press, New York, 2000.
63. Sagan, H., *Introduction to the Calculus of Variations*, McGraw-Hill, New York, 1969.
64. Schechter, M., *Modern Methods in Partial Differential Equations, An Introduction*, McGraw-Hill, New York, 1977.
65. Selengut, D.S., Variational analysis of multidimensional systems. Rep. HW-59129, HEDL, Richland, WA, 1959, 97 pp.
66. Selengut, D.S., On the derivation of a variational principle for linear systems, *Nucl. Sci. Eng.,* 17, 310, 1963.
67. Shapiro, S.S. and Gross, A.J., *Statistical Modeling Techniques*, Marcel Dekker, New York, 1981.
68. Smith, D.L., *Probability, Statistics, and Data Uncertainties in Nuclear Science and Technology*, American Nuclear Society, LaGrange Park, IL, 1991.
69. Smith, L., *Linear Algebra*, Springer-Verlag, New York, 1984.
70. Stacey, W.M., Jr., *Variational Methods in Nuclear Reactor Physics*, Academic Press, New York, 1974.
71. Tomovic, R. and Vucobratovic, M., *General Sensitivity Theory*, Elsevier, New York, 1972.
72. Usachev, L.N., Perturbation theory for the breeding ratio and for other number ratios pertaining to various reactor processes, *J. Nucl. Energy* A/B, 18, 571, 1964.
73. Vainberg, M.M., *Variational Methods for the Study of Nonlinear Operators*, Holden-Day, San Francisco, 1964.
74. Varga, R.S, *Matrix Iterative Analysis*, Prentice-Hall, Englewood Cliffs, NJ, 1962.
75. Weinberg, A.M. and Wigner E.P., *The Physical Theory of Neutron Chain Reactors*, University of Chicago Press, Chicago, 1958.
76. Wacker, H., Ed., *Continuation Methods*, Academic Press, New York, 1978.
77. Wigner, E.P., Effect of small perturbations on pile period, Manhattan Project Report CP-G-3048, 1945.
78. Yosida, K., *Functional Analysis*, Springer-Verlag, Berlin, 1971.
79. Zemanian, H., *Distribution Theory and Transform Analysis*, McGraw-Hill, New York, 1965.

80. Zhou, X. et al., An adjoint sensitivity study of blocking in a two-layer isentropic model, *Mon. Wea. Rev.,* 121, 2833, 1993.
81. Zwillinger, D., *Handbook of Differential Equations*, 2nd ed., Academic Press, New York, 1992.

INDEX

Absolute continuity, 26, 32
Absolute convergence, 56
Absolute measurement errors, 106-109,113,127
Absolute value of error, 106
Absolutely constant
    errors, 112,119,127
Accuracy,
    of measurements, 102, 106-107, 109, 114, 116, 118, 127
Addition,
    of matrices, 3-4, 8
    of operators, 12
Adjoint,
    of bounded linear operator, 28
    continuous linear
        operator, 27-28
    of differential operator, 26
    formal, 26, 32, 134, 150, 177, 183, 203
    formally self-adjoint, 26
    function, 145
    in Hilbert space, 31
    of unbound linear operator, 32
Adjoint matrix, 29
Adjoint Sensitivity Analysis
    Procedure (ASAP),
        for algebraic operators, 209
        for linear opeators, 134
        for nonlinear operators, 202, 221
Adjoint sensitivity equation, 145, 179, 243
A posteriori estimation of
    errors, 118-119
A posteriori probability, 50
A priori estimation of errors, 107, 118-119
A priori probability, 50

    determination of, 50-51
    in Bayesian analyses, 50-51
Arguments,
    measured, 119
Arrangement(s), 42
Average value(s), 18,
    of parameter, 117
    of weak derivative, 27

Banach space, 18, 22
    of bounded linear operators, 27
$\mathscr{L}_p$, 22
Basis(es), 1, 13, 18-20, 29-30, 139, 141
    in Hilbert space, 19
    orthonormal, 19-20, 28-30, 139, 203
Bayes theorem, 40, 46, 50
Bayesian
    interpretation of probability, 49
    statistics, 50
Bernoulli distribution, 166
    definition of, 70-71
    mean value of, 71, 167
    trial(s), 71-73
    variance of, 71, 167
Best linear unbiased estimator, 90
Beta distribution, 83-84
    definition of, 83
    mean value of, 83
    variance of, 83
Bias, 89, 94, 97
Binomial coefficient(s), 58
Binomial distribution,
    definition of, 71
    mean value of, 71
    moment generating
        function for, 71
    variance of, 72
Bifurcation, 247, 252, 256, 258,

262, 264, 266
Bijective operator, 22
Bilinear form,
    associated with, 27, 134, 140, 151
Block, of matrix, 8
Block-multiplications of matrices, 8-9
Blunders, 110-111
Boiling Water Reactor model, 264
Boltzmann transport equation, 182-183
Boson(s), 43
Boundary,
    of a set, 14
Bounded linear operators, 24
    adjoint, 28
    Banach space of, 27
    spectrum, 24

Calibration, 116
Cartesian coordinates, 63
Cauchy distribution, 66
    definition of, 82-83
    mean value of, 82
    variance of, 82
Cauchy-Schwartz inequality, 60
Cauchy sequence, 18
Cayley-Hamilton theorem, 11
Cell, of matrix, 8
Center of gravity, 58
Central limit theorem, 85
    Gaussian distribution and, 77
    proof of, 86
    random error(s) and, 77,
    statement of, 77
Central moments 57-59, 123, 125
    reduced, 64
Chain,
    Markov, 165
Chain rule(s), 34-35
Characteristic
    equation, 11
    function, 24, 65-66, 77-78, 95
    polynomial, 11

value, 10-11, 24, 30-31
vector, 10, 29-30
Chebyshev,
    limit, 45
    polynomials of, 20, 203
    theorem, 64-65
        confidence and, 98
Chi-square distribution 66, 80, 91, 97
    definition of, 84
    degrees of freedom for, 84
    mean value of, 84
    variance of, 84
Climate model with feedback, 205
Closed
    set, 13
    subspace, 13
Closure of set, 13
Coefficient of variation, 58
Column matrix, 2
Combined measurements, 108
Complete (orthonormal), 21
Complete set, 19
Condition number of square
    matrix, 14
        geometrical interpretation of, 15
Conditional probability,
    Bayes' theorem,
        definition of, 45
        Kolmogorov axioms in, 46
Conditionally constant errors, 112
Confidence,
    formal definition of, 98
    indicator associated with
        probability function, 59
    intervals of, 87, 98-100, 117, 119-120
    methods of constructing, 98
    probability, 98
Conjugate index, 16
Consistent estimate, 94
Constrained optimization, 263
Continuity, 16, 34, 35
Continuous functions, 16

Continuous linear functional, 28
Continuous linear operators, 35, 36
    adjoint, 27-28
Continuous spectrum, 24
Continuous uniform distribution,
    definition of, 70
    mean value of, 70
    moment generating function
        for, 70
    variance of, 70
Convergence, 18
    absolute, 56
Convolution,
    Fourier, 54
    Mellin, 54
Coordinate transformation,
    orthogonal, 20
    unitary, 20
Correlation(s), 59-63, 79-80
    meaning of, 63
    parameter(s), 59
Correlation coefficient, 123
Correlation matrix(ces), 60, 196, 197
        covariance matrix(ces), relation to, 59-61
        definition of, 60
        properties of, 60
Covariance, 59
Covariance matrix(ces),
    citations to in literature, 60
    definition of, 59
    symmetry of, 59
Crank Nicolson procedure, 219-221, 229
Critical point, 247, 249-252
Cumulative Distribution Function (CDF), 52

Data,
    adjustment, 205
    consistency of, 60
    fitting of, 97
Dead band, 107, 127
Deduction,
    statistical, 55
Degeneracy, 61
    measures of, 62
Degenerate distribution,
    definition of, 69
    mean value of, 69
    moment generating
        function for, 69
    variance of, 69
Degrees of freedom, 83
DeMoivre-Laplace theorem, 76-77
Determinant of square matrix, 5
Deviations,
    standard, 57
Diagonal matrix, 2
Differential operator, 25
    formal, 25
Dimension,
    of matrix, 1
Dirac delta functional, 69-70
Direct effect term, 136, 148, 223
Direct measurements, 108, 116
    estimate the
        uncertainties of, 118
    multiple, 117-119
    single, 116-117
Directional continuity, 34
Dirichlet problem,
    generalized, 27
Discrete Adjoint Sensitivity
    Analysis Procedure (DASAP), 159, 209, 216, 224, 238
Discrete Adjoint Sensitivity
    Equations, 161, 241-242
Discrete Forward Sensitivity
    Analysis Procedure (DFSAP), 212, 238
Discrete response sensitivity, 161
Discrete Sensitivity System, 159, 240
Discrete uniform distribution,
    definition of, 69
    mean value of, 70
    moment generating function
        for, 70

variance of, 70
Discretized Adjoint Sensitivity
　　Equations, 216
Discretized Forward Sensitivity
　　Equations, 157, 215
Discriminant threshold, 108
Dispersion,
　　measures of, 55, 58, 63
Domain of operator, 22
Drift, 108, 127
Dual spaces, 27
Dynamic characteristics of
　　measuring instruments, 108

Eigenfunctions
　　of linear operator, 10
　　of a matrix, 10-11
　　of an unbounded
　　　　linear operator, 24
Eigenspace, 24
Eigenvalue,
　　of a matrix, 10-11
　　of a (unbounded) linear
　　　　operator, 24
Elements of matrix, 1
Elimination method of Gauss, 9-10
Elimination of systematic
　　errors, 110, 116
Error(s),
　　absolute, 106
　　absolutely constant, 112
　　additional instrumental, 110
　　approximate estimation of, 107
　　conditionally constant, 112
　　dynamics, 111
　　elementary, 111, 127
　　estimation of, 107
　　fractional or relative
　　　　or percent, 58
　　gross, 110
　　instrumental, 110
　　intrinsic, 110, 127
　　mean square, 89
　　of measuring instruments, 110
　　methodological, 110
　　outlying, 110
　　periodic, 117
　　personal, 109
　　progressing, 117
　　propagation of, 120
　　purely-random, 112
　　quasi-random, 113
　　random, 110, 128
　　relative, 106, 128
　　repeatability, 107
　　residual, 110-112, 117
　　setup, 116
　　static, 111
　　systematic, 110, 128
Error propagation,
　　method of, 120
Estimate, 113
Estimation of errors,
　　a posteriori, 107
　　a priori, 107
Estimator(s),
　　best linear unbiased, 90
　　biased, 89
　　consistent, 89
　　definition of, 89
　　interval, 99
　　maximum likelihood, 92
　　point, 97
　　sufficient, 89
　　unbiased, 89
Euler's one-step method, 154-165
Event(s),
　　complement(s) of, 42
　　distinguishable feature(s) of, 42
　　intersection or product of, 41
　　multi-variable, 47
　　mutually exclusive, 42
　　null, 42
　　single-variable, 47
　　statistically independent, 46
　　subevent, 42
　　sum or union of, 41
　　universal, 42
Expansion(s),
　　P3-Legendre angular, 185

series, 86
Taylor series, 37, 85, 121
Expectation,
   concept of, 55
   definition of, 56
Exponential distribution,
   definition of, 74
   mean value of, 74
   moment generating
      function for, 74
   variance of, 74

F-distribution,
   definition of, 83
   degrees of freedom for, 83
   mean value, 83
   variance of, 83
Failure rate analysis, 75, 166
Feedback, 201, 205
Fermion(s), 43
Finite difference method(s), 260
Fluctuations, 115
Form, bilinear, 27, 134, 140, 151
Formal
   differential operator, 25
      bilinear form associated
         with, 27
      ordinary, 25
      partial, 25
Formal adjoint, 26
Formal self-adjointness, 26
Forward Sensitivity Analysis
   Procedure (FSAP) 133
Forward Sensitivity Equations, 133
Fourier
   coefficients, 19, 21
   convolution, 54
   series, 19, 21
   transform, 54
Fréchet
   differentiable operator, 35
   differential, 35
      total, 35
      partial differential, 36
Fréchet derivative,

   of continuous operator, 35
Fredholm alternative
   theorem, 31-32
Fredholm integral equation, 23
Function(s),
   continuity of, 16
   cumulative distribution, 52
Functional, 22
Fundamental parameters,
   of probability functions, 55
Fundamental principle
   of error estimation, 113-114

Gamma distribution,
   definition of, 82
   mean value of, 82
   moment generating function
      for, 82
   variance of, 82
Gamma function, 82
Gamma radiation transport, 182
Gâteaux,
   derivative, 32
   differentiable operator, 34
   differential, 35
      total partial differential, 35
   variation, 33
Gauss elimination method, 9-10
Gaussian curve, 77
Gaussian distribution,
   central limit theorem and, 77
   definition of, 75
   mean value of, 75
   variance of, 75
Gaussian probability distribution
   function, 79
Gaussian random variable, 77,
Generalized Dirichlet problem, 27
Generalized least-square
   method, 97
Geometric distribution,
   definition of, 72-73
   mean value of, 73
   moment generating function
      for, 73

variance of, 73
Global optimization, 249
Global sensitivity analysis, 245
    definition of, 247
Gosset, W. ("Student"), 83
Gross errors, 110
Group,
    unitary, 30

Hamilton-Cayley theorem, 11
Hermite polynomials, 20
Hermitian
    matrix, 3, 30
    operator, 30-31
Hermitian matrix, 3
Hilbert space, 18
    $\mathscr{L}_2$, 22
    Sobolev, 26-27
Hölder's inequality,
    ℓ_p-norm, 14, 16
Homotopy, 257
Hyperbolic partial differential
    equation, 238
Hypothesis(ses),
    testing or verification
        of, 117-118

Identity operator, 23
Image, 22
Image space(s),
Improved Polygon Method, 231
Inaccuracy of measurements, 100-101, 106, 111, 118, 127
Index of nilpotence, 5
Indirect effect term, 136, 148
Indirect measurements, 119, 127
Injective operator, 22
Inner product, 18
Inner product space, 18
Instability of measuring
    instruments, 108
Instruments, measuring, 105
Interior
    of a set, 14
    point, 14

Interval estimator, 99
Inverse
    matrix, 7
    of a transformation, 7

Jacobian matrix, 251
Jacobian of a transformation, 55

k-tuple(s),
    definition of, 42
    distinguishable, 42
    formed with replacement, 42
    formed without
        replacement, 42
    indistinguishable, 42
    occupancy model(s), relation
        to, 42
    ordered, 42
    permutation(s) of, 42
    unordered, 42
Kernel,
    hermitian, 31
    symmetric, 31
Knowledge,
    a priori or prior, 97
    merging of new and old, 97
Kolmogorov's axioms, 43-44
Kronecker symbol, 29
Kurtosis, 64

Lagrange functional, 249
Lagrange multiplier(s), 249, 251, 253-256, 259
Laplace's approach of probability
    theory, 40
Laplace's degree of belief, 43
Laplace's transform method, 174
Law of large numbers, 94
Least square method,
    generalized, 97
Lebesgue measure, 22
Legendre polynomial(s), 20
Leptokurtic distribution(s), 64
Likelihood(s), 50
Likelihood function, 92

Index **279**

Lindeberg condition, 86-87
Linear algebraic models, 142
Linear (in)dependence of
    vectors, 13
Linear transformation, 3
Linear operators, 23
    bounded, 24
    continuous, 23
    definition of, 23
    normal, 29
    self-adjoint operators, 32
    unbounded, 24
Lipschitz condition, 136, 138
Location,
    measure(s) of, 58
Log-normal distribution,
    definition of, 81
    mean value of, 81
    random errors and, 81
    variance of, 81
ℓ_p-norm, 14, 16
$\mathcal{L}_p$, 21-22

Marginal probability
    density, 52
    distribution function, 53
Markov chain, 165
Matrices,
    adjoint, 29
    blocks of, 8
    cells of, 8
    characteristic polynomial of, 11
    column, 2
    commuting, 4
    determinant of, 5,6
    diagonal, 2
    dimension of, 2
    elements of, 2
    fractional error, 62
    function of, 11
    hermitian, 3
    ill-conditioned, 16
    inverse of, 7
        generalized, 12
    minor of, 5
    principal, 5
    multiplication of,
        by matrix, 3-4
        by number, 3
    nilpotent, 5
    non-singular, 7
    norm, 14
    order of, 1
    orthogonal, 20
    partitioned, 8
    permutability of, 4
    permutable, 4
    polynomial, 5
    power of, 5
    quasi-triangular, 9
    rank of, 12
    relative covariance, 62
    row, 2
    singular, 7
    skew-hermitian, 3
    skew-symmetric, 3
    square, 1, 5, 7
    symmetric, 3
        condition number of a
            square, 15
    trace of, 4
    transpose of, 2
    triangular, 2
        lower, 2
        upper, 2
    unit, 2
    unitary, 20
Matrix addition, properties of, 3
Maximum entropy principle, 51
Maximum likelihood
    estimator, 92
    method, 92
Mean value(s), 57
    calculation with moment
        generating function(s), 67
    expectation and, 57
    measure(s) of location, role
        in, 58
    multivariate probability and, 57
    nuclear data files of, 63

univariate probability and, 57
Mean vector, 79
Measure,
  of degeneracy, 62
Measurement(s), 105
  accuracy of, 107, 127
  classification of, 109
  combined, 108
  direct, 108, 116
  dynamic, 109
  equation, 119
  inaccuracy of, 106
  indirect, 108, 119
  multiple, 108, 116, 119
  preliminary, 107
  result(s) of, 105
  single, 108, 116, 119
  static, 109
Measuring instruments, 105
  discrimination threshold, 108
  drift, 108
  instability, 108
  resolution, 108
  sensitivity, 108
Mellin convolution, 54
Mesokurtic distributions, 64
Minimum variance theorem, 58
Minkowski's inequality, 16
  in ℓ_p, 16
Minor, 5
  principal, 5
Model(s), 40
Modified Euler method, 231
Moment(s),
  about the mean value(s) or central, 57
  about the origin, 57
  crude or raw, 57
  mixed, 56
  order k about a point, 56
  propagation equation, 122
  propagation method, 120
  reduced or scaled, 64
  zeroth order, 56
Moment generating function(s)

(MGF), 66-68
  uniqueness property of, 67
Multinomial distribution,
  definition of, 72
  mean values of, 72
  moment generating function for, 72
  variance of, 72
Multiple occupancy of states, 43
Multivariate normal distribution,
  applications for, 79-80
  covariance matrix for, 79
  definition of, 78
  geometrical features of, 80
  mean values of, 79
  moment generating functions for, 79
Multivariate quadratic form, 79

Negative Binomial (Pascal) Distribution
  definition of, 73
  mean value of, 73
  variance of, 73
Neutron diffusion, 146
Neutron transport, 182
Newton's method, 258
Noise,
  generation-recombination, 115
  Schottky, 115
  shot, 115
  thermal, 115
Norm,
  of square matrix, 14
  of vector, 14
  spectral, 15
  sup, 17
Normal linear operator, 29
Null hypothesis, 100
Null space, 24
Null vector, 263
Number of degrees of freedom, 84

Open set, 13-14
Operator (linear),

Index **281**

adjoint of, 28
Gâteaux differential of, 34
hermitian, 30
identity, 23
matrix corresponding to, 28-29
normal, 29
positive definite, 30
positive semidefinite, 30
self-adjoint, 31
spectrum of, 24
unitary, 30
Order of matrix, 1
Ordinary differential operator
    formally, 25
Orthogonal
    matrix, 20
    vectors, 19
Orthogonal complement, 29
Orthonormal basis, 19-20, 29-30, 139, 203

P_3-Legendre angular expansion, 185
Parameter(s),
    location or translation, 61
    of a model, 41
    scaling, 61
Parameter estimation,
    maximum likelihood
        condition in, 92-93
    weakest condition of, 64
Parseval's equality, 21
Particle(s), distinguishable, 43
Particle(s), indistinguishable, 43
Pascal distribution,
    definition of, 73
    mean value of, 73
    moment generating
        function for, 73
    variance of, 73
Pauli exclusion principle, 43
Pearson's curves, 83-84
Platykurtic distribution(s), 64
Poincare theorem, 44
Point estimator, 97

Point spectrum, 24
Poisson distribution,
    definition of, 73
    mean value of, 74
    moment generating function
        for, 74
    variance of, 74
Polynomial(s),
    of Legendre, 20
Polynomial matrix, 5
Population(s), 41
Positive definite, 30
Power of matrix, 5
Precision, 114
Pre-Hilbert space, 18
Preimage, 22
Probability,
    additive property of, 44
    a posteriori or posterior, 50
    a priori or prior, 50
    as relative frequency, 48
    axiom(s) or postulate(s) of, 44
    conditional, 45
    cumulative or integrated, 64
    definition of, 44
    discrete, 77
    distribution(s) of,
        degenerate, 61
        non-degenerate, 61
    existence of, 44
    indicator function(s), role in, 59
    inferred, 55
    normalization of, 44
    postulated, 43
    rational degree of belief, as a
        measure of, 43
    subjective, 49-50
    theorems of, 44
Probability density,
    definition of, 52
    function(s) for, 45, 51
    univariate, 84
Probability function(s), 45, 51
    confidence indicator associated
        with, 59

degenerate, 61
localized characteristics of, 65
moment(s) of, 66-67
multivariate, 56-57
    measure of degeneracy, 62
non-degenerate, 61
univariate, 56
Product, inner, of vectors, 18
Progressing systematic errors, 117
Propagation of higher order
    moments equations, 124
Proper
    value, 10
    vector, 10

Quadratic form(s), 80
    multivariate, 79
Quadrature, 156-159
Quantile, 98

Radioactivity,
    exponential distribution, role
        in, 75
Random error(s),
    definition of, 110
    log-normal distribution and, 81
    purely, 112
        distribution of, 112
    quasi, 112
Random events, 43
Random process(es), 43
Random trial(s), 43
Random variable(s),
    cardinality or dimensionality
        of, 47
    definition of, 47
    discrete, 47
    finite, 47
    fully-anti-correlated, 62
    fully-correlated, 62
    infinite, 47
    nondenumerable or
        uncountable, 47
    orthogonal, 64
    parent, 85, 88

stochastically independent, 54
unaccountable, 47
vector(s) of, 47
Random variable function(s),
    cumulative distribution, 52
    indicator, 59
    probability density, 52-53
    transformations and
        statistics, 54
Random vector(s), 47
Range, 22
Rank of matrix, 12
Relative covariance matrix(ces),
    definition of, 62
    error propagation and, 122
    fractional error(s) and, 62
Reliability, 75
Repeatability of measurements, 110
Reproducibility of
    measurements, 110
Residual errors, 110-112, 117
Residual spectrum, 24
Resolution, 108
Resolvent, 24
Resolvent set, 24
Response, 40
Results of measurements, 105
Riemann integral, 37
Riesz representations theorem, 31
Row matrix, 2
Runge-Kutta method, 231

Sample(s),
    asymptotic behavior of, 85
    central moment(s) of
        order k, 90
    definition of, 41
    mean value(s) of, 90
    moment(s) of, 90
    space(s) of, 41
    standard deviation(s) of, 91
    variance(s) of, 91
Sampling,
    points, 41
    process, 88

Index

random, 88-89
random variable(s) in, 89
Scalar multiplication, 3
Scaling and translation theorem, 61
Scatter coefficient(s), 62
Schottky noise, 115
Schur, formulas of, 10
Self-adjoint operators,
    of bounded linear operators, 31
    of unbounded linear
        operators, 32
Sensitivity,
    analysis,
        global, 245-266
        local, 129-243
    coefficient(s) of, 182
    of measuring instruments, 108
    of the response to the parameter
        variations, 122
Sensitivity matrix(ces), 124
Sensitivity vector, 124
Separability of $\mathscr{L}_p$, 22
Sequence space, 13
Series, convergence of, 20-21, 139, 203, 246
Significance level, 99
Significance, test(s) for, 99
Similarity of matrices, 10
Simple eigenvalue, 11
Simultaneous algebraic
    equations, 23
Skewness, of probability
    distributions, 64
Sobolev space,
    definitions, 26-27
Spectral norm, 15
Spectrum, 24
Standard deviation(s),
    confidence and, 59
    definition of, 57
    dispersion, measure(s) of, 55
    equivalence to error(s), 58
Standard normal distribution,
    definition of, 75
    mean value of, 76

moment generating function
    for, 75
    properties of, 76
    variance of, 76
Statistical,
    deduction, 55
    inference, 55
    model, 41-42
    physics, 115
    tolerance intervals, 99
Statistics,
    applied, 75
    Bayesian, 50
    classical, 50
    counting laws of, 43
    definition of, 39-40
    Fermi-Dirac, 43
    Maxwell-Boltzmann, 43
    weak laws of, 64
Step function, 153, 265
Strictly positive operator, 30
Student's t distribution, 83
    Cauchy distribution, relation
        to, 83
    definition of, 83
    degrees of freedom in, 99
    mean value of, 83
    variance of, 83
Subevent, 42
Sup norm, 17
Support of a function, 17
Surjective operator, 22
Symmetric kernel, 31
Symmetric operators,
    and ordinary differential
        operators, 37
Systematic error(s),
    accuracy and, 110
    combination of, 111
    confidence and, 117
    elimination of, 117
    estimation of, 117
    method associated with, 117
Systems of vectors,
    bi-orthogonal, 29

orthogonal, 19

Taylor expansion
    functional, 121
    operator, 37
Thermal-hydraulics, 217
Tolerance range, 100
Trace of a matrix, 4
Transform variable, 67
Transformation, linear, 3, 7
    of coordinates, 19
    orthogonal, 20
    unitary, 20
Transpose of a matrix, 2
Transposition properties, 2
Trial(s),
    Bernoulli, 70
    independent, 71
    random, 43, 70
    sampling and, 41
Triangle inequality, 14
True value(s),
    of a measurable quantity,
        106, 128

Unbiased estimator, 88-92, 94-95, 97
Unbound linear operators,
    adjoint, 32
    definition of, 24
    eigenvalue of, 24
    resolvent, 24
        set of, 24
    self-adjoint, 32
    spectrum, 24
Uncertainty(ies), 100
    confidence and, 101
    covariance matrix(ces)
        and, 60-62
    equivalence to error(s), 100
    file(s) of, 63
    variance(s) and, 60-62
Uniform boundedness principle, 86
Uniform continuity, 16
Unit

matrix, 2
operator, 30
Unitary,
    group, 30
    linear operator, 30
    matrix, 20
    transformation, 20
Unpredictable behavior, 43

Value(s),
    characteristic, 10-11, 24, 30-31
    latent, 10
    proper, 10
Variance(s), 57-61, 69-71, 73-75,
        77-78, 81-84, 86, 87, 90,
        91, 93, 115, 121, 123, 143,
        166, 167, 169, 170, 172,
        174, 182, 192
    Bayesian methodology and,
        definition of, 57
    generalized, 61
    measure(s) of dispersion, 58
    minimum, condition or
        theorem of, 58
    random variable(s) and,
Variance-covariance
    matrix(ces), 59
Variance generalized, 61
Vector(s),
    bi-orthogonal, 29
    inner product of, 18, 20, 22,
        27, 32
    latent, 10
    linear dependence of, 13
    linear independence of, 13
    norm of, 14
    orthogonal, 19
    orthogonalization of
        sequence, 19
    proper, 10
    scalar product of, 21
    systems of, bi-orthogonal, 29
Vector space, 12
    basis of, 12
    complex, 13

Index

  dimension of, 13
  finite-dimensional, 13
  infinite-dimensional, 13
  real, 13
Vector subspace, 19, 22, 24, 29, 33

Weak derivatives, 26-27